THE FAMILY TRICHODORIDAE: STUBBY ROOT AND VIRUS VECTOR NEMATODES

Developments in Plant Pathology

VOLUME 6

The Family Trichodoridae: Stubby Root and Virus Vector Nematodes

by

W. DECRAEMER

Koninklijk Belgisch Instituut voor Natuurwetenschappen,
Brussels, Belgium
Guest Professor at Gent University

Published with the financial support of the Fondation Universitaire – Universitaire Stichting, and the Fonds Francqui, Brussels, Belgium

Springer-Science+Business Media, B.V.

Library of Congress Cataloging-in-Publication Data

Decraemer, W. (Wilfreda)
The family Trichodoridae : stubby root and virus vector nematodes / by W. Decraemer.
p. cm. -- (Developments in plant pathology ; v. 6)
Includes bibliographical references (p.) and index.

1. Trichodoridae. 2. Plant nematodes. 3. Plant nematodes as carriers of disease. 4. Virus diseases of plants. I. Title. II. Series.
SB998.T75D435 1995
589.1'82--dc20 95-39767

ISBN 978-90-481-4632-1 ISBN 978-94-015-8482-1 (eBook)
DOI 10.1007/978-94-015-8482-1

Coverpictures:
Anterior end of *Trichodorus cedarus*, male and
tail region of *T. paracedarus*, male.

Printed on acid-free paper

Originally published by Kluwer Academic Publishers in 1995.
Softcover reprint of the hardcover 1st edition 1995

CONTENTS

FOREWORD

The family Trichodoridae was established by Thorne in 1935 but it remained of limited taxonomic interest until 1951 when Christie and Perry associated *Trichodorus christie* (now *Paratrichodorus minor*) with a "stubby root disease" that affected certain crops in Florida, USA and interest further increased from 1960 when *P. pachydermus* was implicated as a vector of the economically important tobacco rattle virus. Such discoveries gave an impetus to studies on many aspects of the biology of trichodorid nematodes with a consequent proliferation of new species described from different parts of the world. The number of species increased from twelve in 1957 to a present day total exceeding eighty. This rapid expansion in the number of species was accompanied by several reappraisals of the taxonomy and systematics of the family, the most notable changes being the creation of the genus *Paratrichodorus* by Siddiqi in 1974 and the recognition of the monodelphic genera *Monotrichodorus* and *Allotrichodorus* respectively by Andrássy in 1976 and Rodriguez-M, Sher and Siddiqi in 1978.

Reliable and unambiguous identification of taxa is dependent on the known reliability of a range of characters, which remains of paramount importance in taxonomy even when biochemical and molecular methodology is available. In this book each of the species, including synonyms and *species inquirenda*, that comprise the family Trichodoridae is re-described from meticulous microscopic examination of type specimens, material from official nematode collections and populations in the field. The descriptions are supported by detailed drawings and micrographs and in many instances new characters of taxonomic significance have been identified. Attention to detail and keen observation have enabled the author to provide a considered and authoritative opinion on the taxonomy and systematics of the Trichodoridae. Further, the evaluation and selection of specific characters clearly demonstrates the importance of reliable identification of species when undertaking investigations on geographical distribution, ecology and other facets of the biology of individual species, particularly those of economic importance. It is inevitable that in an actively investigated taxon, such as the Trichodoridae, there will be a continuing addition of new species which will in turn encourage further expression of taxonomic opinion. If that proves to be so, Dr. Decraemer's book will provide a well established foundation for any new opinion and will remain as a standard reference to the Trichodoridae as well as a substantial contribution to the science of nematology.

C.E. TAYLOR

FOREWORD

Trichodorids constitute a group of plant parasitic nematodes, together with tylenchs and longidorids. It is the smallest one in number of genera, species and also individuals in the soil. However, trichodorids are not at all devoid of interest on both economical and zoological points of view, and their study by agronomists and taxonomists is receiving an increasing attention.

The first trichodorid was described by de Man in 1880, as *Dorylaimus primitivus*, the genus *Trichodorus* being created more than 30 years later by Cobb (1913). It was necessary to wait for nearly 40 more years to demonstrate the parasitism and the pathogenicity of representatives of the genus (Christie & Perry, 1951), and 10 other years to prove the trichodorids are vectors of plant viruses (Sol & Seinhorst, 1961). So, research progress is slow with trichodorids as they progress themselves in the soil.

Trichodorids are important in strictly agronomical research, as they are able to cause direct damage to the root system of various crops; they are the agent of the "docking disorder" of the sugar beet, a severe disease in Europe. They are also able to transmit various viruses to a series of important crops. These viruses are different from those transmitted by the longidorids and also are - as opposite to longidorids- not strictly associated to one or to a restricted number of species. Trichodorids constitute therefore a potential active group of direct and indirect predators.

The taxonomic aspects of the study of trichodorids are interesting also, as reflected by the diverging opinions of the specialists not only for the number and definition of genera and subgenera but also, what is not frequent, for the higher taxonomic categories (order, suborder) in which the trichodorids have to be placed.

The only conclusion to draw at present is that the trichodorids constitute a very original group of nematodes. Their study needs to be continued and even extended to clarify both their role in plant disorder and their position among the other nematodes.

In that aim, the present book written by one of the best specialists of the group constitutes a very helpful basis for any nematologist intending to work on trichodorids. It produces an updating of data concerning the morpho-anatomy of all the species of the group. Moreover, the credit of the book is to having been based, not on literature as too much often, but on personal observations from the author. Every nematologist can take profit from Dr. Decraemer's book, and so she has to be complimented for having done such an important contribution to the nematological literature.

Michel LUC

FOREWORD

Frieda Decraemer started her nematology carrier with the study of marine nematodes from the Great Barrier Reef. Especially her contribution on Desmoscolecida (since 1974), on Epsilonematidae and Draconematidae (from 1986 onward) are wellknown reference works.

In 1977 she became involved with Trichodoridae when she was employed as a researcher in several projects dealing with the occurrence of these nematodes in potential seed potato plots in Belgium. Since then she has divided her research between marine nematodes on the one hand and plant parasitic nematodes on the other.

Whatever the kind of nematodes, Frieda Decraemer's descriptions are always based on detailed and accurate observations and are beautifully illustrated.

This book on Trichodoridae is the result of her personal experience with this family. It would have been so much easier to make a compilation of the information available in the literature. However, although a lot of this information stems from her previous work, it is not her style. She has again verified, amended and corrected the data where she felt it was necessary. The result is an authoritative contribution which allows successful identification of these economically important, plant parasitic and virus transmitting nematodes.

Correct identification is the basis for all other investigations and therefore this book should help and stimulate not only further taxonomic and faunistic studies, but also more general research on soil ecology, on plant parasite relationships and on soil born virus transmission. I hope it will also enhance further studies to solve the still problematic position of the family inside the Nematoda.

August COOMANS

PREFACE

The present monograph is the result of nearly twenty years of experience in the field of stubby root and virus vector nematodes (Trichodoridae), a study which started in 1977 with a survey for Trichodoridae in Flanders (Belgium) in plots designated for the production of seed potatoes. This research was to be used as a basis for more extensive research on the significance of virus-transmitting nematodes in agricultural crops.

Dealing with the taxonomy of the Trichodoridae, it soon became apparent that there was a requirement for illustrated keys to the species of this family. Therefore the systematics of the family was revised and the validity of the genera and subgenera discussed. New identification keys were prepared (Decraemer, 1980). In addition to the creation of a genus database for the genera *Trichodorus* and *Paratrichodorus* in collaboration with Dr. R. Fortuner and his program NEMYSIS, the taxonomic value and intraspecific variability of the characters used for species identification were studied in detail (Decraemer, 1988; 1989).

The present work is a compilation of the systematics of the family Trichodoridae. Included are all currently accepted species and morphological data, elaborate species diagnoses, and illustrations mainly based upon type material are provided. Keys to the genera and species are included.

The morphology of the family is treated in detail, with emphasis on the diagnostic features and intraspecific variability.

Some aspects of the ecology of Trichodoridae are provided and their rôle on the transmission of plant viruses is reviewed.

In presenting my understanding of the family Trichodoridae, I hope to help students, nematologists, and plant pathologists especially with the identifications to species level and to inspire at least some of them to start further research on the ecology, molecular biology and all aspects of transmission of plant viruses by the stubby root nematodes.

I thank Drs. D. J. Brown, A. Coomans, E. Geraert and C. Taylor for critical reading of the manuscript and for their constructive discussions. I take the opportunity to thank all my colleagues, friends and co-nematologists, too many to mention by name, who generously helped me by sending type-specimens on loan, reprints of papers, specimens of nematodes, photographs, permissions to reproduce printed work (Marcel Dekker Inc., Fundamental and applied Nematology, Nematologica, Journal of

Nematology, Bulletin Koninklijk Belgisch Instituut voor Natuurwetenschappen).

My thanks also go to the members of the reading committee of the Fondation Universitaire - Universitaire Stichting for their careful reading, constructive comments and financial support.

I am grateful to Dr. D. Cahen, Director of the Royal Belgian Institute of Natural Sciences for a generous grant towards the cost of the publication of this book.

I wish to thank Dr. J. Van Goethem, Head of the Department of Invertebrates and my colleagues for their support and for the stimulating atmosphere they provide. I also am grateful to the technicians Ing. C. Lauwens, H. Van Paesschen and the photographer, T. Hubin for their help.

Last, but not least, I would like to express my deep sense of gratitude to my mother and daughter for their support and their willingness to let me devote most of my attention to research during the preparation of this book.

I dedicate this contribution to my father, the late Wilfried Decraemer, who throughout his work, was very devoted to nematology and provided me with much valuable advice concerning the art of drawing.

W. DECRAEMER

CHAPTER 1

INTRODUCTION

Trichodorid species, in general, are root ectoparasites, usually aggregating at the root tip. They possess a long stylet, called an onchiostyle, with which they pierce plant cells. They feed by repeated thrusts of the onchiostyle into epidermal cells and, root hairs, with a preference for the meristemic tissue near the tip of the growing root.

The group first received wide attention in 1951 by the discovery of their plant pathogenic role. Christie and Perry (1951) demonstrated that the species, *Trichodorus christiei*, now known as *Paratrichodorus minor*, caused damage to vegetable crops (e.g. bean, celery, cowpea, pea, pepper, tomato, etc.) in Florida. The symptoms produced on host plants were abnormally stunted roots, caused by cessation of root elongation and an overall reduction in root development, a condition descriptively called "stubby root". Since then, trichodorid nematodes have been referred to as "stubby root" nematodes.

Interest in the group increased rapidly from the discovery in the early 1960s, that several trichodorid species were vectors of plant viruses. Sol, Heuven and Seinhorst (1960) established that *Paratrichodorus pachydermus* could transmit Tobacco Rattle Virus (TRV). Pea early-browning virus (PEBV) was reported to be transmitted by *P. pachydermus* and *Paratrichodorus teres* by van Hoof in 1962. About ten years later, Salomao (1973) found a third virus, Pepper Ringspot (PRV), to be transmitted by *Paratrichodorus minor*. The economic consequence of crop losses due to trichodorids, both as causal agents of direct damage to plants and as vectors of plant virus diseases, is a worldwide problem (see chapters 8 and 9).

The Trichodoridae is a small family with four genera: *Trichodorus* Cobb, 1913, *Paratrichodorus* Siddiqi, 1974, *Monotrichodorus* Andrassy, 1976 and *Allotrichodorus* Rodriguez-M, Sher & Siddiqi, 1978 (Fig. 1.1). Siddiqi (1974) proposed three subgenera within the genus *Paratrichodorus viz.*: *Paratrichodorus*, *Atlantadorus* and *Nanidorus*, which he raised to genus level in 1980.

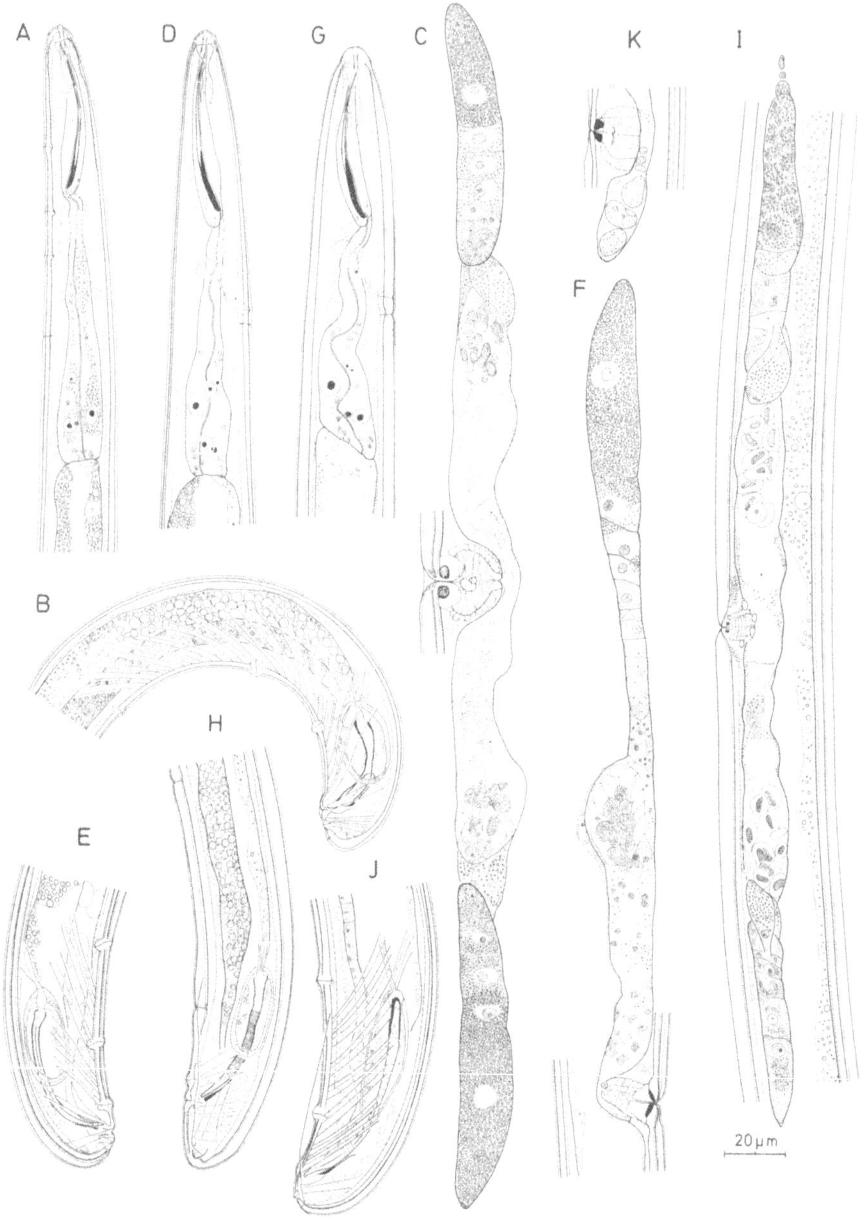

Figure 1.1. The genera of the Trichodoridae. Anterior body region in male: (A) *Trichodorus velatus* (paratype), (D) *Monotrichodorus m. vangundyi*, (G) *Paratrichodorus weischeri* (paratype). Male posterior body region: (B) *T. velatus* (paratype), (E) *M. vangundyi*, (H) *P. pachydermus*, (J) *Allotrichodorus campanulatus*. Female reproductive system: (C) *T. taylori* (holotype), (F) *M. vangundyi*, (K) *A. campanulatus*, vulva region, (I) *P. weischeri* (paratype)(after Decraemer, 1991, Courtesy of Marcel Dekker Inc.).

The genera *Atlantadorus* Siddiqi, 1974 and *Nanidorus* Siddiqi, 1974 subsequently were rejected by several reviewers, but recently were reinstated at the subgeneric level by Ahmad (1989) and Jairajpuri & Ahmad (1992). This view point is not followed (see chapter on taxonomy) and the Trichodoridae are considered to be comprised of only four genera.

The two largest genera *Trichodorus* (47 valid species) and *Paratrichodorus* (31 species) are didelphic and occur worldwide. The two remaining genera, *Monotrichodorus* Andrássy, 1976 and *Allotrichodorus* Rodriguez-M, Sher & Siddiqi, 1978, have fewer species (7 nominal species each), are monodelphic, and thus far have only been recorded from South and Central America.

CHAPTER 2

GENERAL MORPHOLOGY

2.1. Habitus

Trichodoridae are rather short and relatively plump nematodes, varying in length from about 350 to 1800 µm. Their body is largely cylindrical, slightly tapered anteriorly to a more or less rounded head end and posteriorly with a short, usually blunt, tail. They often look cigar-shaped when moribund, especially in typical straight females and males in the genera *Paratrichodorus* and *Allotrichodorus*. Males of the genera *Trichodorus* and *Monotrichodorus* have the posterior body region clearly curved ventrally.

2.2. Structure of the body cuticle

The cuticle of most trichodorids tends to swell in response to chemical stimuli, including fixatives, detergents and water when heat killed. In particular, representatives of the genera *Paratrichodorus* and *Allotrichodorus* frequently exhibit a markedly swollen condition of the body cuticle.

With the light microscope, a multilayered structure of the body cuticle is visible. It is usually described as a three-layered cuticle consisting of a thin outer layer (cortical zone) , followed by a thicker middle (median zone) and a thinner inner layer (basal zone) (Fig. 2.1.A) or sometimes as a two-layered structure. Often, more than two or three layers (representing subdivisions) may be distinguished. The body cuticle looks smooth but a pseudoannulation of the basal layer can be discerned.

Few data are available on the ultrastructure of the body cuticle. In *Paratrichodorus allius*, TEM (transmission electron microscopy) revealed an eight-layered cuticle, with a triple-layered epicuticle (Raski *et al.*, 1969). Hirumi *et al.* (1969) described the body

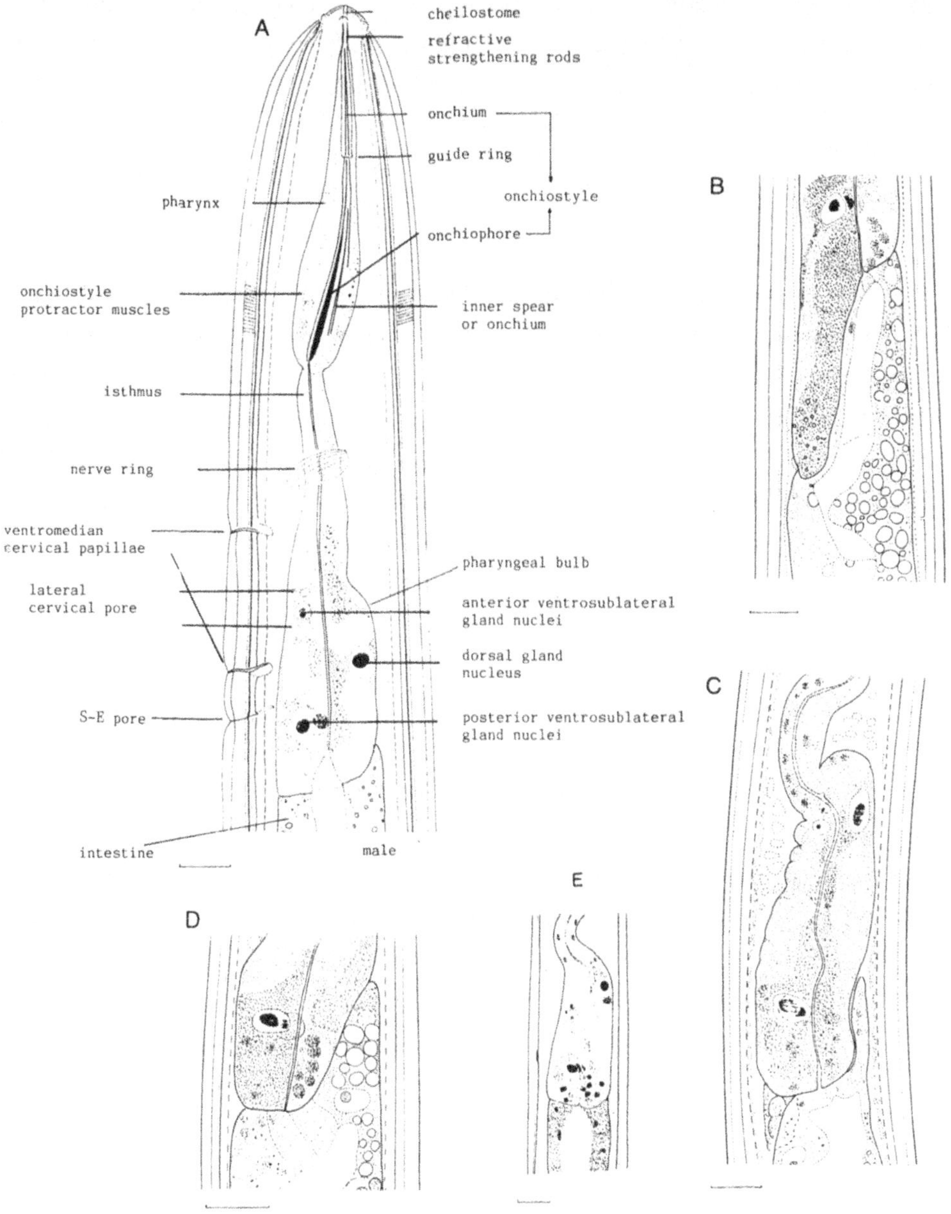

Figure 2.1. Basic structure of anterior region of a trichodorid nematode. (A) *Trichodorus petrusalberti*, pharyngeal region; (B-E) *Paratrichodorus teres*, subventral overlap pharyngeal glands (B), anterior dorsal overlap of intestine (D), both kind of overlap together (C) and an offset bulb (E) respectively (scale= 10 µm).

wall in the cephalic region of *Trichodorus christiei* (= *P. minor*) as being composed of two main layers, loosely connected and further subdivided; the outer layer being surrounded by a three-zone marginal layer (= epicuticle).

2.3. Somatic musculature

The somatic musculature in trichodorids is of the basic type of musculature in nematodes. The muscle cell is termed *platymyarian*, and the cells are grouped into a *meromyarian* arrangement. The somatic muscle cells are arrayed longitudinally, forming a single muscular layer just beneath the epidermis. More muscle cells are located in the anterior half of the body than in the posterior half (Hirumi *et al., 1971).* These authors considered the muscle cells in *P. minor* (=*T. christiei*) to be very primitive, capable of performing only simple sluggish movements.

2.4. Cephalic region

The lip region is dome-shaped with amalgamated lips. The sensilla are arranged in two circles: an inner circle of 6 labial papillae (2 subdorsal, 2 subventral, 2 lateral) around a triradiate oral opening, and an outer circle of 10 papillae (6 external labial papillae and four cephalic papillae). The four cephalic papillae with the 2 subdorsal and 2 subventral outer labial papillae form "double papillae"; both single lateral outer labial papillae are anterior to the level of the "double papillae" (Sher *et al.*, 1977) . Two or three receptors are found in the inner and outer labial sensilla, each arising from a single dendrite (Coomans & De Grisse, 1981). The lip region may appear to be offset from the neck region by the elevation caused by the outer ring of 10 papillae (Fig. 2.2.).

2.5. Amphids

These paired lateral sensory organs are postlabial and have large, transverse slit-like apertures, often with an exudate (= corpus gelatum). The aperture leads to a cup-shaped pouch or amphidial fovea, separated posteriorly by a constriction from the sensillum

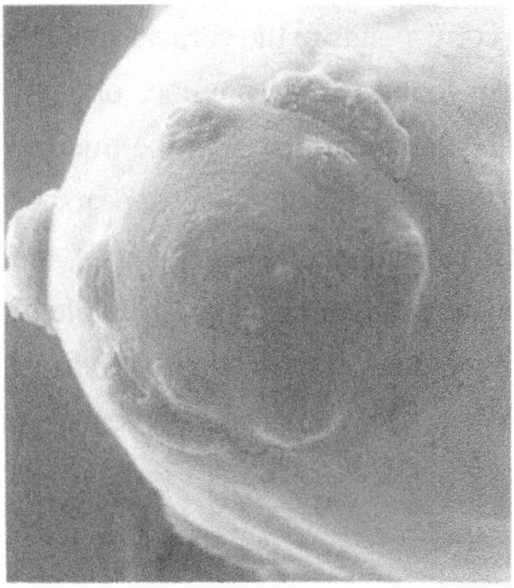

Fig. 2.2. Scanning electron microscopic micrograph of *en face* view of a male of *Trichodorus intermedius* (scale x 5000) (courtesy of the University of California, Riverside and *Journal of Nematolgy*)

pouch or fusus. The amphids are the main chemoreceptor organs and possess the largest number of receptors in contact with the external environment (23 amphidial receptors in *Trichodorus* according to Coomans & De Grisse, 1981).

2.6. Buccal cavity

The buccal cavity is situated between the triradiate mouth opening and the pharynx. Until recently, the buccal cavity was considered to consist of five regions, which from the anterior to posterior were defined as the cheilostom, prostom, mesostom, metastom, and telostom (Bird & Bird, 1991). In ultrastructure studies of trichodorid nematodes the buccal cavity was only described as stoma (Hirumi *et al.*, 1968; Raski *et al.*, 1969). De Ley *et al.* (1995) revised the buccal terminology based upon the ultrastructure of the stoma in Rhabditida, and divided the stoma into six regions (cheilostom, gymnostom, pro-, meso-, meta- and telostegostom). The cross sections illustrated in the earlier ultrastructure studies on trichodorids, do not allow the determination of the different stoma regions. Provisionally, the buccal cavity in trichodorids can be described as follows: the cheilostom is lined by invaginated labial cuticle , comprising epicuticle (layer

1), external cuticle (layer 2) and inner cuticle (layers 3, 4 in Raski *et al.*, 1969). Probably the external layer corresponds to the cortical zone, the inner layer to the median zone. No evidence has been found for the presence of a gymnostom which is defined as "a part surrounded by arcade epidermis". At the middle of the stoma, the external cuticle becomes thin and the pharyngeal tissue appears between the "external" and "internal" cuticle, probably indicating the beginning of the stegostom. In the most anterior region of the pharynx, strengthening rods, running longitudinally, support the cuticular wall of the pharyngeal lumen; their anterior ends are fused into the base of the cheilostom (Robertson & Wyss, 1983). These may serve as muscle attachments.

2.7. Pharynx

The term pharynx is used here instead of the term oesophagus, which is still frequently used. Providing both terms refer to the same structure, no confusion should arise.

Trichodorids have a pharynx with three well-defined regions: (1) an anterior muscular part or corpus that has the feeding apparatus in its lumen, (2) a slender mid-part or isthmus, surrounded at the mid-point by the nerve ring and (3) a basal bulb, accommodating the pharyngeal glands.

The anterior part of the corpus is composed mostly of muscular tissues: anteriorly with three longitudinal muscle cells (one dorsal and two subventral); posteriorly with six longitudinal muscle cells (two subdorsal, two lateral and two subventral) (Hirumi *et al.*,1968).

The onchiostyle varies in length among species from 20 to 188 µm and is curved ventrad. It consists of a stylettiform onchium and an onchiophore. The onchiostyle may contain an inner onchium, in which case the whole structure is described as being composed of an "outer" and an "inner" spear (Fig. 2.1.A). The solid tip (onchium) projects into the stoma lumen and gradually enlarges posteriorly, moving to the dorsal side of the lumen. There is no opening in the "outer" spear, that could connect with the lumen. Posteriad, the onchium fuses with the dorsal wall of the pharynx. At this level, the pharyngeal wall forms a fold which acts as a guiding ring. The posterior half of the onchiostyle, the onchiostyle extension or onchiophore, actually forming the dorsal wall of the pharyngeal lumen consists of solid cuticle. The histological structure of the cuticle lining the onchiostyle extension is similar to that of the onchium and continuous with it.

Pharyngeal muscles are attached to the dorsal surface of the outer spear. Just posteriad to the fusion of the onchiostyle with the dorsal wall of the pharynx, an oblique opening occurs at the dorsal side of the spear through which is inserted the apical portion of the 'inner spear'. The tip of the inner spear is of solid cuticle, posteriorly its diameter increases and has a central hole into which the cytoplasm of the pharyngeal tissue extends (Fig. 2.1.A). The inner spear has been observed with the light microscope only in adults of a few species: *Paratrichodorus minor* (in Hirumi *et al.*, 1968), *P. nanus* (in Seinhorst, 1970), *P. porosus* (in Bird, 1971) and *Trichodorus petrusalberti* (in Decraemer & Marais, 1993) and *T. nanjingensis* (in Decraemer & Cheng, 1994).

The inner spear is of the same structure as observed in juveniles at the same location. It presumably replaces the anterior end of the onchiostyle (= onchium) but formation nor shedding of the spear has yet been observed during molting (Bird & Mai, 1968; Morton & Perry, 1968; Kuiper, 1977). The terminology of the different parts of the stylet differs in the literature; here the anterior solid part is defined as onchium, the posterior part as onchiophore and the whole stylet as onchiostyle.

Muscles associated with the alimentary tract have been studied in *P. allius* (Raski *et al.*, 1969). In the cheilostome region there are two short sets of muscles (four labial muscles or buccal dilators and three perpendicular pharyngeal muscles) for protraction and contraction of the anterior part of the pharynx with feeding apparatus and a third set appear to act on the food canal in the region of the strengthening rods, possibly to break the seal on the feeding tube when the nematode leaves the feeding site (Robertson & Wyss, 1983). No retractor muscles have been reported, and apparently retraction of the onchiostyle results from turgor pressure provided in that region.

The isthmus, or narrow mid-portion, of the pharynx enlarges posteriorly to form a pyriform to elongate bulb accommodating the pharyngeal glands. According to Bird & Bird (1991) the basal bulb in *P. porosus* is non-muscular. The basal enlarged part of the pharynx has a triradiate lumen and is composed of dorsal and ventrosublateral glandular sections; there are five pharyngeal gland nuclei present, one dorsal nucleus and two pairs of ventrosublateral nuclei. The dorsal gland extends over the full length of the dorsal sector and opens into the lumen of the pharynx near the anterior end of the bulb; its nucleus varies considerably in position within a species from opposite the anterior ventrosublateral gland nuclei to the level of the posterior ventrosublateral gland nuclei (Decraemer & De Waele, 1981). The two posterior ventrosublateral glands open into the lumen near their nuclei; the latter vary in position within a species from opposite the dorsal nucleus to the pharyngo-intestinal junction. The two smaller anterior ventrosublateral glands have small nuclei in the anterior half of the bulb, and the orifices of these glands are difficult to observe.

The pharyngeal bulb may be (1) offset from the intestine as in the majority of the *Trichodorus* species or (2) the pharyngo-intestinal junction is characterized by a ventral to subventral pharyngeal overlap by the ventrosublateral glands: present in about half of the species of the genus *Paratrichodorus*, less frequent in the other genera, or (3) there may be a dorsal intestinal overlap, or (4) both the pharyngeal and intestinal overlap may occur together as in several *Paratrichodorus* species. The dorsal intestinal overlap occurs less frequently and is usually shorter compared to the pharyngeal gland overlap. All types of pharyngo-intestinal junctions may occur together within a single species (Fig.2.1.B-E, *P. teres*).

Siddiqi (1974) used the position of the gland nuclei and the type of pharyngo-intestinal junction to separate genera (*Trichodorus*/ *Paratrichodorus*) as well as subgenera (*Paratrichodorus*, *Atlantadorus*, *Nanidorus* in the genus *Paratrichodorus*).

The pharynx is linked to the intestine by a triangular pharyngo-intestinal valve.

The intestine is a long straight tube, without posterior differentiation into a prerectum. Hirumi (1970) distinguished two different intestinal regions in *P. minor*: (1) an anterior region with a lumen surrounded by three cells (a dorsal and two ventrosublaterals), its wall provided with small irregular microvilli protruding into the lumen, and (2) a posterior region without a lumen. However, these findings do not correspond with previous data in Hirumi & Hung (1969) where the intestine was described as having a granular anterior part and a microvillous posterior region. Since then, no other TEM observations have been made to resolve this point. The rectum is a cuticle-lined invagination; an intestinal-rectal valve has not been clearly observed with light microscopy.

2.8. Secretory-excretory system

Bird & Bird (1991) proposed the use of the term *secretory-excretory* (S-E) *system*, the assignment of the name *excretory system* being based on morphological criteria, without unequivocal evidence for an excretory function. The nematode S-E system opens to the exterior through the S-E pore. The terminology of Bird & Bird (1991) well suits the

Trichodoridae, where no clear evidence of an excretory function is available.

In trichodorid nematodes, there is a minute duct leading inward from the pore (Fig. 2.1.A). The latter was regarded as an excretory pore because of its position, usually close to the nerve ring, which is a common position of the S-E pore in many nematodes. The S-E pore in trichodorids is located mid-ventrally ranging from halfway along the pharynx to the pharyngo-intestinal junction, rarely occurring within the anterior intestinal region. The extreme posterior position of the S-E pore, which is uncommon in trichodorids, has been considered a diagnostic feature at the subgenus/genus level (see chapter on taxonomy). A small renette cell associated with the secretory-excretory system in trichodorids, referred to in Hunt (1993), has only been observed in *Allotrichodorus guttatus* (Rodriguez-M *et al.*, 1978). After study of type material this observation could not be confirmed. In *A. loofi*, a similar small cell with nucleus is present opposite the anterior end of the pharyngeal bulb. However, other comparable cells are present more anteriorly along the pharynx.

2.9. Caudal pores

All species of *Trichodorus, Allotrichodorus, Monotrichodorus* and most species of *Paratrichodorus* have one pair of caudal pores (except for females of *P. weischeri* with two pairs), located terminally or subterminally (Figs 2.3.E, G). In some *Paratrichodorus* species, they are obscure (interpreted as a single pore by some authors) or absent. In *Monotrichodorus* and in a few *Trichodorus* species both pores lie in close proximity to one another and can be difficult to distinguish as separate units.

2.10. Males

2.10.1. REPRODUCTIVE SYSTEM

Trichodorids are monorchic; they possess a single outstretched testis. The germinal zone is relatively short, followed by a longer growth zone in which the germ cells increase in size. Spermatids are stored in the vesicula seminalis at the posterior end of the testis. The vesicula seminalis leads to the long, glandular vas deferens, which is not differentiated posteriorly, into a muscular ejaculatory duct (Fig. 2.3.B).

In trichodorids sperm appear to be ameboid-like cells, as in most nematodes. Diversity in sperm structure has been observed within the genus *Paratrichodorus* and may be a useful taxonomic character (see chapter 3).

In *Paratrichodorus*, up to five types of sperm cells can be distinguished (1) large subcylindrical cells with a granular sausage-shaped nucleus as in *P. anemones*, (2) medium-sized cells with rounded nucleus as in *P. mirzai*, (3) small oval to rounded cells with small nucleus as in *P. allius* and present in most *Paratrichodorus* species (cells exceptionally with fibrillar appearance as in *P. lobatus*), (4) very long (fusiform) sperm with elongated nucleus as in *P. macrostylus* or (5) minute thread-like structures without distinct nucleus as in *P. acutus* (Decraemer & Chaves, 1988).

Most *Trichodorus*- and *Monotrichodorus* species have oval cells with sausage-shaped nucleus e.g. *T. cottieri*, although some have globular sperm cells with a globular nucleus as in *T. dilatatus*. In *Allotrichodorus*, most species have well developed rounded sperm cells with a nearly globular nucleus. In *Trichodorus*-species, sperm usually show a fibrillar inner structure (microtubules ?).

Copulatory apparatus

The spicules consist of a head (manubrium or capitulum), a shaft (calamus) and a flattened lamina which may have a membranous extension (the velum). The shaft is tubular with a central cytoplasmic core. In some trichodorid species, the cuticular tube is surrounded by an additional sheath formed by scales, and the shaft possesses an inner septum. Under the light microscope, the distal ends of the scales of the covering sheath sometimes appear separated from the spicule body and are interpreted as "bristles" (setae, spines). These bristles are often difficult to observe in retracted spicules. The opening of the spicules is terminal or ventro-subterminal and the spicule tip may be bifid (Rodriguez-M & Bell, 1978).

The size (22 to 87 µm) and shape of the spicules usually varies between species; both are used as diagnostic characteristics in taxonomy (see chapter on taxonomy). *Paratrichodorus* species have mainly straight spicules with transverse striae, no bristles (except Italian populations of *P. tunisiensis* have spicules with small bristles; not checked). *Trichodorus* males display a more pronounced differentiation in spicule shape (- spicule, cephalated or not; - spicule shaft, with or without a constriction, at different levels; - spicule straight or ventrally curved; curvature totally or only in specific regions)

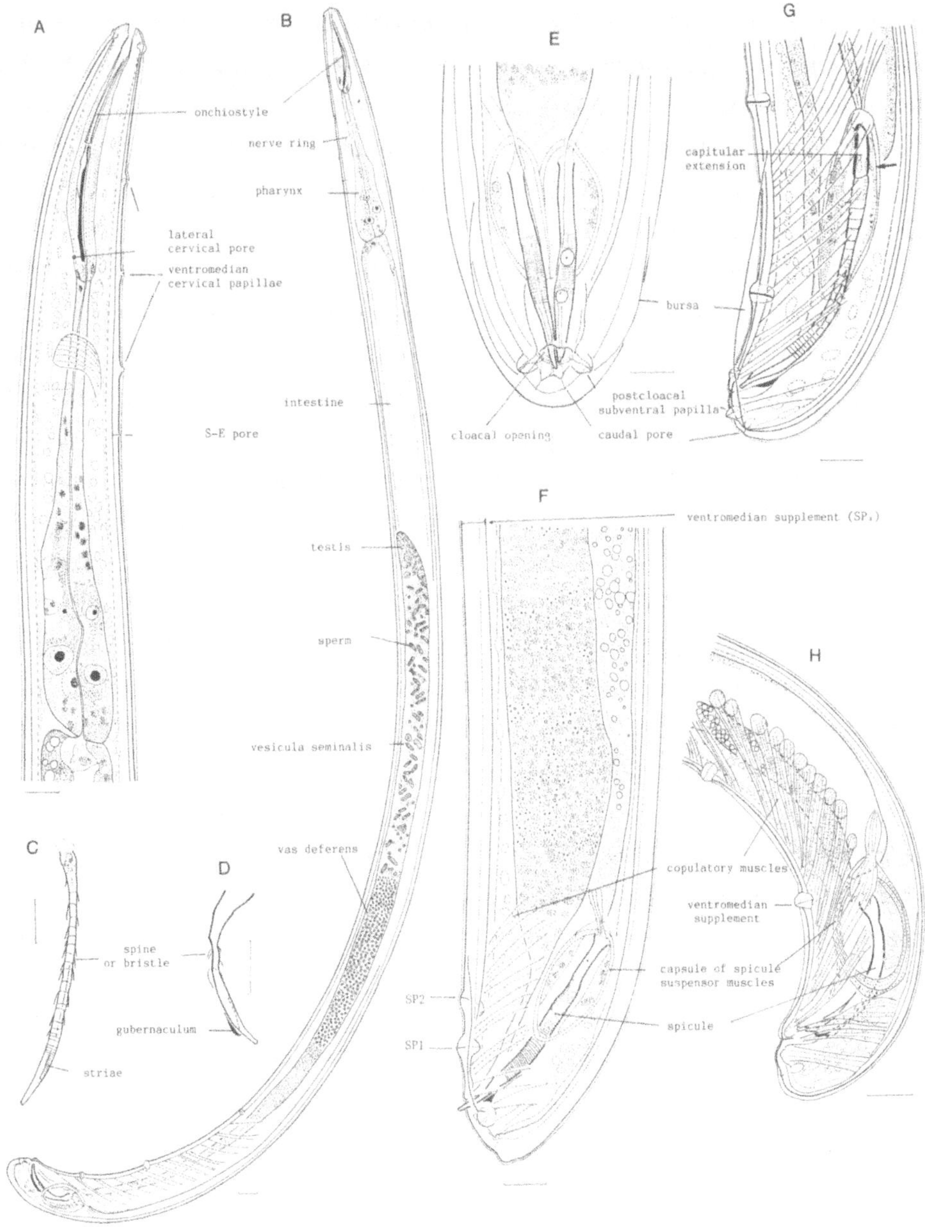

Figure 2.3. Basic structure of a trichodorid male. *Trichodorus*: (A) anterior region, lateral; (B) total male, lateral; (D) copulatory apparatus; (H) posterior region, lateral. *Paratrichodorus*: (E-F) posterior region, respectively ventral and lateral. *Allotrichodorus*: (C) spicule; (G) posterior region, lateral (scale = 10 μm).

and a more diverse ornamentation (transverse striae, bristles or spines, a ventral flange or without ornamentation i.e. smooth). *Monotrichodorus* males have long, slender, arcuate spicules (little variation among species), with transverse striae and bristles. *Allotrichodorus* males (Fig. 2.3.C) have long, slender spicules, less arcuate than in *Monotrichodorus* and with transverse striae. The spicules of two *Allotrichodorus*-species (*A. campanullatus* and *A. guttatus*) have a capitular extension additional to the spicule head (Fig. 2.3.G).

The spicule protractor muscles form a capsule of longitudinal muscles (termed the suspensor muscles by Siddiqi, 1974). They enclose the proximal half of each spicule and the spicular pouch; they are attached distally to the cloacal wall in the region of the proximal half of the gubernaculum. Proximally, the protractors are not attached directly to the head of the spicule but to a polar point from where retractor muscles connect the capsule to the subdorsal body wall. Two spicule retractor muscles extend towards the head of the spicule. The capsules of suspensor muscles show some differentiation in the degree of development between the genera. They are generally less developed in *Paratrichodorus* and *Allotrichodorus* species (Fig. 2.3.E-F and G respectively).

The gubernaculum is merely the cuticular thickening of the dorsal wall of the cloaca, consisting of a thin corpus with a keel-like thickening at its distal end and with a thin ventro-medial plate (cuneus) projecting between the spicules. This structure of the gubernaculum is similar in most trichodorids.

Copulatory muscles and bursa

The copulatory muscles show differences in degree of development according to genera. They are well developed and extend far anteriorly in *Trichodorus* (Fig. 2.3.H), poorly developed and restricted to the bursa region in *Paratrichodorus* (Fig. 2.3.F) and moderately developed in both monodelphic genera (Fig. 2.3.G). Upon contraction of the copulatory muscles (for example under the influence of fixatives), the posterior part of the body is ventrally bent. Dilator muscles of the cloaca run dorsally towards the body wall.

Caudal alae or bursa are found in the posterior region of some males. The bursa is usually narrow and rather weakly developed and unlikely to assist in copulation. It is always present in the genera *Paratrichodorus* and in *Allotrichodorus*, with various degrees of development. Sometimes the bursa is at most a small thickening of the ventrosublateral cuticle in the cloacal region. A bursa is absent in *Trichodorus* species (except for *T. cylindricus* and *T. paracedarus*), and in *Monotrichodorus* species. In several *Trichodorus* species (e.g. *T. vandenbergae*), the ventrosublateral body wall slightly protrudes from just anterior to the cloacal aperture to the level of the caudal pores, flanking the cloacal lips and is reminiscent of a rudimentary bursa (see chapter on taxonomy).

Bernard (1992b) observed narrow but distinct caudal alae on living, active specimens of *T. elefjohnsoni*, but they were obscure or indistinct in fixed or processed individuals. During movements of the males, the alae tended to flatten, then reappear. According to this author, it seems likely that the alae in *Trichodorus* species are formed by temporary muscular activity and are not fixed cuticular expansions, as is the case in many Tylenchida. From observations on fixed *Trichodorus* specimens, some individuals had the ventral caudal wall indented (slightly to extremely) probably due to contraction of the caudal muscles as a reaction upon fixation. These muscles run dorsoventrally and are not attached to the lateral body wall. Upon contraction, the lateral sides appear expanded, resembling caudal alae. In Hunt (1993) TEM transverse sections through the spicule region of *Paratrichodorus pachydermus* and *Trichodorus primitivus* show respectively the presence of subventral bursal wings, and lateral thickenings and a right subventral thickening of the body cuticle. No comments were given as to the presence of these cuticular differentiations in *Trichodorus*.

Supplementary pre- and postcloacal papillae

Precloacal supplements. The number of precloacal ventromedian supplements (SP) is generally constant within a species. They are usually described as papillae although they are not typical papilliform. Species of the genus *Trichodorus* usually have 3 SP (exceptionally 2, 4 or 5).

When three or more supplements are present, the anteriormost is usually the least developed (may be pore-like as in *Paratrichodorus* and *T. paracedarus*) and not protruding. The number and position of the precloacal supplements in relation to the retracted spicules is considered a diagnostic feature at species level; Jairajpuri & Ahmad (1992) used it also at subgenus level (see chapter 3). *Trichodorus* males usually have 1SP, *Paratrichodorus* and *Monotrichodorus* have 2SP and *Allotrichodorus* males have 3SP within the region of retracted spicules.

Postcloacal papillae. With few exceptions, all trichodorid males have only one pair of postcloacal subventral papillae, however, some species of the genus *Paratrichodorus* have two pairs (e.g. *P. atlanticus*, *P. mirzai*, *P. tunisiensis* (only Italian specimens) and *P. pachydermus* (in Seinhorst, 1954; Allen, 1957; Kuiper, & Loof, 1962; Baujard, 1980). According to Sturhan (1985) *P. pachydermus* possesses only one pair of postcloacal papillae; but often small pores may be observed just posterior to the cloacal opening.

2.10.2. VENTROMEDIAN CERVICAL PAPILLAE (CP)

Most trichodorid males possess ventromedian cervical papillae. Cervical papillae are absent in a few *Paratrichodorus* species (*P. acaudatus*, *P. delhiensis*, *P. lobatus*, *P. minor*, *P. nanus*, *P. teres*, exceptionally also in some specimens of *P. meyeri* and *P. rhodesiensis*). In *Trichodorus*, cervical papillae are lacking in *T. obscurus* and exceptionally in some specimens of *T. eburneus* and *T. nanjingensis*, and in *Allotrichodorus* they are absent in *A. brasiliensis* and *A. guttatus*.

In *Trichodorus*, the number of ventromedian cervical papillae varies between zero and four (the extremes of the range being exceptions). In the majority of species they are anterior to the S-E pore; even reaching to the onchiostyle region in a few species. The ventromedian cervical papilla(e) are usually relatively prominent and well differentiated from the S-E pore.

In *Paratrichodorus*, the majority of the species have a single ventromedian cervical papilla. The distinction between this papilla and the S-E pore can be difficult (Siddiqi,1962 for *P. mirzai*). Some specialists prefer to omit mention of the S-E pore when a ventral cell is not observed. The difference between S-E pore/CP may be associated with the presence or absence of a sensory cilium (Van de Velde, personal communication).

The species of the genus *Monotrichodorus* each have one ventromedian cervical papilla. In the genus *Allotrichodorus*, five out of seven species have one ventromedian cervical papilla, the other two species having none.

2.10.3. LATERAL CERVICAL PORES

Most trichodorid males have a pair of lateral cervical pores, situated either along the posterior third of the onchiostyle region, or between the nerve ring and the pharyngeal bulb, or along the pharyngeal bulb. Exceptionally, lateral cervical pores can be observed at the base of the amphids, as for example in *P. acaudatus*. They have not been observed in a few species of *Paratrichodorus* and *Trichodorus*.

Two receptors have been reported for the lateral pores of the genus *Trichodorus* (Hirumi *et al.*, 1970).

2.10.4. LATERAL BODY PORES

Their presence was observed only in male specimens of two *Paratrichodorus* species: *P. weischeri* (Sturhan, 1985) and *P. anemones* (Loof, 1965).

2.11. Females

2.11.1. REPRODUCTIVE SYSTEM

All species of the genera *Trichodorus* and *Paratrichodorus* are didelphic, amphidelphic: they have two genital tubes and their uteri are opposed; the vulva is situated near mid-body. The species of the genera *Monotrichodorus* and *Allotrichodorus* are monodelphic and prodelphic: they have one anteriorly directed genital tube; the vulva is situated at 60-90% of the total body length.

A female reproductive branch consists of ovary and gonoduct. The latter is made up of the oviduct and the uterus, which may be simple or subdivided into uterus and spermatheca. The vagina opens to the exterior through the vulva.

The uterus is a uniform sac (*Paratrichodorus*, *Monotrichodorus*, *Allotrichodorus*) or may be differentiated into a uterine sac and the uterus proper (*Trichodorus*). In the monodelphic genera a postvulvar uterine sac is usually present. The dorsal wall of the uterine sac in *T. variopapillatus* consists of seven cells, the uterus proper of two groups of four large, finely granular cells (Fig. 2.5.C,D). In *T. primitivus* there are approximately twenty uterine cells. In *Paratrichodorus*, the uterus wall consists either of several small and obscure cells (*P. pachydermus*, Fig. 2.5.E) or of very large cells (*P. weischeri*) (Geraert *et al.*, 1980). Large, typical sclerotized tubular structures (Fig. 2.5.C) may be present over the entire uteri in *Trichodorus* (Geraert *et al.*, 1980). They have been observed in *T. variopapillatus*, *T. elefjohnsoni* and *T. minzi* (chapter 4: Fig. 4.13).

All species of *Trichodorus*, *Monotrichodorus*, *Allotrichodorus* (except *A. westindicus*) and *Paratrichodorus* (partim) are bisexual species and possess a spermatheca

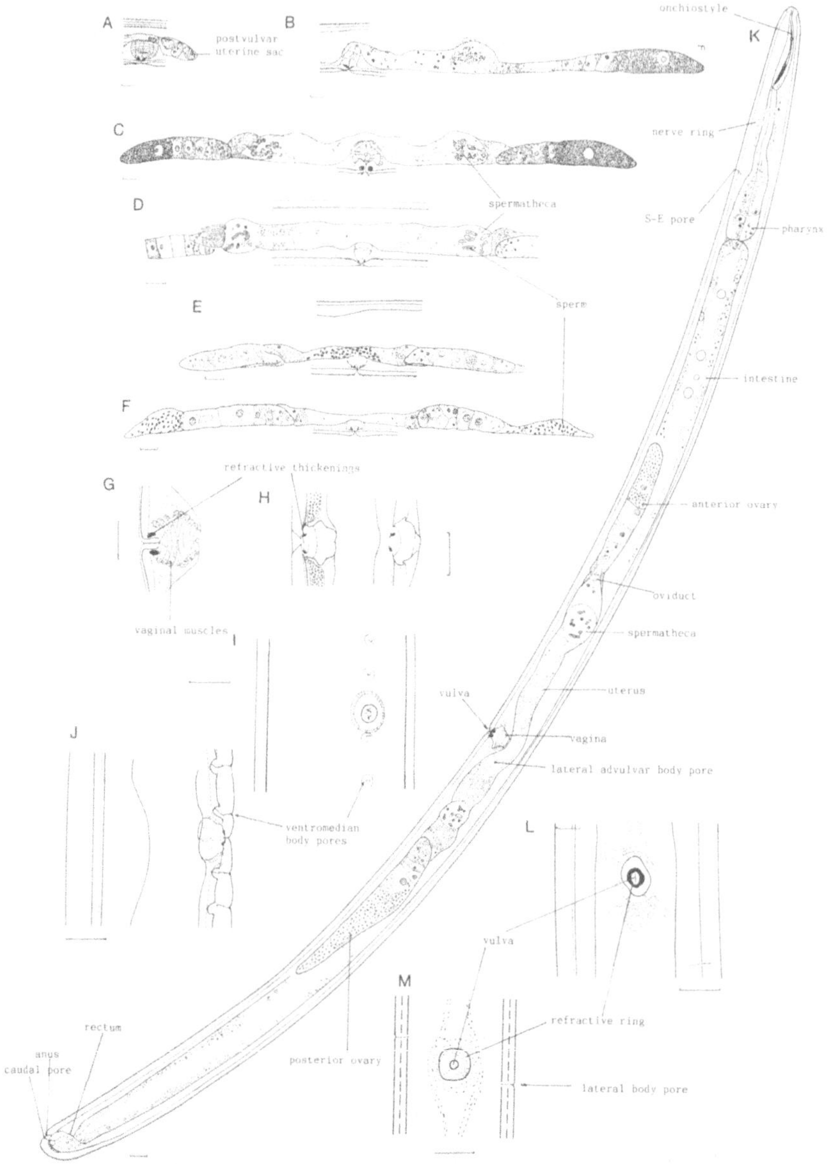

Figure 2.4. Basic structure of a trichodorid female. Monodelphic genital system: (A) *Allotrichodorus*: vulvar region; (B) *Monotrichodorus*: anterior branch. Didelphic genital system: (C) *Trichodorus*; *Paratrichodorus*: (D) with spermathecae; (E-F) without spermathecae, respectively with sperm in uterus and at top of ovary. Vagina and sclerotized pieces. *Trichodorus*: (G) lateral view, (M) ventral view; *Paratrichodorus*: (H) lateral, (L) ventral; *P. porosus* with ventromedian body pores: (I) ventral view, (J) lateral view. *Trichodorus*: total female in longitudinal optical section, lateral view (scale = 10 μm).

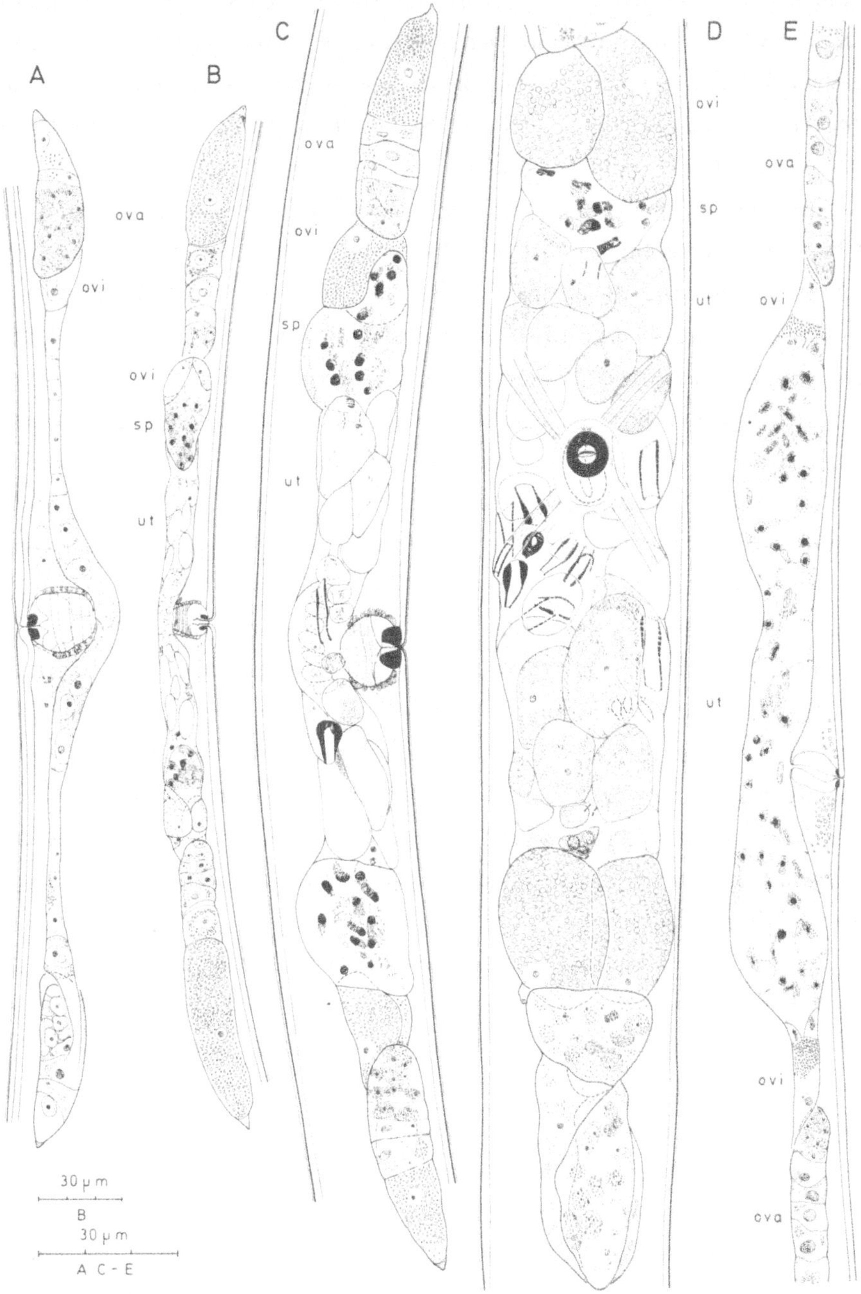

Figure 2.5. Female reproductive system. (A) *Trichodorus cylindricus*, young female, lateral view; (B) *T. primitivus*, lateral view; (C-D) *T. variopapillatus*, respectively lateral and ventral view; (E) *Paratrichodorus pachydermus*, lateral view (after Geraert *et al.*, 1980)(Courtesy *Nematologica*).

near the junction of the oviduct and uterus. The spermatheca is usually filled with sperm and is clearly visible but, in the absence of sperm, the spermatheca may be very poorly developed, difficult to discern and may be recorded as being absent (e.g. *T. dilatatus*, *T. intermedius* in Rodriguez-M & Bell, 1978). *T. nanjingensis* was described without spermathecae but with sperm-filled uteri.

In *Paratrichodorus*, part of the uterus near the oviduct may serve as a sperm storage chamber and may be slightly differentiated (= widened) or not e.g. *P. weischeri*, *P. anemones*, *P. anthurii*. A spermatheca has been observed in both branches of nine species: *P. anemones*, *P. anthurii*, *P. atlanticus*, *P. hispanus*, *P. lobatus* (only in original description), *P. macrostylus*, *P. paraporosus*, *P. tunisiensis*, *P. weischeri*; while in the other species, sperm cells were dispersed throughout the uterus: *P. acutus*, *P. allius* (=*P. tansaniensis*), *P. catharinae*, *P. grandis*, *P. faisalabadensis*, *P. lobatus*, *P. meyeri*, *P. minor*, *P. nanus*, *P. orrae*, *P. pachydermus*, *P. paramirzai*, *P. porosus*, *P. psidii*, *P. queenslandensis*, *P. rhodesiensis*, *P. sacchari* and *P. teres*. Several of the latter species e.g. *P. minor*, were reported to reproduce parthenogenetically (Winfield & Cooke, 1975). Unusual sperm location in the genital tract was observed in *P. allius*, *P. nanus* and *P. rhodesiensis*: numerous sperm cells were present at the distal end of each branch of the genital system where the ovary is reflexed (Decraemer, 1988; Sturhan, 1989a). According to Sturhan (1989a) *P. allius* and *P. nanus* reproduce by automixis (syngonic hermaphroditism). However, from fixed specimens and without investigation of the nuclear fusion-meiosis cycle, it is not possible to determine if these species reproduce by automixis (fusing nuclei being products of the same meiosis) or by autogamy (fusing nuclei of different meiosis).

The oviduct consists of two large cells with a distinct nucleus towards the ovary and usually situated laterally, one on the right and one on the left side. Next to these two cells, a narrow duct runs laterally along the ovary apparently joining the ovary near the flexure of the genital branch. The ovary is always reflexed in all trichodorid nematodes; at the end of the growing zone a small cap is present (Geraert *et al.*, 1980).

Bivulval females have been observed in *Trichodorus viruliferus* (Coiro *et al* ., 1988), *T. velatus* (Almeida, personal communication) and *Paratrichodorus mirzai* (Nasira & Maqbool, 1994). Both vulvae and vaginae were fully developed and apparently functional. They were interconnected by the uterus. Both the anterior and posterior genital branches were normally and equally developed.

In ventral view, the vulva varies in shape between species but is constant within a species. Three different types may be distinguished: a pore, a transverse slit or a longitudinal slit. The vulva shape has been used as a diagnostic feature for the (sub)genera *Paratrichodorus* (a small longitudinal slit), *Atlantadorus* (pore-like) and *Nanidorus* (a small transverse slit) (Siddiqi, 1980a). Each type is found in about the same number of species in the genus *Paratrichodorus*. The vulva shape is of restricted taxonomic use at genus level, but Jairajpuri & Ahmad (1992) considered this character to be important at subgenus level. Currently, only the genus *Allotrichodorus* is characterized by one type of vulva shape in ventral view: a transverse slit; the other genera have two or three different types.

At the junction of the vulva with the vagina, a refractive (sclerotized) ring is present and in lateral optical section, the refractive "pieces" have a characteristic shape. They vary from inconspicuous dots to large quadrangular or rod-shaped pieces. In general, they are least developed in the genus *Paratrichodorus*. The shape of the vaginal sclerotizations in lateral view is a very important diagnostic character at species level.

The vagina is well developed in *Trichodorus*, *Monotrichodorus* and *Allotrichodorus* species, usually about half a body width long (except in *A. westindicus* and a few *Trichodorus* species e.g. *T. elefjohnsoni*), and is surrounded by well developed sphincter muscles. In *Paratrichodorus* species, the vagina is clearly smaller and surrounded by a weakly developed muscular system. The shape and length of the vagina may vary within a species, for example between relaxed specimens or specimens with contracted vaginal constrictor muscles (e.g. due to fixation).

2.11.2. LATERAL/SUBLATERAL/SUBVENTRAL OR MEDIOVENTRAL BODY PORES

Nearly all *Trichodorus* species possess a pair of lateral postadvulvar body pores i.e. behind but within one body diameter from the vulva; rarely in subventral position as in *T. eburneus* (De Waele & Carbonell, 1983) and *T. cedarus* (Shishida, 1979). Prevulvar lateral body pores (one or two pairs) were found in one third of *Trichodorus* species.

In *Paratrichodorus*, lateral body pores are rarely located advulvar; they vary in number and position intraspecifically (e.g. in *P. anemones* from 0 to 6 pores in total), and also among species. In species with one lateral body pore on each side, the pore is mainly postvulvar. Lateral body pores are not always at the same level on both sides of the body and are not readily observed in lateral view; they are much more easily observed in dorsal or ventral view.

P. porosus and *P. paraporosus* have no lateral body pores, but possess medioventral pores, respectively pre- and postvulvar (*P. porosus*) or restricted to the postvulvar region (*P. paraporosus*). Lateral advulvar body pores are absent in *Allotrichodorus* species. One pair of lateral advulvar body pores is present in *Monotrichodorus* species.

2.11.3. LATERAL CERVICAL PORES IN FEMALE

Paired lateral cervical pores situated immediately behind the amphidial apertures were observed in *M. vangundyi*. A pair of lateral cervical pores on each side in the region of the onchiostyle were described for *P. pachydermus* (De Waele *et al.*, 1985) and some specimens of *P. teres* (De Waele & Kilian, 1992). The lateral cervical pores in *Trichodorus* are less obvious than the lateral body pores correlated with the vulva region.

2.12. Embryology.

Little information is available on the embryogenesis of trichodorid nematodes. Bird *et al.* (1968) made several observations on the embryology of *T. christiei* (= *P. minor*). In this species, the eggs (oval, 60-70 µm long and 38-42 µm wide) were laid in the single cell stage. The first cleavage was transverse to the longitudinal axis of the egg and produced two blastomeres of about equal size. The cleavage of each of the blastomeres was also transverse to the longitudinal axis, producing four blastomeres of similar size. These cleavages occurred during the first 24 hours after the egg was laid. A third set of cleavages occurred 24-30 hours after the egg was laid, forming eight blastomeres. Subsequent mitotic divisions occurred during the next 18 hours, resulting in an oval mass of cells. The blastula and gastrula stages occurred between 48 and 72 hours after the egg was laid, resulting in a vermiform embryo. This stage continued to elongate and soon became capable of occasional movement. Bird *et al.* (1968) recognized the first-stage juvenile within the egg 96 hours after the egg was laid; it had a well-developed stomatal armature. The pharynx and intestine developed rapidly after the 96th hour, and the juvenile emerged from the egg between 100 and 120 hours after the egg was laid. No molt was observed in the egg. The stomatal armature was flexible and could be protruded within the egg. The time for hatching was in agreement with the observations of Morton & Perry (1968), who reported that the first-stage juvenile emerged from the egg four days after it was laid.

Morton & Perry (1968) observed that the first-stage juvenile molted to second stage shortly after emerging from the egg. The third juvenile stage was observed seven days after inoculating a gravid female to tomato seedlings, the fourth juvenile stage on the tenth day, and the adult in two weeks. During the process of molting, a break developed in the old cuticle circling the body at the base of the spear. As a result, a "hood" was formed, which usually remained attached to the old cuticle at a single point on the dorsum. The stage of the life cycle was determined on the basis of the size of the genital primordium. Zoom (personal communication) distinguished four juvenile stages in *P. teres* based upon differences in body length.

To study the growth and development of *T. christiei* (= *P. minor*) Bird & Mai (1968) made an analysis of morphometric and allometric characters. They did not observe first-stage juveniles, and referred to a similar observation for *T. christiei* by Rohde & Jenkins (1957). They presumed that the first-stage juveniles molt shortly after emerging from the egg. Four life cycle stages (second-, third-, fourth-stage juvenile, or adult) were differentiated by body length and degree of maturation of the reproductive system. The magnitude, location and time of morphometric growth was determined for fourteen characters. Morphometric growth was calculated as a percentage of the total morphometric magnitude for four time intervals, determined as follows:

Time interval 1 (% morphometric growth to second life cycle stage)=

$$\frac{\text{second-stage character}}{\text{adult character}} \times 100$$

Time interval 2 (% morphometric growth between the second and third-stage juveniles)=

$$\frac{(\text{third-stage character} - \text{second-stage character})}{\text{adult character}} \times 100$$

Time interval 3 (% morphometric growth between the third and fourth-stage juveniles)=

$$\frac{(\text{fourth-stage character} - \text{third-stage character})}{\text{adult character}} \times 100$$

Time interval 4 (% morphometric growth between the fourth-stage juveniles and adults)=

$$\frac{(\text{adult character} - \text{fourth-stage character}) \times 100}{\text{adult character}}$$

With the exception of the stylet length, body width (both increasing more in the second than in the third and fourth interval), and characters pertaining to the reproductive system, morphometric characters increased in size more in the second and fourth than in the third developmental time interval. It was during this retardation in the growth of the total body length, and pharynx length, that the greatest increase in ovary length occurred. They also observed that the length of the onchiostyle in *P. minor* increased more in the second time interval than in the third and fourth.

During a study of the occurrence and behavior of *P. teres* in a polder, reclaimed from the Zuiderzee in the Netherlands, Kuiper (1977) investigated the embryology of this species. The life cycle was completed after 36 to 40 days at temperatures below 20° C. The different juvenile stages were determined based upon the length of the remaining hood after molting and on the body length . Kuiper (1977) observed only three juvenile stages, and no remains of the hoods of the first stage. He presumed that possibly the hood of the first and second stages could not be distinguished. However, the few hoods he observed contradict this possibility.

Apart from the former studies, juvenile specimens of the Trichodoridae have hardly been studied morphologically or morphometrically. The identification to species level appears almost impossible and juvenile specimens are seldom included in species descriptions.

Juvenile specimens were studied in three species (*T. primitivus*, *T. similis*, *P. pachydermus*) from mixed populations from seed-potatoes fields in Belgium (Decraemer, 1979). Based upon morphometric and morphological studies the juveniles could be identified to species level and the different juvenile stages could be distinguished. As in previous studies only three juvenile stages were observed. The onchiostyle length is a useful diagnostic character for differentiating *T. primitivus* and *T. similis* for the different developmental stages (except first stage which was not observed) (Fig. 2.6.). The juvenile stages (except the first) of these species were identified morphologically and morphometrically based upon the length of the genital system and the relation between onchiostyle length and length genital system. The morphometric growth was determined for *T. primitivus* and for *T. similis* and revealed that the body length and onchiostyle length increased more in the second and the fourth time interval than in the third time interval (Decraemer, 1979).

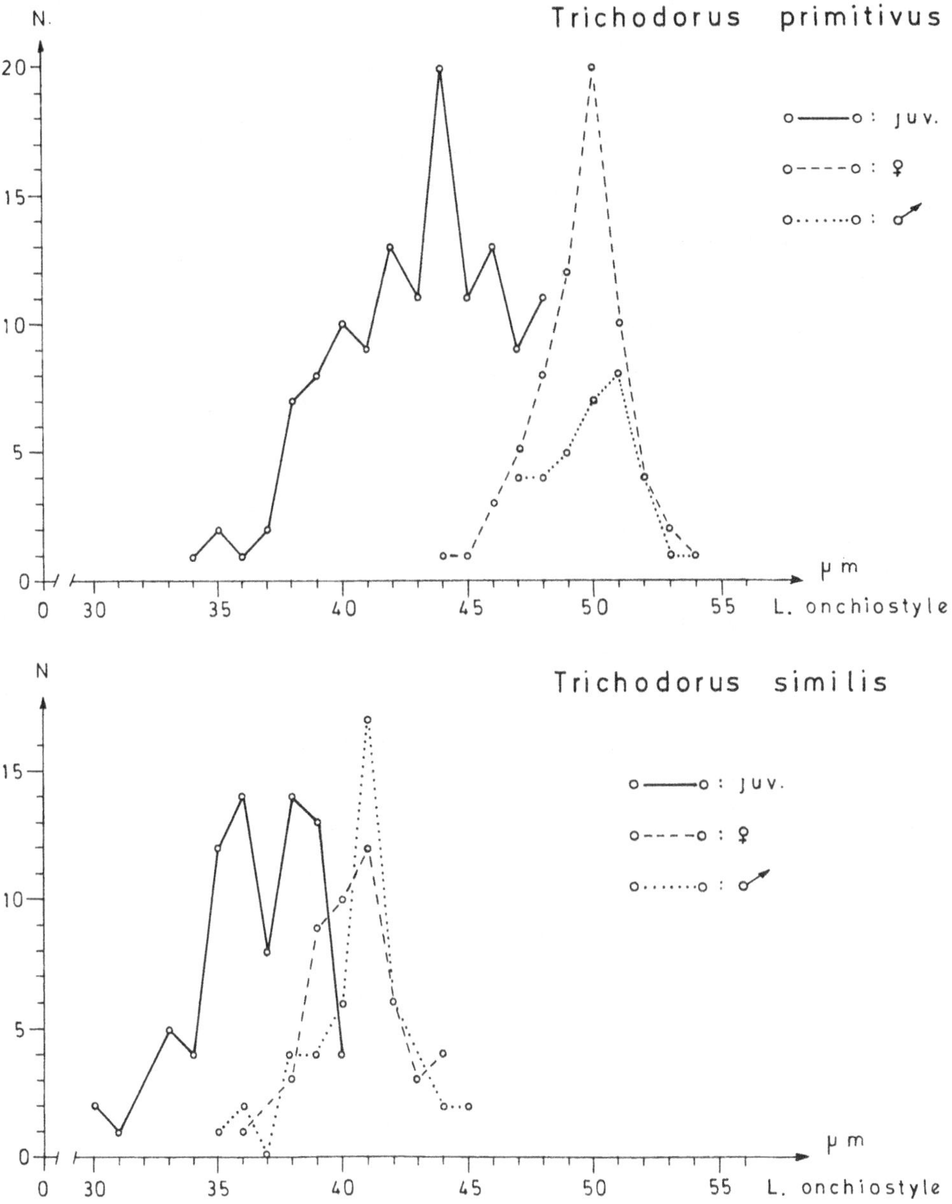

Figure 2.6. Relation between length of onchiostyle and body length in *T. primitivus* and *T. similis* (after Decraemer, 1979).

CHAPTER 3

SYSTEMATICS

Trichodoridae (Thorne, 1935) Clark, 1961

Order Triplonchida Cobb,1920

Suborder Diphtherophorina Coomans & Loof, 1970

Superfamily Trichodoroidea (Thorne, 1935) Siddiqi, 1974

Family Trichodoridae (Thorne, 1935) Clark, 1961

Type genus: *Trichodorus* Cobb, 1913

3.1. Diagnosis of superfamily Trichodoroidea

Body cigar-shaped. Onchiostyle relatively long, ventrally curved without basal knobs; guiding ring simple, surrounds anterior end of onchiostyle. Nerve ring about halfway the pharynx, at isthmus region. Pharynx with swollen posterior glandular bulb; five gland nuclei present: one posterior ventrosublateral pair, one small anterior ventrosublateral pair, and a single large dorsal nucleus; dorsal nucleus position varies within same species between anterior and posterior ventrosublateral pairs. Male with one testis, outstretched; precloacal supplements well developed; poorly developed caudal alae present or absent; spicules straight to ventrally curved, with or without ornamentation. Spicule protractor muscles form capsule of suspensor muscles around the spicules. Female reproductive system didelphic-amphidelphic or monodelphic-prodelphic, ovary (ovaries) reflexed; oviduct consisting of two finely granular cells; spermatheca(e) present or absent; rectum almost parallel to longitudinal body axis; anus subterminal. Tail short, maximum length one anal body width.

Diagnosis of family Trichodoridae: same as superfamily.

3.2. Differential diagnosis of the genera and species lists

3.2.1. GENUS *TRICHODORUS* COBB, 1913

Diagnosis

Trichodoridae. Cuticle usually not strongly swollen when fixed; basal layer with pseudoannulation. Pharynx usually with offset bulb, but also with ventral overlap of pharyngeal glands or dorsal intestinal overlap. Secretory-excretory pore usually at level of isthmus or anterior region of the pharyngeal bulb. Female reproductive system didelphic-amphidelphic; spermathecae present (exc. *T. nanjingensis*); vagina well developed, length usually about one-half body width; vaginal constrictor muscle well developed; vaginal sclerotization conspicuous, small to large, slightly variable in shape. Vulva a pore, or a transverse slit, or, rarely, a longitudinal slit. One to four lateral body pores present on each side, rarely absent; one advulvar pair within one body width posterior to the vulva. Males common with one to three ventromedian cervical papillae; rarely none or four. Lateral cervical pores usually present, rarely absent; usually one pair near onchiostyle base or just behind nerve ring. Sperm with large sausage-shaped or rounded nucleus. Spicules ventrally arcuate, rarely straight; smooth or with ornamentation of bristles, striations or velum. Capsule of spicule suspensor muscles usually well developed. Oblique copulatory muscles well developed, extending far anterior to spicular region, causes ventral curvature of posterior end in male upon death, caudal alae absent or at most rudimentary, except in *T. cylindricus* and *T. paracedarus* where present. Three precloacal ventromedian supplements, rarely two, four or five, more or less evenly spaced, usually at least one within the region of retracted spicules. Tail with one pair of postcloacal subventral papillae and one pair of caudal pores.

Type species: *Trichodorus obtusus* Cobb, 1913

syn. *Trichodorus proximus* Allen, 1957, n.syn.

List of Trichodorus *species*

T. aequalis Allen, 1957

T. aquitanensis Baujard, 1980

T. azorensis Almeida, De Waele, Santos & Sturhan, 1989

T. beirensis Almeida, De Waele, Santos & Sturhan, 1989

T. borai Rahman, Jairajpuri & Ahmad, 1985

T. borneoensis Hooper, 1962

T. californicus Allen, 1957

T. carlingi Bernard, 1992

T. cedarus Yokoo, 1964: syn. *T. kurumeensis* Yokoo, 1966; syn. *T. longistylus* Yokoo, 1964

T. complexus Rahman, Jairajpuri & Ahmad, 1985

T. coomansi De Waele & Carbonell, 1983

T. cottieri Clark, 1963

T. cylindricus Hooper, 1962

T. dilatatus Rodriguez-M & Bell, 1978

T. eburneus De Waele & Carbonell, 1983

T. elefjohnsoni Bernard, 1992

T. elegans Allen, 1957

T. giennensis Decraemer, Roca, Castillo, Pena-Santiago & Gomez-Barcina, 1993

T. hooperi Loof, 1973

T. intermedius Rodriguez-M & Bell, 1978

T. kilianae Decraemer & Marais, 1993

T. lusitanicus Siddiqi, 1974

T. magnus Decraemer & Marais, 1993

T. minzi De Waele & Cohn, 1992

T. nanjingensis Liu & Cheng, 1990

T. obscurus Allen, 1957: syn. *T. primitivus* apud Thorne, 1939 and Goodey, 1951

T. obtusus Cobb, 1913: syn. *T. proximus* Allen, 1957, n.syn.

T. orientalis De Waele & Hashim, 1984

T. pakistanensis Siddiqi, 1962: syn. *T. litchi* Edward & Misra, 1970

T. paracedarus Xu & Decraemer, 1995

T. parorientalis Decraemer & Kilian, 1992

T. paucisetosus Bernard, 1992

T. persicus De Waele & Sturhan, 1987

T. petrusalberti De Waele, 1988

T. philipi De Waele, Meyer & Van Mieghem, 1990

T. primitivus (de Man, 1880) Micoletzky, 1922: syn. of *Dorylaimus primitivus* de Man, 1880; syn. *T. castellanensis* Arias Delgado, Jiminez Millan & Lopez Pedregal, 1965; n.syn. *T. mirabilis* Ivanova, 1977

T. rinae Vermeulen & Heyns, 1984

T. sanniae Vermeulen & Heyns, 1984

T. similis Seinhorst, 1963

T. sparsus Szcygiel, 1968

T. taylori De Waele, Mancini, Roca & Lamberti, 1982

T. tricaulatus Shishida, 1979

T. vandenbergae De Waele & Kilian, 1992

T. variopapillatus Hooper, 1972

T. velatus Hooper, 1972

T. viruliferus Hooper, 1963
T. yokooi Eroshenko & Teplyakov, 1975
T. henanica Wang & Wu, 1991: nom.nud.

Discussion on the taxonomic status of T. obtusus

Cobb (1913) described a new species *Trichodorus obtusus* and also introduced a new genus *Trichodorus*. The description of *T. obtusus* was brief with illustrations of the head and onchiostyle region and of the copulatory apparatus and tail of a male specimen only.

Micoletzky (1922) synonymized *T. obtusus* with *T. primitivus* (de Man, 1880). However, the identity of *T. primitivus* is unambiguous due to the illustrations by de Man (1884) of the copulatory apparatus in the male and of the vaginal sclerotizations in lateral view of the female. Also, reference can be made to the type specimens which have been preserved (Loof, 1975). Thus, *T. obtusus* and *T. primitivus* are two distinct species.

Thorne (1939) accepted the synonymy by Micoletzky (1922) and redescribed *T. primitivus* (syn. *T. obtusus*) based on numerous specimens collected by Cobb at Arlington, Virginia (type locality of *T. obtusus*) and from populations collected by himself from cultivated and virgin soil at various localities in Utah, Colorado and California, U.S.A. He also gave illustrations of the onchiostyle region, the tail of the female and tail and copulatory apparatus of the male. The illustration of the male copulatory apparatus and tail is not that of *T. primitivus* nor that of *T. obtusus*. The shape of the spicules differ from both species as well as the position of the posterior precloacal supplement (clearly within the region of retracted spicules *vs* not in the spicule region but at level of the spicule head in *T. obtusus* and *T. primitivus*).

Allen (1957) assumed that Cobb could have described *T. obtusus* from specimens of *T. primitivus* collected in Virginia since *T. primitivus* had been found in the U.S.A. by Jenkins in 1955. He, therefore, concluded that Micoletzky (1922) was correct in placing *T. obtusus* in synonymy with *T. primitivus*.

Allen (1957) redescribed *T. primitivus*, but as Seinhorst (1963) subsequently reported, Allen's description of *T. primitivus* referred to two species: the females were the real *T. primitivus*, the males represented an undescribed species now known as *T. similis* Seinhorst, 1963. Therefore, as Seinhorst (1963) and Loof (1975) correctly concluded, *T. obtusus* and *T. primitivus* are two distinct species.

Allen (1957) redescribed the specimens collected by Thorne in 1931 from the type locality of *T. obtusus* as a new species: *T. obscurus*, close to *T. californicus*.

Thorne (1974) gave additional information on *T. obtusus* Cobb,1913, including illustrations of a female specimen. On this occasion he contradicted his earlier report (Thorne, 1939) and did not treat *T. obtusus* as a synonym of *T. primitivus*. He considered *T. obtusus* to most closely resemble *T. proximus* but different from the latter species by having a shorter onchiostyle (40-45 µm in *T. obtusus vs* 48-65 µm(♀), 49-70 µm (♂) in *T. proximus)*, and by the females possessing one postadvulvar body pore on both sides. Former differences between both species are inconsequential: the length of the onchiostyle of *T. obtusus* falls within the range known for *T. proximus*; a pair of postadvulvar body pores is also present in *T. proximus* and the more anterior prevulvar body pore in *T. proximus* may have been overlooked in *T. obtusus* or was absent as in the population described in Norton *et al.* (1982). Moreover, both species have a similar spicule shape and arrangement of the ventromedian precloacal supplements. They also have the same host preference.

Loof (1975) calculated the length of the onchiostyle in *T. obtusus* type specimen from the formula by Cobb as 75 µm, the spicule length as 56 µm. Both the onchiostyle length and spicule length are similar to those of *T. proximus* (spicule length 48-65 µm). Therefore, *T. proximus* is considered to be a synonym of *T. obtusus*.

Siddiqi (1974) gave an up-to-date key to the species of *Trichodorus*. He omitted four species, amongst them *T. obtusus* which according to Loof (1975) Siddiqi considered identical with *T. primitivus*. However, Siddiqi (1974) did not make this statement.

Ahmad (1989) indicated *T. primitivus* as type species of the genus but also included *T. obtusus* as a valid species in the identification key.

Hunt (1993) considered *T. obtusus* Cobb, 1913 as a *species inquirenda* and indicated *T. primitivus* Cobb, 1913 as type species of the genus, although *T. primitivus* is clearly a different species (see also Loof, 1994).

3.2.2. GENUS *PARATRICHODORUS* SIDDIQI, 1974

syn. *Atlantadorus* Siddiqi, 1974

syn. *Nanidorus* Siddiqi, 1974

Diagnosis

Trichodoridae. Cuticle usually strongly swollen when heat-killed or treated with acid fixative. Ventral pharyngeal and/or dorsal intestinal overlap usually present, pharyngeal bulb rarely offset. Female reproductive system didelphic-amphidelphic; spermathecae present, or absent with sperm throughout uterus or rarely sperm at proximal end of genital branch. Lateral body pores present in almost 50% of species, rarely located advulvar. Vagina short (clearly less than half the corresponding body diameter), constrictor muscles inconspicuous, vaginal sclerotization small to inconspicuous. Vulva a pore, or a longitudinal or a transverse slit. Males rare or unknown in about 26% of species. Males ventromedian cervical papillae absent or, if present, usually one papilla near secretory-excretory pore, exceptionally with two. Lateral cervical pores absent or, if present, usually one pair near onchiostyle base or secretory-excretory pore. Male tail straight in death; caudal alae present, obscure to distinct. Sperm cells large, subcylindrical with large sausage-shaped nucleus, or small with small, oval to rounded nucleus, or very long fusiform with elongated nucleus, or medium-sized with rounded nucleus, or thread-like and nucleus obscure. Oblique copulatory muscles poorly developed, usually restricted to caudal alae region; capsule of spicule suspensor muscles inconspicuous. Spicules usually straight; corpus striated. One to three ventromedian precloacal supplements, exceptionally four; usually two supplements present within caudal alae region, third supplement less developed, anterior to caudal alae; or some species with only one supplement present. Usually one pair, rarely two, subventral postcloacal papillae. A pair of caudal pores present, rarely absent.

Type species: *Paratrichodorus tunisiensis* (Siddiqi, 1963) Siddiqi, 1974
syn. *Trichodorus tunisiensis* Siddiqi, 1963

List of Paratrichodorus *species*

P. acaudatus (Siddiqi, 1960) Siddiqi, 1974: syn. *T. acaudatus* Siddiqi, 1960; syn. *P. (P.) acaudatus* (Siddiqi, 1960) Siddiqi, 1974

P. acutus (Bird, 1967) Siddiqi, 1974: syn. *T. acutus* Bird, 1967; syn. *P. (N.) acutus* (Bird, 1967) Siddiqi, 1974; *Nanidorus acutus* (Bird, 1967) Siddiqi, 1974

P. alleni (Andrássy, 1968) Siddiqi, 1974; syn. *T. alleni* Andrássy, 1968; syn. *P. (P.) alleni* (Andrássy, 1968) Siddiqi, 1974

P. allius (Jensen, 1963) Siddiqi, 1974: syn. *T. allius* Jensen, 1963; syn. *P. (P.) allius* (Jensen, 1963) Siddiqi, 1974; syn. *P. tansaniensis* Siddiqi, 1974

P. anemones (Loof, 1965) Siddiqi, 1974: syn. *T. anemones* Loof, 1965; syn. *P. (A.) anemones* (Loof, 1965) Siddiqi, 1974; syn. *Atlantadorus anemones* (Loof, 1965) Siddiqi, 1974

P. anthurii Baujard & Germani, 1985 (*P. anthuriae* in Hunt, 1993)

P. atlanticus (Allen, 1957) Siddiqi, 1974: syn. *T. atlanticus* Allen, 1957; syn. *P. (A.) atlanticus* (Allen, 1957) Siddiqi, 1974; syn. *A. atlanticus* (Allen, 1957) Siddiqi, 1974

P. catharinae Vermeulen & Heyns, 1983: syn. *P. (A.) catharinae* Vermeulen & Heyns,1983, n.syn.

P. delhiensis (Khan, Saha & Lal, 1993) n. comb.: syn. *Atlantadorus delhiensis* Khan, Saha & Lal, 1993: n.syn.

P. faisalabadensis Nasira & Maqbool, 1994

P. grandis Rodriguez-M. & Bell, 1978: syn. *P. (A.) grandis* Rodriguez-M & Bell, 1978; syn. *Atlantadorus grandis* (Rodriguez-M & Bell, 1978) Siddiqi, 1974, n.syn.

P. hispanus Roca & Arias, 1986

P. lobatus (Colbran, 1965) Siddiqi, 1974: syn. *T. lobatus* Colbran, 1965; syn. *T. clarki* Yeates, 1967; syn. *P. (P.) lobatus* (Colbran, 1965) Siddiqi, 1974

P. macrostylus Popovici, 1989

P. meyeri De Waele & Kilian, 1992

P. minor (Colbran, 1956) Siddiqi, 1974: syn. *T. minor* Colbran, 1956; syn. *P. (N.) minor* (Colbran, 1956) Siddiqi, 1974; syn. *N. minor* (Colbran, 1956) Siddiqi, 1974; syn. *T. christiei* Allen, 1957; syn. *P. (N.) christiei* (Allen, 1957) Siddiqi, 1974; syn. *N. christiei* (Allen, 1957) Siddiqi, 1974; syn. *T. obesus* Razjivin & Penton, 1975; syn. *P. (N.) obesus* (Razjivin & Penton, 1975) Rodriguez-M & Bell, 1978

P. mirzai (Siddiqi, 1960) Siddiqi, 1974: syn. *T. mirzai* Siddiqi, 1960; syn. *P. (P.) mirzai* (Siddiqi, 1960) Siddiqi, 1974; syn. *T. musambi* Edward & Misra, 1970

P. nanus (Allen, 1957) Siddiqi, 1974: syn. *T. nanus* Allen, 1957; syn. *P. (N.) nanus* (Allen, 1957) Siddiqi, 1974; syn. *N. nanus* (Allen, 1957) Siddiqi, 1974

P. orrae Decraemer & Reay, 1991

P. paramirzai Siddiqi, 1991

P. pachydermus (Seinhorst, 1954) Siddiqi, 1974: syn. *T. pachydermus* Seinhorst, 1954; syn. *P. (A.) pachydermus* (Seinhorst, 1954) Siddiqi, 1974; syn. *A. pachydermus* (Seinhorst, 1954) Siddiqi, 1974

P. paraporosus Khan, Jairajpuri & Ahmad, 1989

P. porosus (Allen, 1957) Siddiqi, 1974: syn. *T. porosus* Allen, 1957; syn. *P. (A.) porosus* (Allen, 1957) Siddiqi, 1974; syn. *A. porosus* (Allen, 1957) Siddiqi, 1974; syn. *T. bucrius* Lordello & Zamith, 1958

P. psidii Nasira & Maqbool, 1994

P. queenslandensis Decraemer & Reay, 1991

P. renifer Siddiqi, 1974: syn. *P. (N.) renifer* Siddiqi, 1974; syn. *N. renifer* Siddiqi, 1974

P. rhodesiensis (Siddiqi & Brown, 1965) Siddiqi, 1974: syn. *T. rhodesiensis* Siddiqi & Brown, 1965; syn. *P. (P.) rhodesiensis* (Siddiqi, & Brown, 1965) Siddiqi, 1974

P. sacchari Vermeulen & Heyns, 1983: syn. *P. (A.) sacchari* Vermeulen & Heyns, 1983, n.syn.

P. teres (Hooper, 1962) Siddiqi, 1974: syn. *T. teres* Hooper, 1962; syn. *P. (P.) teres* (Hooper, 1962) Siddiqi, 1974; syn. *T. flevensis* Kuiper & Loof, 1962

P. tunisiensis (Siddiqi, 1963) Siddiqi, 1974: syn. *T. tunisiensis* Siddiqi, 1963; syn. *P. (P.) tunisiensis* (Siddiqi, 1963) Siddiqi, 1974

P. weischeri Sturhan, 1985: syn. *P. (P.) weischeri* Sturhan, 1985, n.syn.

Discussion on the taxonomy of the genus Paratrichodorus

Ahmad (1989) reinstated *Atlantadorus* and *Nanidorus* at the subgeneric level, characterizing the subgenera by the diagnoses of Siddiqi (1980a). No comment or discussion was provided on the reintroduction of the three subgenera in *Paratrichodorus* in Ahmad (1989) nor in Jairajpuri & Ahmad (1992).

Ahmad (1989) gave a taxonomic revision of the Trichodoridae and added a table with the differentiating characters of the subgenera of *Paratrichodorus*. The following characters were considered: lateral cervical pores (presence/absence); ventromedian cervical "pore" (presence/absence); lateral and caudal pores in female (presence/absence); secretory-excretory pore (position in relation to nerve ring and to base of pharynx); ventrosublateral pharyngeal gland (apparently concerning the position of the nuclei of the posterior ventrosublateral glands in relation to pharyngo-intestinal junction and to the dorsal nucleus); pharyngeal overlap (presence or absence of pharyngeal or intestinal overlap); vulva shape in ventral view and number of precloacal supplements.

Remarks. The position of the secretory-excretory pore, the position of the posterior ventrosublateral gland nuclei, and the presence of pharyngeal and intestinal overlaps are variable characters within the same species and between species (Baujard & Germani, 1985; Decraemer & De Waele, 1981; Decraemer, 1989b).

The ventral vulva shape is a reliable diagnostic character but difficult to observe since specimens in ventral view are needed.

The number of precloacal supplements differs between *Paratrichodorus* / *Atlantadorus* and *Nanidorus* (3 *vs* 1). However, the subgenus *Nanidorus* contains only four valid species, two of which are described only from females (*P. acutus*, *P. renifer*), and two (*P. minor*, *P. nanus*) in which males are rare.

Lateral pores in females are described as being absent in the four *Nanidorus* species and caudal pores are either absent or obscure (e.g. *P. minor*). Moreover, lateral pores are often difficult to observe.

The other characters such as lateral cervical pores and ventromedian cervical papillae in males are variable characters within the subgenus *Paratrichodorus*.

Taking these remarks into account, the earlier statement on the synonymization of *Atlantadorus* and *Nanidorus* with *Paratrichodorus* is confirmed and herewith, only four genera in the family Trichodoridae are recognized (Decraemer, 1980a; Decraemer & De Waele, 1981).

3.2.3. GENUS MONOTRICHODORUS ANDRÁSSY, 1976

Diagnosis

Trichodoridae. Cuticle usually not swollen when fixed. Usually no pharyngeal or intestinal overlaps. One pair of caudal pores. Female reproductive system monodelphic-prodelphic; spermatheca present; postvulvar uterine sac present, minute to large. Vagina well developed, length about half to more than one-half body width and anteriorly directed. Vaginal sclerotization small to medium-sized, often weakly sclerotized. One pair of lateral advulvar body pores present. Vulva a transverse or a longitudinal slit. Males with one ventromedian cervical papilla and one pair of lateral cervical pores. Spicules long, slender; without a spicule capitular extension, shaft striated, with or without bristles. Caudal alae absent or at most rudimentary as in *M. sacchari.* Oblique copulatory muscles moderately developed, extending to slightly anterior to the retracted spicules. Three medioventral precloacal supplements present. One pair of large subventral postcloacal papillae.

Type species: *Monotrichodorus monohystera* (Allen, 1957) Andrássy, 1976
syn. *Trichodorus monohystera* Allen, 1957
syn. *M. acuparvus* Siddiqi, 1991, n.syn.
syn. *M. parvus* Siddiqi, 1991, n.syn.
syn. *M. proporifer* Siddiqi, 1991, n.syn.

List of Monotrichodorus *species*

M. monohystera monohystera (Allen, 1957) Andrássy, 1976: syn. *T. monohystera* Allen, 1957, n.syn.; *M. acuparvus* Siddiqi, 1991, n.syn.; syn. *M. parvus* Siddiqi, 1991, n.syn.; syn. *M. proporifer* Siddiqi, 1991, n.syn.

M. monohystera vangundyi Rodriguez-M, Sher & Siddiqi, 1978: syn. *M. vangundyi*, n.grad.

M. muliebris Andrássy, 1989

M. sacchari Baujard & Germani, 1985

The validity of the species within *Monotrichodorus* is discussed under the species descriptions.

3.2.4. GENUS *ALLOTRICHODORUS* RODRIGUEZ-M, SHER & SIDDIQI, 1978

Diagnosis

Trichodoridae. Cuticle variable when fixed: not swollen, or moderately to distinctly swollen. Anteriorly directed dorsal intestinal overlap present but short, or pharyngeal bulb offset. One pair of caudal pores, but absent in *A. westindicus*. Female reproductive system monodelphic-prodelphic; spermatheca present (except *A. westindicus*; postvulvar uterine sac present, minute to large. Vagina well developed, length about one-half body width, except in *A. westindicus* where weakly developed. Vaginal sclerotization small to conspicuous. Lateral advulvar body pores absent. Vulva a transverse slit. Males usually without ventromedian cervical papillae, or one papilla near secretory-excretory pore. One pair of lateral cervical pores near nerve ring. Spicules long, slender; with or without spicule capitular extension. Caudal alae present, minute to well developed. Oblique copulatory muscles restricted to caudal alae region or just anterior to it. Capsule of spicule suspensor muscles moderately to poorly developed. Three to four medioventral precloacal papillae. One pair of subventral postcloacal papillae.

Type species: *Allotrichodorus campanullatus* Rodriguez-M, Sher and Siddiqi, 1978

List of Allotrichodorus *species*

A. brasiliensis Rashid, De Waele & Coomans, 1986
A. campanullatus Rodriguez-M, Sher & Siddiqi, 1978
A. guttatus Rodriguez-M, Sher & Siddiqi, 1978
A. longispiculis Rashid, De Waele & Coomans, 1986
A. loofi Rashid, De Waele & Coomans, 1986
A. sharmae Rashid, De Waele & Coomans, 1986
A. westindicus (Rodriguez-M, Sher & Siddiqi, 1978) Rashid, De Waele & Coomans, 1986: syn. *P. (N.) westindicus* Rodriguez-M, Sher & Siddiqi, 1978; syn. *N. westindicus* Rodriguez-M, Sher & Siddiqi, 1978.

Discussion on the taxonomic position of A. westindicus

A. westindicus is considered a species *incertae sedis* by Hunt (1993). According to this author, apart from a single functional genital tract, *A. westindicus* shares little else with the other nominal members of *Allotrichodorus*: the secretory-excretory pore is located posterior to the pharyngeal base as opposed to being near the nerve ring; the vulva is considerably more anterior (60-66% compared to 81-89%); the vaginal musculature is small and weakly developed *vs* well developed in *Allotrichodorus* species; the female tail is conoid, with rounded to subdigitate end, relatively long and with the rectum at an oblique angle to the body axis (*Allotrichodorus* has the tail virtually nonexistent and the rectum parallel to the body axis as it leads to the subterminal anus); there are no lateral or caudal pores (present in *Allotrichodorus*); short body length (smaller than 500 μm as opposed to longer than 500 μm); males are lacking in all three described populations as opposed to present in the six other species of *Allotrichodorus* and the distribution is largely allopatric (one population described from Brazil, the other two from Martinique and Trinidad). Hunt (1993) considered the species to be closer to *Paratrichodorus* in general respects, but for the time being tentatively placed near *Allotrichodorus* because of monodelphy, but nevertheless as a *species incertae sedis*.

A. westindicus is in several aspects "apart" from the other species of the genus, and takes a somewhat intermediate position between the genera *Allotrichodorus* and *Paratrichodorus*. However, some of the characters treated in earlier discussions on the taxonomic status of the species are not relevant at genus level. For example: the body length (varies intra- and inter-specifically in the other genera of the family), the tail shape (similar exceptions are found within the genus *Paratrichodorus*), the more allopatric distribution (but restricted to South America as are the other monodelphic species), to a lesser extent the position of the secretory-excretory pore (similar as in three species of *Paratrichodorus*, but showing variation as in some specimens of e.g. *P. anthurii* where the position of the S-E pore shows a wide intraspecific variability), the absence of caudal pores (exceptions are also found in *Paratrichodorus*). Finally the absence of lateral body pores is a feature which is similar to *Allotrichodorus* and not different from it as mentioned in Hunt (1993).

3.3. Discussion and appraisal of differentiating characters at genus level

Although trichodorids are a small group, their identification to genus and to species level can pose some problems. Therefore, a study of the morphometric variability and value of the characters used for species differentiations is very important.

A list of all the diagnostic characters which have been used for defining genera in the family Trichodoridae is presented here (Table 3.1.). This also includes the differential features of *Atlantadorus* and *Nanidorus*, two genera erected by Siddiqi (1980a), considered at subgenus level by Ahmad (1989) and Jairajpuri & Ahmad (1992) but not accepted by several authors who have revised the family Trichodoridae (Decraemer, 1980a; Decraemer & De Waele, 1981; Sturhan, 1985; Hunt, 1993).

3.3.1. PRIMARY IMPORTANT DIAGNOSTIC FEATURES (with * in Table 3.1.)

The most important diagnostic characters at generic level are:

Female:

- **number of genital branches and relative position of the vulva**: differentiation of didelphic and monodelphic genera,
- **length of vagina as a proportion of corresponding body diameter**: differentiation of the didelphic genera; vagina much shorter than half the corresponding body width in *Paratrichodorus*, usually longer i.e. about half the body width in *Trichodorus*,
- **shape and dimensions of vaginal sclerotizations**: the shape of the sclerotized ring at the junction vulva/vagina in lateral optical section is a primary diagnostic feature at species level. The degree of development may have, to some extent, taxonomic value at generic level. In *Trichodorus*, the vaginal sclerotizations are usually well developed and clearly differentiated in size and shape, although some species possess only small to minute vaginal sclerotizations (Figs 3.1; 3.2). In *Paratrichodorus*, the vaginal sclerotizations are small to inconspicuous (Figs 3.12; 3.13). In *Monotrichodorus* (Fig. 3.18) and *Allotrichodorus* (Fig. 3.20), the vaginal sclerotizations are usually well developed, varying from small to large, and conspicuous.
- **presence of lateral advulvar body pores**: this feature is one of the few differential characters between the monodelphic genera: *Monotrichodorus* females possess a pair of lateral advulvar body pores (Fig 1.1.F), in *Allotrichodorus* females they are absent. In all *Trichodorus* species, except for *T. beirensis*, *T. eburneus* and *T. cedarus*, female specimens possess a pair of postadvulvar lateral body pores. In *Paratrichodorus* the lateral body pores are rarely located postadvulvar (i.e. within one body diameter from the vulva).

Male

- **presence of caudal alae (=bursa)**: absent in *Trichodorus* (except for *T. cylindricus*, *T. paracedarus*) and in *Monotrichodorus*, always present in *Paratrichodorus* and *Allotrichodorus*.

Remark: Caudal alae are present in all species of *Paratrichodorus* but are usually narrow and rather weak (Figs 3.15- 3.16). They are present in *Allotrichodorus* species in various degrees of development (Fig. 3.20); exceptionally, they appear rudimentary.

Table 3.1. Review of all diagnostic characters of the genera in the family Trichodoridae from the literature; the taxonomically most important features are indicated with *.

GENERAL MORPHOLOGY

Body

- habitus
- body length
- swelling of cuticle upon fixation *
- caudal pores:presence/number/position/development
- presence of incisures in lateral field

Tail

- shape
- length

Secretory-excretory pore

- Position in relation to pharyngeal bulb

Digestive system

- length onchiostyle
- ventral pharyngeal overlap:presence/length
dorsal intestinal overlap:presence/length
presence offset pharyngeal bulb
- dorsal pharyngeal nucleus:position in relation to pharyngeal bulb/position in relation to ventrosublateral gland nuclei/size
- ventrosublateral pharyngeal gland nuclei:position in relation to pharynx - intestine junction/position in relation to dorsal pharyngeal gland nucleus/size
- presence of an inner onchium in adults

CHARACTERS OF FEMALES

Reproductive system

- number of genital branches *
- vulva:ratio V*/ shape in ventral view
- Vagina:length as a proportion of body diameter*/ orientation/shape and dimension of vaginal sclerotization*/ development of vaginal constrictor muscles
- presence of spermatheca
- sperm cells:shape/ position
- postvulvar uterine sac: length
- lateral advulvar body pores: presence */number
- lateral body pores:presence/number
- position of anus
- position of lateral cervical pores

CHARACTERS OF MALES

- males:presence / abundance
- spicules: shape/length/ornamentation/ presence of capitular extension
- development of capsule of suspensor muscles*
- gubernaculum: development
- precloacal supplements/ number/ number in region of retracted spicules or bursa region/position of precloacal supplements /distance between SP2-SP3/ position of anteriormost supplement in relation to retracted spicules/distance of SP2 from cloacal opening/development of anteriormost supplement
- copulatory muscles:extension / development
- ventromedian cervical papillae: presence/ number/ position in relation to S-E pore/ position in relation to onchiostyle
- lateral cervical pores:presence/position in relation to nerve ring/position in relation to S-E pore
- presence of caudal alae (=bursa) *
- postcloacal papillae:presence/development/position in relation to cloacal opening

Similar observations were made for *Monotrichodorus sacchari* (Decraemer, 1986) and some *Trichodorus*- species e.g. *T. sparsus* (Loof, 1973), *T. elefjohnsoni*, *T. nanjingensis* and *T. vandenbergae* with protruding ventrosublateral cuticle in the cloacal region.

- **degree of development of copulatory muscles and capsule of spicule suspensor muscles**: both well developed and copulatory muscles extending far anteriorly in *Trichodorus* (Fig. 1.1.B); both poorly developed and copulatory muscles restricted to bursa region in *Paratrichodorus* (Fig. 2.3.F); copulatory muscles mediumly developed in the monodelphic genera (Fig. 1.1.E, J);

In both sexes:

- **the reaction of the body cuticle upon fixation or heat killing**: mainly used to distinguish *Trichodorus/Monotrichodorus* (non-swollen) from *Paratrichodorus /Allotrichodorus* (clearly swollen).

Remark: So far, the degree of swelling of the body cuticle as a reaction to fixation or heat killing is included in the diagnoses of the genera of the Trichodoridae. The genus *Paratrichodorus* (Fig. 1.1.G, H, I) with the body cuticle usually much swollen (in various degrees), the genera *Trichodorus* (Fig. 1.1.A, B) and *Monotrichodorus* (Fig. 1.1.D, E, F) with the cuticle generally not swollen (except in badly fixed specimens) and the genus *Allotrichodorus* (Fig. 1.1.J, K) with the body cuticle varying from not swollen, to moderately swollen or distinctly swollen upon fixation.

3.3.2. SECONDARY DIAGNOSTIC CHARACTERS

Some of the characters that have been used in the diagnoses of the genera are not suitable as diagnostic features due to the large variability displayed, for example: the position of the pharyngeal gland nuclei, body length, position of the secretory-excretory pore (Decraemer & De Waele, 1981; Decraemer, 1989a, b). Other variable features are useful but have a more limited taxonomic value at generic level:

Habitus

This feature can to some extent help to differentiate males of the genera *Trichodorus* (Fig. 2.3 B)/ *Monotrichodorus* (posterior body region clearly curved ventrally) from those of *Paratrichodorus* (straight body, Fig 2.3.F)/ *Allotrichodorus* (usually straight body). However, the habitus is susceptible to fixation and typical straight females and straight males of *Paratrichodorus* and *Allotrichodorus* may also be observed as arcuate specimens or specimens with a slightly ventrally curved posterior body region, especially in the latter genus.

The presence or absence of anterio-dorsal intestinal or posterio-ventral/subventral pharyngeal gland overlap

Although the pharynx-intestine junction in *Trichodorus* may vary inter- and intra-specifically (flat, with ventral pharyngeal overlap, or with anterio-dorsal intestinal overlap or rarely with both pharyngeal and intestinal overlap), the majority of the species have an offset (=flat) pharyngeal bulb. In *Paratrichodorus*, all types of overlapping may be observed, but the ventral pharyngeal overlap occurs most frequently and is also more pronounced (Fig. 2.1.B-D). Currently, in *Allotrichodorus* an offset pharyngeal bulb or a short anterio-dorsal intestinal overlap is present, while in *Monotrichodorus* an offset bulb or, at most, a negligible ventral pharyngeal overlap has been observed.

Presence of spermatheca(e)

In *Trichodorus* (Fig. 1.1.C), except *T. nanjingensis* , *Monotrichodorus* (Fig. 1.1.F) and *Allotrichodorus*, except *A. westindicus*, all species possess a spermatheca; in *Paratrichodorus*, spermathecae or sperm concentration near the oviduct have been observed in nine species (Fig. 2.4.D), while in other species sperm is dispersed throughout the uterus (Fig. 2.4. E). Therefore, this character is of restricted value for differentiating the genus *Paratrichodorus* from the other trichodorid genera.

Shape of sperm cells

Trichodorus (Fig. 3.11.) and *Monotrichodorus* species have large sperm cells, usually with sausage-shaped nucleus. In *Paratrichodorus* (Fig. 3.17.), most species possess small sperm cells, although, some species have sperm as large as in the former genera. In *Allotrichodorus*, most species have well developed sperm cells with more or less globular nucleus. Therefore, this character is of restricted value but still helpful for differentiating several species of *Paratrichodorus* from the other trichodorid genera.

Spicule shape and ornamentation

The spicule shape is an important diagnostic character for species differentiation. At generic level, its value is more restricted. It may be used to distinguish males of *Paratrichodorus* with mainly straight spicules and with transverse striae (Fig. 3.14.) from *Trichodorus* males displaying a more pronounced differentiation in spicule shape together with a more diverse ornamentation (transverse striae, bristles, ventral flange, smooth) (Figs 3.3.-3.4.), from *Monotrichodorus* males having long slender arcuate spicules with transverse striae and with or without bristles (Fig. 3.16.Q-S) and from *Allotrichodorus* males with long slender spicules, less arcuate than in *Monotrichodorus,* and with transverse striae, rarely with bristles (Fig. 3.20.). The presence of a capitular extension of the spicules is restricted to two species of *Allotrichodorus.*

Number of precloacal supplements (SP)

This character is not a useful taxonomic feature at generic level. It can be used only to distinguish *Paratrichodorus* from the other genera, by those species with 1 (2) precloacal supplement(s). *Trichodorus* species mainly have 3 SP (exceptionally 2, 4 or 5) (Fig. 3.5.- 3.7.); *Paratrichodorus* species have 1 to 4 SP but usually 3 (exceptionally 4) (Fig. 3.15.- 3.16.); *Monotrichodorus* have 3 SP and *Allotrichodorus* usually have 3, exceptionally 4.

Position of precloacal supplements in relation to the retracted spicules

This character is susceptible to fixation and is of very restricted value at generic level. *Trichodorus* males usually have 1 SP, *Paratrichodorus* and *Monotrichodorus* males have 1-2 SP and *Allotrichodorus* males have 3 SP within the 'spicular' region.

Vulva shape (ventral view)

This feature was used as a diagnostic feature for the characterization of the (sub)genera of *Paratrichodorus* (Siddiqi, 1974, 1980a; Ahmad, 1989; Jairajpuri & Ahmad, 1992) but is of no taxonomic value at the genus level. Presently, only the genus *Allotrichodorus* is characterized by having one type of vulva shape (a transverse slit), whereas the other genera have 2 or 3 different types of vulva shape (a pore, a transverse slit or a longitudinal slit).

Number of ventromedian cervical papillae (CP)

In *Trichodorus* species the number of CP varies between 0 and 4, with an approximately equal distribution between the species of 1CP, 2CP and 3CP; 0 and 4CP are rare. The majority of *Paratrichodorus* species have 1CP, *Monotrichodorus* species have 1CP and *Allotrichodorus* species have either 1 or 0 CP. The number of ventromedian cervical papillae is of restricted taxonomic importance at generic level, permitting the differentiation of only the genus *Trichodorus* from the other genera when more than 1CP is present (exc. *P. faisalabadensis*, *P. mirzai* and *P. psidii* with 2CP).

Presence of males/mode of reproduction

In *Trichodorus*, *Monotrichodorus* and *Allotrichodorus* (except *A. westindicus*) all species are bisexual. In *Paratrichodorus* most species are bisexual. Three species (*P. acutus*, *P. renifer* and *P. anthurii*) with males unknown and five species (*P. allius*, *P. minor*, *P. nanus*, *P.porosus* and *P. teres*) with males usually rare and sperm degenerated, were reported to reproduce parthenogenetically, except *P. allius*, *P. nanus* and *P. anthurii* which would reproduce by syngonic hermaphroditism and *P. porosus* with well developed sperm and reproduction mode not determined.

3.5. General remark

Although the genera *Trichodorus* and *Paratrichodorus* are usually well characterized, their differentiation may appear rather confusing in some species , either in male or in female. A few species of the genus *Trichodorus* have some of their characters in one of the sexes similar to those of the genus *Paratrichodorus* which makes their classification less distinct.

3.5.1. MALES

T. cylindricus males have a bursa, little developed copulatory muscles and capsule of suspensor muscles and a habitus with a straight posterior body region as in *Paratrichodorus*. *T. cylindricus* was classified within the genus *Trichodorus* because of the non-swollen body cuticle in well fixed specimens, the structure of the spicules (shape and ornamentation typical of *Trichodorus*) and females with the diagnostic characters typical of *Trichodorus*.

T. paracedarus males have a small bursa, straight spicules and a habitus with a straight posterior body region typical for *Paratrichodorus*. They were identified as *Trichodorus* based on the ornamentation of the spicules (with bristles as in many *Trichodorus* species, absent in *Paratrichodorus*) and the combination of the following characters: males with three ventromedian cervical papillae, an offset pharyngeal bulb and thickened terminal tail cuticle, and females with spermathecae, a long vagina and well developed vulvar constrictor muscles, distinct, small vaginal sclerotized pieces in lateral view and one pair of lateral postadvulvar body pores.

A few *Trichodorus* species (*T. elefjohnsoni*, *T. nanjingensis*, *T. orientalis*, *T. parorientalis*, *T. petrusalberti*, *T. sparsus* and *T. vandenbergae*) may have the subventral body wall slightly swollen, indicating a rudimentary bursa. The other characters in males and females are typical of the genus *Trichodorus*, although the vaginal sclerotizations are small, they are usually well marked.

3.5.2. FEMALES

T. nanjingensis females have (1) no spermathecae and sperm dispersed throughout the uteri and (2) a vagina which is shorter than half the corresponding body diameter, two features which are characteristic for many *Paratrichodorus* species. *T. nanjingensis* is classified within the genus *Trichodorus* because of the non-swollen body cuticle, well developed vulvar constrictor muscles and vaginal sclerotizations and the presence of one pair of postadvulvar lateral body pores in the females and males with the diagnostic characters typical of *Trichodorus* species.

A number of *Trichodorus* species have small to minute vaginal sclerotizations and a few species have additionally a short vagina (about 1/3 rd of the corresponding body diameter e.g. *T. elefjohnsoni*). Both features are typical for *Paratrichodorus* species. However, these species are assigned to *Trichodorus* when considering a combination of other characters as e.g. sperm shape, presence of spermathecae, an offset pharyngeal bulb, lateral postadvulvar body pores and typical *Trichodorus* males.

Table 3.2. Differentiating characters between the genera of the Trichodoridae: ♂ characters

♂ characters	*Trichodorus*	*Paratrichodorus*	*Monotrichodorus*	*Allotrichodorus*
caudal alae	absent (2 exc.)	present	absent	present
posterior body region	curved ventrad	straight	curved ventrad	straight or curved ventrad
capsule of spicular suspensor muscles	well developed	poorly developed	moderately developed	moderately developed
copulatory muscles	extending far anterior to retracted spicules	restricted to region of retracted spicules	extending to just anterior to retracted spicules	restricted to bursa region or to just anterior to it
number of SP	3 (exc. 2, 4 or 5)	1 to 3 (exc. 4)	3	3 to 4
number of SP in region retracted spicules	0, 1 or 2 (mainly 1)	1 to 2 (mainly 2)	1 or 2	3 or 2
shape of spicules	diverse, usually bent	similar, usually straight	similar long, slender, bent	diverse, slender, bent
ornamentation of spicules	with or without striae, bristles, velum	smooth or with striae	with striae, with or without bristles	with or without striae, bristles, capitular extension
number of CP	1 to 3 (exc. 0 or 4)	0 or 1 (exc. 2)	1	0 or 1
n° lateral cervical pores	1 pair (exc. absent)	1 pair or absent	1 pair	1 pair

Table 3.2. Differentiating characters between the genera: continued with female characters

♀ characters	*Trichodorus*	*Paratrichodorus*	*Monotrichodorus*	*Allotrichodorus*
genital system	didelphic	didelphic	monodelphic	monodelphic
spermatheca(e)	present (1 exc.)	absent or present	present	present (1 exc.)
length vagina	about half the body width (few exc.)	much shorter than half the body width	more than one half body width	one body width (1 exc.)
vaginal sclerotizations	distinct, small to large	little developed, small to minute	distinct, small	distinct, small to large
vaginal constrictor muscles	well developed	poorly developed	well developed	well developed
shape vulva	pore, transverse or longitudinal slit	pore, transverse or longitudinal slit	transverse or longitudinal slit	transverse slit
lateral advulvar body pores	present (exc. absent)	rarely present	one pair	absent
lateral body pores	1 to 2 pairs (exc. absent)	absent or present	absent	absent
n° caudal pores	1 pair	1 pair (exc. absent)	1 pair	1 pair (1 exc.)

Table 3.2. Differentiating characters between the genera: continued with male and female characters

genera	swelling body cuticle upon fixation	sperm	pharyngo-intestinal junction
Trichodorus	not swollen	never minute; large sausage-shaped or rounded nucleus	generally with offset bulb
Paratrichodorus	swollen	sperm cells minute to large, or thread-like,nucleus large sausage-shaped, small oval	usually with pharyngeal or/and intestinal overlaps
Monotrichodorus	not swollen	large cells, usually rounded nucleus	offset bulb (exc. minute ventral overlap)
Allotrichodorus	variable swelling or not swollen	large cells, rounded nucleus	offset bulb or short dorsal intestinal overlap

3.4. Key to the genera of the Trichodoridae

Females

1. - Genital system monodelphic, prodelphic 2
   - Genital system didelphic, amphidelphic 3
2. - 1 pair of lateral advulvar body pores, cuticle non-swollen *Monotrichodorus* (Fig. 1.1.F)
   - No lateral advulvar body pores, cuticle non-swollen to strongly swollen *Allotrichodorus* (Fig. 1.1.K)
3. - Body cuticle non-swollen, vagina usually up to half a body width long, vaginal sclerotizations in lateral view well developed, small to large; pharyngeal bulb usually offset *Trichodorus* (Fig. 1.1.C)

- Body cuticle swollen, vagina always much shorter than half a body width long, vaginal sclerotizations very small; pharyngo-intestinal junction usually with overlaps *Paratrichodorus* (Fig 1.1.I)

Males

1. - Bursa present, body cuticle usually well swollen, copulatory muscles restricted to spicule region 2
   - Bursa absent (few exceptions), cuticle non-swollen, copulatory muscles extending shortly to far beyond region of retracted spicules 3
2. - Body cuticle pronounced swollen, spicules short to long, mainly straight, copulatory muscles and capsule spicular suspensor muscles poorly developed, sperm generally small; pharynx mostly with ventral overlap of its glands *Paratrichodorus* (Figs 1.1.G, H)
   - Swelling of body cuticle variable; spicules long slender, with or without capitular extension, copulatory muscles and capsule spicular suspensor muscles moderately developed, sperm large, well developed; pharynx mostly with small dorsal intestinal overlap *Allotrichodorus* (Fig. 1.1.J)
3. - Spicules mostly curved, rarely elongated, diverse in shape; copulatory muscles extending far anteriorly beyond spicule region; 1- 3 (exc. 0, 4) ventral cervical papillae (CP); usually 1 precloacal supplement (SP) in region of retracted spicule *Trichodorus* (Figs 1.1.A, B)
   - Spicules slender, curved; copulatory muscles extending to just anterior to the retracted spicules; 1 CP; 1-2 SP in region of retracted spicules *Monotrichodorus* (Figs 1.1.D, E)

3.6. Discussion of differential characters at species level

For species diagnoses/differential diagnosis/relationships (different headings treating the characterization of a new species) up to 48 characters have been used for *Paratrichodorus* species, 42 for *Trichodorus* species, 21 for *Monotrichodorus* species and 10 for *Allotrichodorus* species. Half of the most frequently used diagnostic characters (i.e characters appearing in descriptions of 20% or more species per genus) are common to both *Trichodorus* and *Paratrichodorus*, with several having restricted taxonomic value. A quarter of these characters are also present in the monodelphic genera. The data for the genus *Monotrichodorus* are those for seven nominal species i.e. before synonymization (see species description).

List of diagnostic features used in about 20% of the species of the genera of the Trichodoridae with the percentage occurrence per genus (adapted from Decraemer, 1988)

Characters shared by *Trichodorus* and *Paratrichodorus* (also by *Monotrichodorus* and *Allotrichodorus*)

- body length: 29.5% / 39.5% (57% /-)
- onchiostyle length: 50.5% / 65% (57% / 57%)
- position of S-E pore: 25% / 29% (-/ -)
- shape of vaginal sclerotizations: 71.5% / 47% (-/ 85.7%)
- spicules: - length: 33.5% / 47% (42% / 57%)
    - shape: 73.5% / 32.5% (-/ 57%)
    - ornamentation: 21% / - (-/-)
- precloacal supplements:- number: 33.5% / 43% (-/-)
    - position: 73.5% / 47% (57% /-)
- ventromedian cervical papillae: - number: 73.5% / 32.5% (-/ 85.7%)
    - position: 46/- (57% /-)

Trichodorus

- shape of vagina: 25%
- number and position of lateral body pores in female: 61%
- position of precloacal papillae in relation to retracted spicules: 61%
- ventromedian cervical papillae:
    - position in relation to onchiostyle: 25%
    - position in relation to S-E pore: 46%

Paratrichodorus

- ventral overlap of pharyngeal glands: 43%
- shape of vulva: 43%
- number of postcloacal papillae: 21.5%

Monotrichodorus

- ratio V: 57%
- shape of vulva: 57%
- length of postvulvar uterine sac: 42%
- presence of lateral cervical pores in female: 28%
- swelling of the body cuticle: 42%
- position precloacal supplements in relation to retracted spicules: 57%

Allotrichodorus

- number of ovaries: 28.6%
- presence of capitular extension of spicules: 57%

3.6.1. PRIMARY DIAGNOSTIC FEATURES

The most important diagnostic features used at the species level for all genera are:

- in females: the shape of the vaginal sclerotizations;
- in males: the shape of the spicules.

Shape of the vaginal sclerotization

This is the most frequently used character for distinguishing females of *Trichodorus* and *Allotrichodorus* species, and the second most important diagnostic feature in *Paratrichodorus*, but so far not important in *Monotrichodorus* (see species description). In *Trichodorus* (Figs 3.1., 3.2.) the vaginal sclerotizations are usually well developed and species differentiation is based on the following features: size, shape (round, rectangular, triangular, quadrangular, oval), orientation (oblique, parallel to the longitudinal body axis or to the vaginal lumen), dimension of the inner diameter of the sclerotized ring (small aperture: pieces close to one another, 1 µm apart; aperture about 2 µm or wider: pieces well separated). About 9 groups of species can be distinguished: (1) vaginal sclerotized pieces in lateral view large, roughly triangular, inner side concave, tips bent toward vagina (*T. californicus*, *T. dilatatus*, *T. intermedius*, *T. elegans*), (2) sclerotized pieces parallel to the longitudinal body axis, large rectangular or small (*T. cottieri*, *T. obscurus* and *T. complexus*), (3) medium-sized, triangular pieces with tips toward vulva (*T. similis*, *T. cylindricus*, *T. yokooi*, *T. velatus*: partim), (4) sclerotized pieces triangular to rounded triangular, rarely oval, large or medium-sized (*T. beirensis*, *T. lusitanicus*), (5) large about quadrangular pieces (*T. variopapillatus*, *T. azorensis*, *T. aquitanensis*), (6) rounded sclerotized pieces (large: *T. taylori*, *T. persicus*, *T. minzi*; medium-sized: *T. hooperi*, *T. paucisetosus*; small, dot-like: *T. elefjohnsoni*, *T. pakistanensis*, *T. philipi*, *T. sanniae*, *T. tricaulatus*), (7) small to minute triangular, rounded triangular to oval sclerotized pieces, with oblique orientation (*T. aequalis*, *T. borai*, *T. borneoensis*, *T. cedarus*, *T. coomansi*, *T. eburneus*, *T. giennensis*, *T. kilianae*, *T. magnus*, *T. nanjingensis*, *T. obtusus*, *T. orientalis*, *T. paracedarus*, *T. parorientalis*, *T. petrusalberti*, *T. rinae*, *T. sparsus*, *T. vandenbergae*), (8) elongated oval pieces, parallel to the vaginal lumen (*T. primitivus*, *T. viruliferus*: may be slightly oblique), (9) small trapezoidal to rectangular sclerotized pieces, oblique (*T. carlingi*, *T. velatus*: partim). Some minor intraspecific variability possible.

Paratrichodorus (Figs 3.12., 3.13.) shows less differentiation in the shape and dimension of the vaginal sclerotizations than the other genera. About seven groups of comparable vaginal sclerotizations can be distinguished, mainly based upon shape: (1) small quadrangular pieces, well separated (*P. acaudatus*), (2) small reniform (*P. renifer*), (3) narrow triangular pieces, well separated and oblique (*P. anemones*, *P. pachydermus*, (4) minute triangular, close (*P. anthurii*, *P. sacchari*), (5) rounded triangular pieces, close (*P. grandis*, *P. hispanus*, *P. nanus*), (6) oval or rod-shaped pieces (parallel to the vaginal lumen: *P. tunisiensis*, *P. weischeri*; parallel to the longitudinal body axis: *P. minor*, *P. porosus* partim; *P. rhodesiensis* partim; oblique: *P. allius*, *P. faisalabadensis*, *P. macrostylus*, *P. orrae*, *P. porosus* partim, *P. queenslandensis*, *P. rhodesiensis* partim), (7) small oval to dot-like pieces (close: *P. acutus*, *P. psidii*, separated: *P. catharinae*, *P. lobatus*, *P. meyeri*, *P. mirzai*, *P. paramirzai*, *P. paraporosus*, *P. teres*). The description of small vaginal sclerotized pieces as small or minute, rounded triangular, oval or dot-like often represents intraspecific variability and depends on the subjective interpretation of the author.

In the monodelphic genera (Figs 3.18., 3.19.), vaginal sclerotizations also differ interspecifically in size and shape, especially in *Allotrichodorus*; so far, *Monotrichodorus* species had essentially similar vaginal sclerotizations (see species description). Some minor intraspecific variability should be taken into account (e.g. in *Allotrichodorus*).

Spicule shape

This feature is an important diagnostic character for species differentiation in *Trichodorus* and *Allotrichodorus*.

In *Trichodorus* (Figs 3.3., 3.4.), about nine groups with similar spicule shape can be recognized: (1-3): spicules with a mid-shaft constriction, respectively with bristles: Fig. 3.3 (1a-f), without bristles: Fig. 3.3 (3a-d), and with a septum in front of the indentation: Fig. 3.3 (2a-b), with or without bristles; (4) spicules proximally curved ventrally, mid-shaft straight: Fig. 3.3 (4a-h); (5) spicules with an enlarged posterior shaft: Fig. 3.3 (5a-f); (6, 7) spicules with a wide, elongated manubrium, respectively offset: Fig. 3.3 (6a-c) or not offset: Fig. 3.3 (7a-e); (8) spicules with a small more or less knob-like manubrium, offset from the shaft: Fig. 3.3 (8a-g) and (9) spicules usually slightly to more ventrally curved, shaft about equally wide : Fig. 3.3 (9a-m').

In *Allotrichodorus* (Figs 3.19., 3.20.), this character can be used to separate the species into two groups: species with males possessing spicules with a capitular extension (Figs 3.19.J, 3.20.A) and those without capitular extension. In both other genera, the spicular shape is less important taxonomically since *Paratrichodorus* species (Fig. 3.14) have mainly straight spicules and show less variation and *Monotrichodorus* species have spicules with more or less the same shape (Fig. 3.16.O-S).

Comments on a possible correlation between spicule structure and vaginal sclerotization in the genus Trichodorus (Figs 3.8.-3.10.).

In *Trichodorus* species, males with slender spicules usually correspond to females with small to minute vaginal sclerotizations, but there are exceptions as for example in *T. pakistanensis*, Fig. 3.8 (4a). Furthermore, slender spicules do not obviously correspond with closely located sclerotized pieces in longitudinal optical section as for example in *T. primitivus*, Fig. 3.9 (7c) and *T. hooperi*, Fig. 3.9 (7e, 8e, e'). However, in the former species, this may be compensated by the invagination of a rather thick body cuticle and/or presence of well developed rod-like pieces parallel to the vagina.

Large, well developed vaginal sclerotizations in females usually correspond with stout spicules in males, but again there are exceptions e.g. in *T. aquitanensis*, Fig. 3.9 (7d), in which the diameter of the sclerotized ring is small and spicules fine.

The grouping of *Trichodorus* species based upon similarities of spicule shape does not completely correspond with the species groups recognized by similarity in shape of the vaginal sclerotizations in lateral view. In particular the grouping of spicules based upon differences in ornamentation (e.g. presence or absence of bristles) or the presence of a mid-shaft constriction, appears rather artificial. Moreover, the distance between the sclerotized pieces in lateral view (as a measure of the inner diameter of the sclerotized ring) does not always correspond to the width of the spicule shaft (e.g. slender spicules

with a narrow vaginal sclerotized ring).
The largest agreement between spicule shape and vaginal sclerotizations in lateral view is present, for example, in the following groups of species: - the Portuguese species (*T. azorensis*, *T. beirensis*, *T. lusitanicus*) and *T. velatus*, - *T. taylori*, *T. minzi* and *T. persicus*.

3.6.2. SECONDARY IMPORTANT TAXONOMIC FEATURES

The secondary important taxonomic characters (i.e. features with some variability which must be taken into account) at species level are: length of onchiostyle, spicule length, position of precloacal supplements and number of ventromedian cervical papillae in males (of restricted use in *Paratrichodorus*).

Onchiostyle length
In *Trichodorus* and *Paratrichodorus*, this character appears relatively constant in populations of a single species (Decraemer, 1988; 1989b). The differences for the mean value of the onchiostyle length within a single *Trichodorus* species is maximally 21.5 µm and 13 µm for *Paratrichodorus* species. The range of the mean value of onchiostyle length within the genus *Trichodorus* is from 40 to 147.5 µm and from 22 to 82.5 µm for *Paratrichodorus*. In *Trichodorus*, however, long-speared and short-speared populations are known to occur in the same species (e.g. in *T. sparsus:* onch.= 48-60 µm but 58-73 µm in two German populations (Loof, 1973) and *T. primitivus*: 40-50 (46) µm in Dutch specimens and 49-55 (52) µm in English specimens (Hooper, 1962). In *Allotrichodorus*, differences were observed between different populations (e.g. in *A. campanullatus*). Within a population of *A. campanullatus*, the maximum difference between females and males for the mean onchiostyle length was 17 µm and 12 µm, respectively. These values are less than reported for *Trichodorus* species. The range of the mean onchiostyle length for *Allotrichodorus* is from 32 µm (*A. westindicus*, females) to 70 µm (*A. guttatus*, males). *Monotrichodorus* species, except *M. monohystera* and *M. vangundyi* (= *M. m. vangundyi*), are only known by their type populations, the mean onchiostyle length within this genus varies from 44 µm (males) (*M. acuparvus* = *M.m. monohystera*)/48.5 µm (females) (*M. monohystera*) to 58.5 µm (males)/59 µm (females) (*M. proporifer* = *M. m. monohystera*).

Spicule length

In both didelphic genera, the spicule length shows little intraspecific variability: maximally 13 µm difference for the mean value of different populations of a single species for *Trichodorus* and 25 µm for *Paratrichodorus*. In *Monotrichodorus*, the mean values of the spicule length vary between 42 µm (*M. acuparvus* = *M. m. monohystera*) and 55 µm (*M. monohystera*). In *Allotrichodorus*, the range of the mean spicule length within the genus is from 49 µm (*A. campanullatus*) to 70 µm (*A. guttatus*). Within a single species, the difference of the mean spicule length is maximally 12 µm (*A. campanullatus*), these values are lower than for the didelphic genera.

Position of precloacal supplements (SP)

The position of SP is given in relation to the caudal alae (*Paratrichodorus*), or the region of the retracted spicules (all genera) and also by the total distance from the cloacal opening in µm, or as a % of spicule length measured from the spicule tip (monodelphic genera). In *Trichodorus*, four groups can be recognized: a) species with one precloacal supplement clearly within the region of the retracted spicules, b) species with the posteriormost supplement near the proximal end of the retracted spicules (taking some variability into account from just anterior to just posterior to the spicule head), c) species with two supplements within the region of the retracted spicules and d) species without precloacal supplements in the region of retracted spicules. *Paratrichodorus* species either have 1 to 4 precloacal supplements; the majority with 2 or 3 supplements can be grouped in : a) species having the two posteriormost supplements opposite the distal half of the spicules and b) species with the posterior two supplements more dispersed (1 - 2 in the region of retracted spicules). As occurs in both didelphic genera, the position of the precloacal supplements in relation to the spicules, shows some variability in some *Allotrichodorus* species, especially the position of SP3 (at the level of the spicule head, just anterior or just posterior to it).

Number of ventromedian cervical papillae (CP)

Monotrichodorus species all have 1 CP in males; in *Paratrichodorus*, the majority of the species also have 1 CP (rare intraspecific variability has been observed only in *P. weischeri* and *P. rhodesiensis*). This feature is of restricted importance in both genera. In *Trichodorus* and *Allotrichodorus*, the number of ventromedian cervical papillae is of more taxonomic importance. In *Trichodorus*, the number of CP varies between zero and four (intraspecific variability rare) and in *Allotrichodorus* this feature separates the species into two groups: species with 1 CP and species without CP, no intraspecific variability has been observed.

3.6.3. ADDITIONAL DIAGNOSTIC CHARACTERS AT SPECIES LEVEL

In *Trichodorus*, spicule ornamentation and shape of vagina together with the arrangement of the vaginal constrictor muscles are of additional help in species identification. Of minor importance are the number and position of lateral body pores in females, and shape and position of the gubernaculum.

In *Paratrichodorus*, structure of sperm cells, presence of spermatheca, shape of vulva are all taxonomically useful features. Tail shape, number of precloacal and postcloacal papillae and presence of male specimens are of minor and/or restricted diagnostic importance.

In *Monotrichodorus*, presence of lateral cervical pores in females, number of precloacal supplements in relation to the retracted spicules, shape of vulva in ventral view and presence of minute caudal alae are all of restricted taxonomic value e.g. characterizing a single species, e.g. lateral cervical pores in females have only been observed in *M. vangundyi* (= *M.m. vangundyi*).

In *Allotrichodorus*, tail shape, body length and presence of caudal pores are of restricted taxonomic value (i.e. differentiating one or a few species from the other species, e.g. absence of caudal pores and body length smaller than 500 µm in *A. westindicus* only).

3.6.4. GENERAL REMARK

A review of Trichodoridae species gives the impression that several species look very similar, and that their taxonomic status can be questioned. Moreover, some species have a more or less similar geographic distribution.

The following examples of the genus *Trichodorus* nevertheless illustrate that most of the closely related species are valid species, differing by features with no or rare intraspecific variability or by a combination of differences of several features, some of them showing intraspecific variability.

1. Typical Portuguese species (*T. lusitanicus*, *T. beirensis*, *T. azorensis*) show few difference in the males (e.g. only in detail of spicule shape and onchiostyle length), but

females differ considerably: differences in shape of vaginal sclerotizations and vulva shape (pore-like in *T. lusitanicus vs* a transverse slit in *T. beirensis* and *T. azorensis*) and in onchiostyle length. *T. azorensis* occurs only in the Azores, while *T. lusitanicus* and *T. beirensis* are recorded from mainland Portugal. However, the latter species have not been found occurring together (Abrantes & Almeida, personal communication). *T. viruliferus*, a species with males comparable to the Portuguese species, has been recorded from Spain and several other European countries but so far not from Portugal. However, *T. viruliferus* females clearly differ from the formerly mentioned Portuguese species by possessing a different type of vaginal sclerotization.

2. In South Africa, eight species of *Trichodorus* were considered endemic (De Waele & Kilian, 1992) and, except for *T. magnus*, the males have a similar spicule shape and the females similar minute vaginal sclerotizations. They differ, however, in more than one character (onchiostyle length, details of spicule shape, shape of vulva, minor differences in vaginal sclerotizations, number of lateral body pores in female and differences in thickening of the terminal cuticle in males) and therefore are considered valid species, although they probably derive from the same ancestral species.

Fig. 3.1. Vaginal region and vaginal sclerotizations in *Trichodorus* (adapted from Decraemer, 1988): A: *T. californicus*; B. *T. dilatatus*; C. *T. intermedius*; D. *T. elegans*; E. *T. cottieri*; F, f. *T. obscurus*; G. *T. complexus*; H. *T. similis*; I, i, i'. *T. cylindricus*; J. *T. yokooi*; K, k, k'. *T. beirensis*; L, l. *T. lusitanicus*; M, m. *T. velatus*; N. *T. variopapillatus*; O. *T. azorensis* (after Almeida *et al.*, 1989); P, p. *T. aquitanensis*; Q, q. *T. taylori*; R. *T. persicus*; S, s. *T. minzi*.

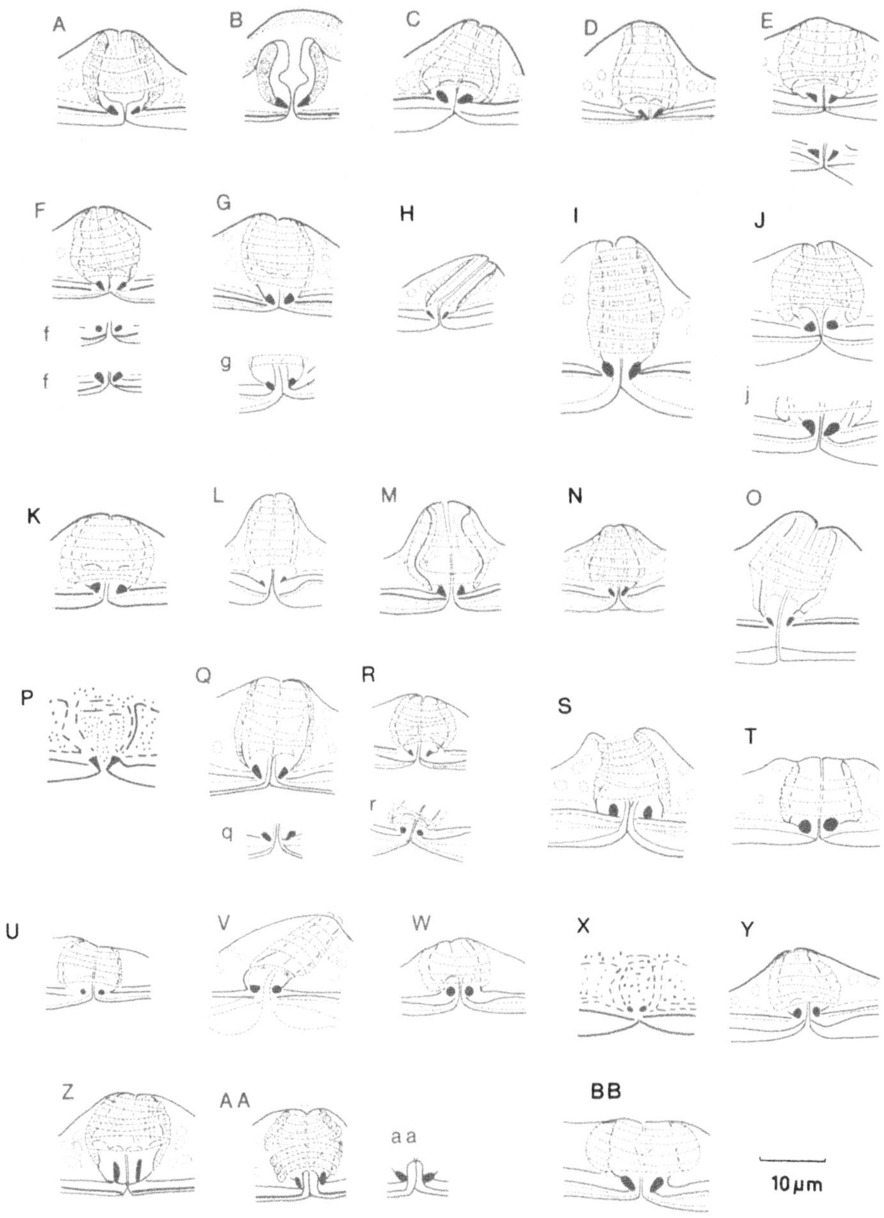

Fig. 3.2. Vaginal region and vaginal sclerotizations in *Trichodorus* (continued): A. *T. aequalis*; B. *T. borai*; C. *T. borneoensis*; D. *T. cedarus*; E, e. *T. coomansi*; F, f, f'. *T. eburneus*; G, g. *T. giennensis*; H. *T. kilianae*; I. *T. magnus*; J, j. *T. nanjingensis*; K. *T. obtusus*; L. *T. orientalis*; M. *T. paracedarus*; N. *T. parorientalis*; O. *T. petrusalberti*; P. *T. rinae*; Q, q. *T. sparsus*; R, r. *T. vandenbergae*; S. *T. hooperi*; T. *T. paucisetosus* (after Bernard, 1992a); U. *T. elefjohnsoni*; V. *T. pakistanensis*; W. *T. philipi*; X. *T. sanniae*; Y. *T. tricaulatus*; Z. *T. primitivus*; AA, aa. *T. viruliferus*; BB. *T. carlingi*.

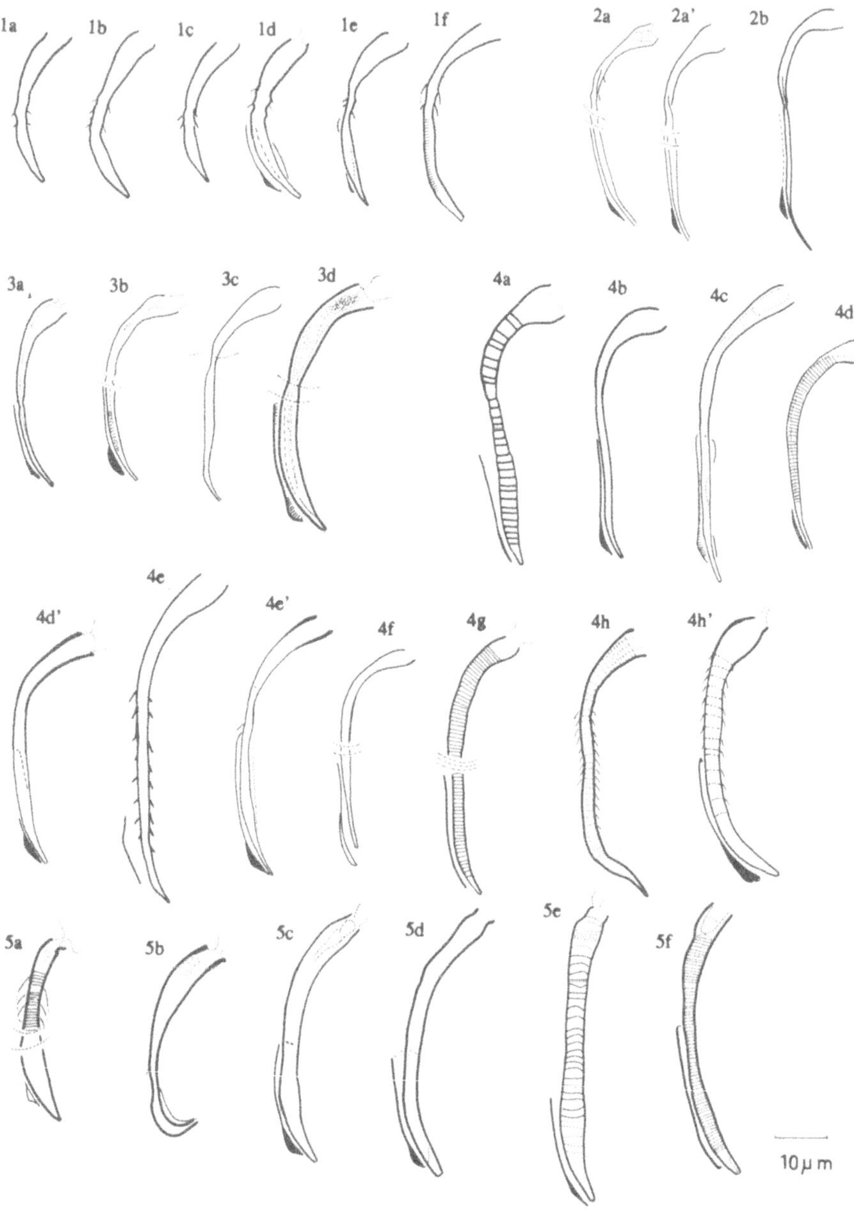

Fig. 3.3. Spicule shape in *Trichodorus* (adapted from Decraemer, 1989a). 1a. *T. beirensis*; 1b. *T. lusitanicus*; 1c. *T. azorensis*; 1d. *T. velatus*; 1e. *T. viruliferus*; 1f. *T. parorientalis*; 2a, a.' *T. kilianae*; 2b. *T. sanniae*; 3a. *T. orientalis*; 3b. *T. parorientalis*; 3c. *T. vandenbergae*; 3d. *T. magnus*; 4a. *T. pakistanensis*; 4b. *T. rinae*; 4c. *T. petrusalberti*; 4d, d'. *T. eburneus* (resp. paratype, from Ivory Coast); 4e, e'. *T. sparsus*; 4f. *T. philipi*; 4g. *T. nanjingensis*; 4h, h'. *T. californicus* respectively from Alaska and California, h after Bernard, 1992a); 5a. *T. cylindricus*; 5b. *T. carlingi*; 5c. *T. cottieri*; 5d. *T. obscurus*; 5e. *T. elegans*; 5f. *T. borneoensis*.

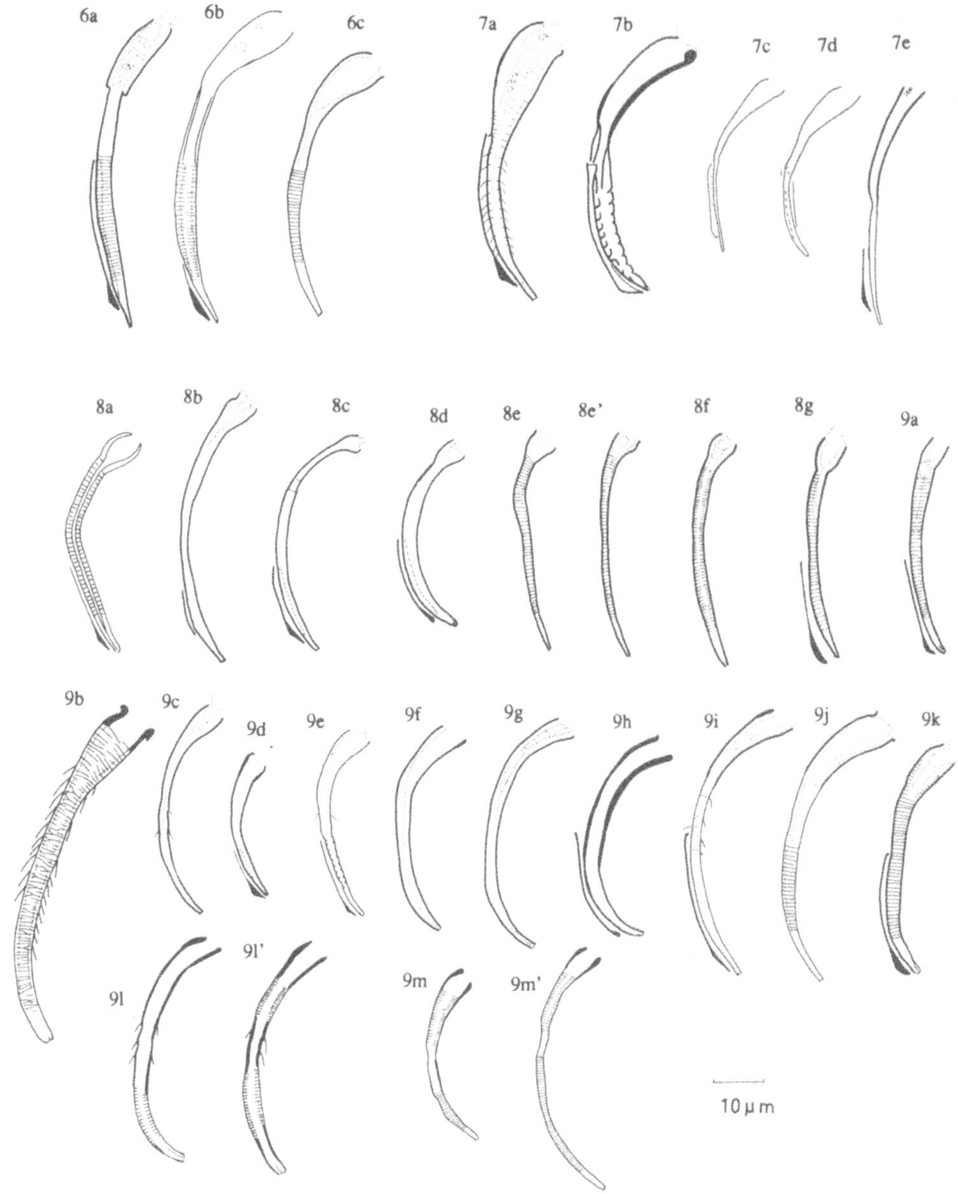

Fig. 3.4. Spicule shape in *Trichodorus* (continued). 6a. *T. taylori*; 6b. *T. minzi*; 6c. *T. persicus*; 7a. *T. dilatatus*, 7b. *T. yokooi*; 7c. *T. primitivus*; 7d. *T. aquitanensis*; 7e. *T. hooperi* (after Loof, 1973); 8a. *T. complexus*; 8b. *T. coomansi*; 8c. *T. variopapillatus*; 8d. *T. similis*; 8e, e'. *T. hooperi*; 8f. *T. cedarus*; 8g. *T. paracedarus*; 9a. *T. cedarus*; 9b. *T. californicus* (after Bernard, 1992a); 9c. *T. giennensis*; 9d. *T. borai* (after Rahman *et al.*, 1985); 9e. *T. tricaulatus*; 9f. *T. aequalis*, 9g. *T. proximus* (= *T. obtusus*), 9h. *T. obtusus* (after Cobb, 1913); 9i. *T. sparsus*; 9j. *T. persicus*; 9k. *T. intermedius*; 9l, l'. *T. paucisetosus* (after Bernard, 1992a); 9m, m'. *T. elefjohnsoni* (after Bernard, 1992b).

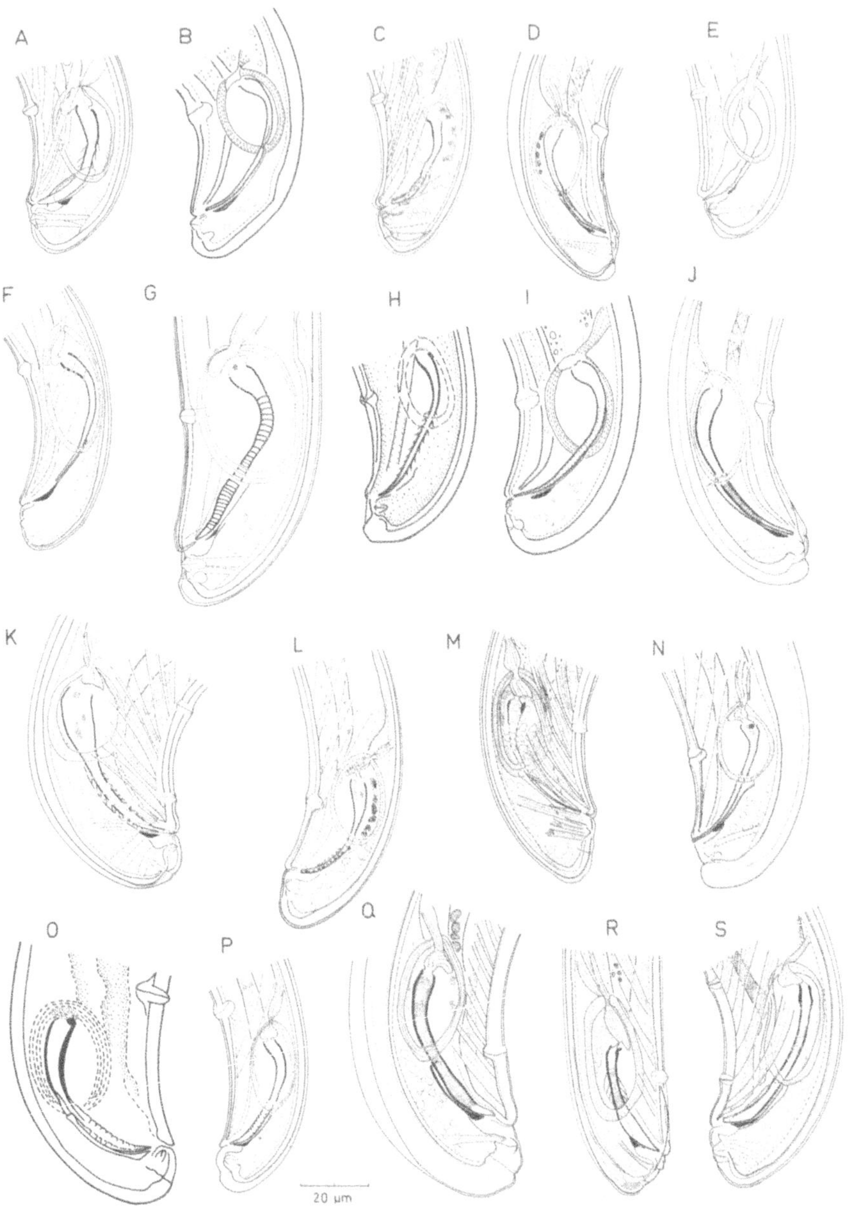

Fig. 3.5. Tail and copulatory apparatus in *Trichodorus* (adapted from Decraemer, 1988). A. *T. velatus*; B. *T. sanniae* (after De Waele, 1988); C. *T. viruliferus*; D. *T. orientalis*; E. *T. lusitanicus*; F. *T. eburneus*; G. *T. pakistanensis*; H. *T. sparsus* (after Loof, 1973); I. *T. rinae* (after De Waele, 1988); J. *T. persicus*; K. *T. dilatatus*; L. *T. aquitanensis*; M. *T. primitivus*; N. *T. hooperi*; O. *T. yokooi*; P. *T. tricaulatus*; Q. *T. cottieri*; R. *T. cylindricus*; S. *T. obscurus* (all figures except A, C, H, M, R, based on type specimens) .

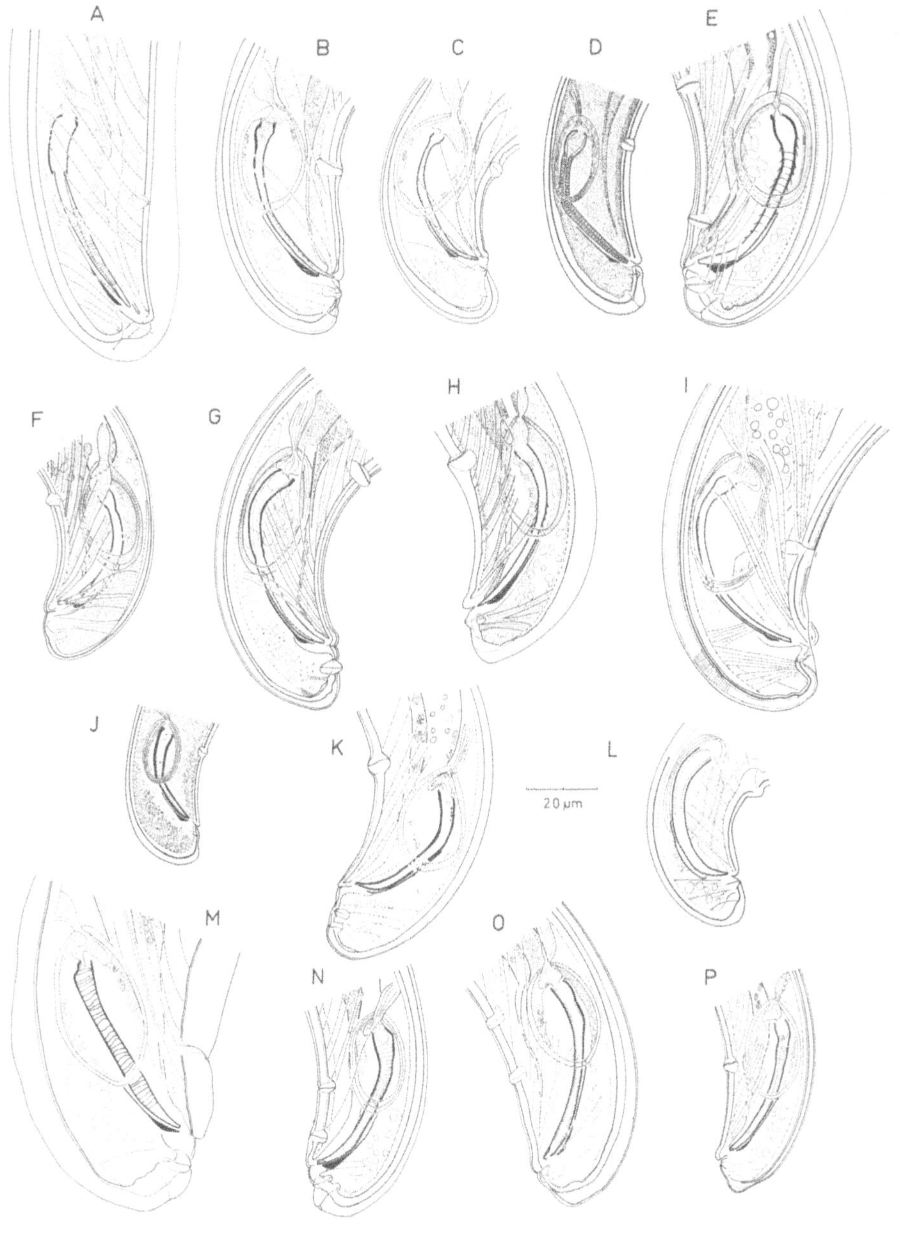

Fig. 3.6. Tail and copulatory apparatus in *Trichodorus* (adapted from Decraemer, 1988). A. *T. taylori*; B. *T. coomansi*; C. *T. variopapillatus*; D. *T. complexus*; E. *T. californicus*; F. *T. similis*; G. *T. proximus* (=*T. obtusus*); H. *T. sparsus*; I. *T. petrusalberti*; J. *T. borai*; K. *T. aequalis*; L. *T. obtusus* (redrawn from Cobb, 1913); M. *T. elegans*; N. *T. intermedius*; O. *T. borneoensis*; P. *T. cedarus* (all figures except C, F, H, K based on type specimens)(Courtesy *Bulletin Koninklijk Belgisch Instituut voor Natuurwetenschappen*).

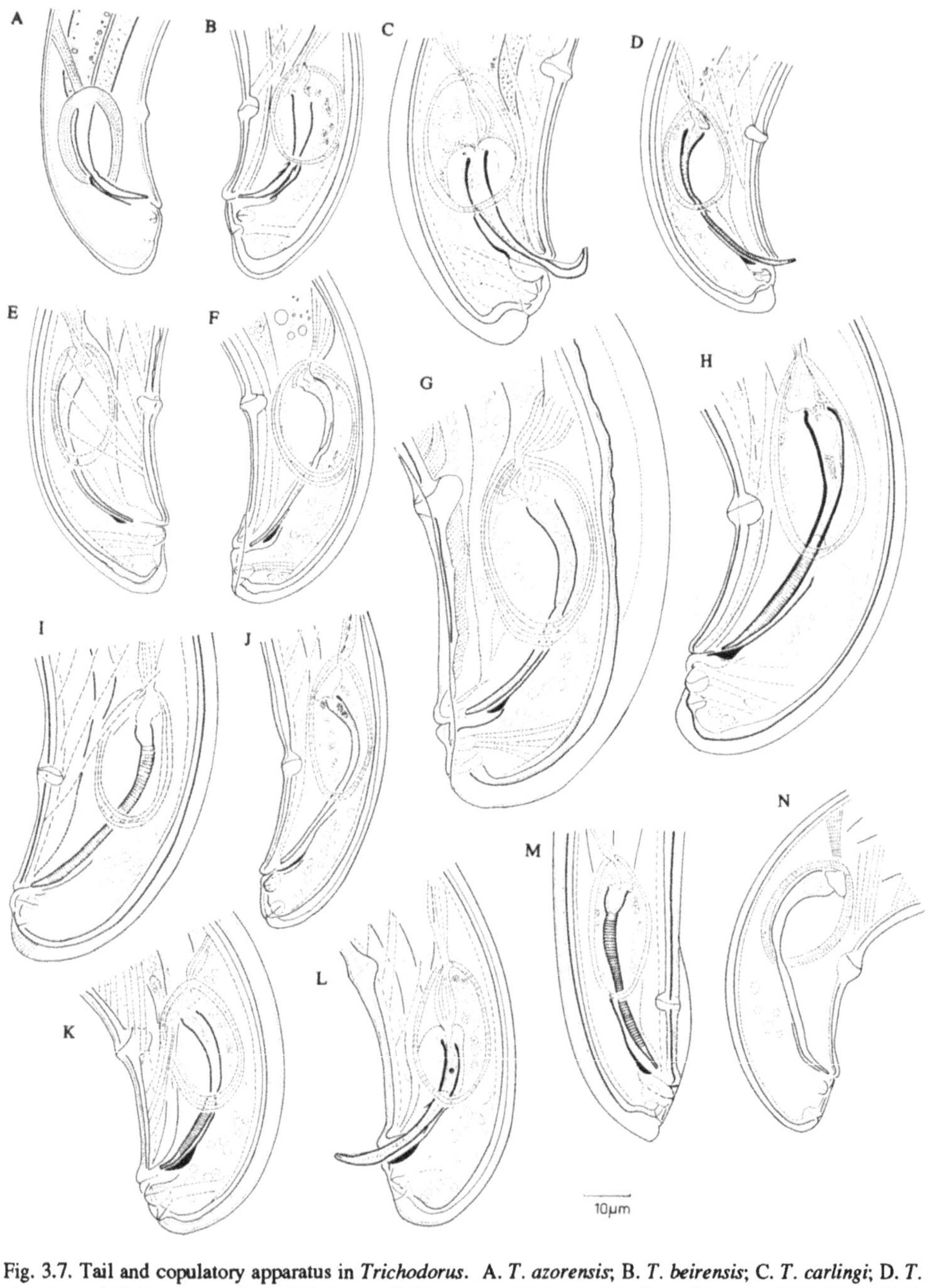

Fig. 3.7. Tail and copulatory apparatus in *Trichodorus*. A. *T. azorensis*; B. *T. beirensis*; C. *T. carlingi*; D. *T. elefjohnsoni*; E. *T. giennensis*; F. *T. kilianae*; G. *T. magnus*; H. *T. minzi*; I. *T. nanjingensis*; J. *T. philipi*; K. *T. parorientalis*; L. *T. paucisetosus*; M. *T. paracedarus*; N. *T. vandenbergae* (all figures based on type material; E-G, K: courtesy of *Fundamental and applied Nematology*).

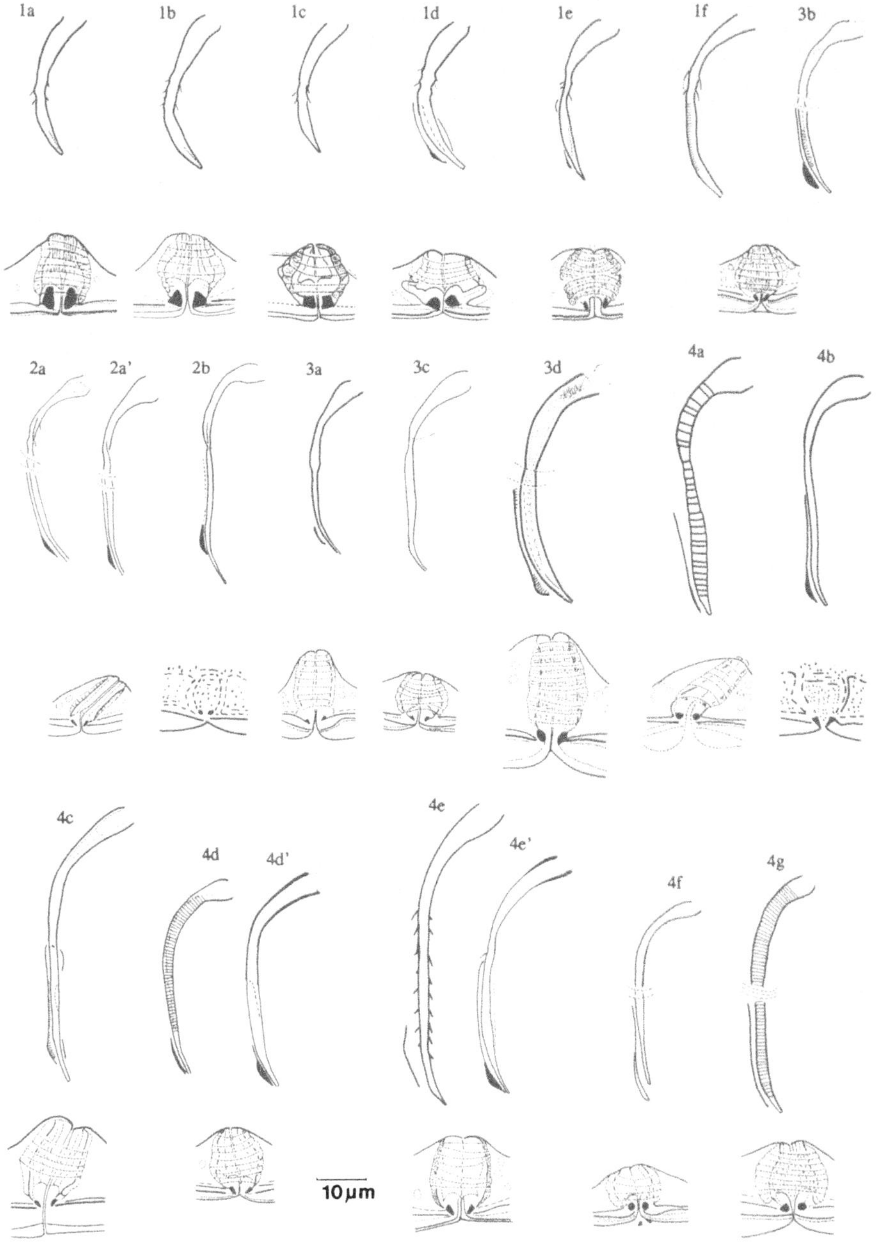

Fig. 3.8. Relation between spicule shape and vaginal sclerotizations in *Trichodorus*. 1a. *T. beirensis*; 1b. *T. lusitanicus*; 1c. *T. azorensis*; 1d. *T. velatus*; 1e. *T. viruliferus*; 1f, 3b. *T. parorientalis*; 2a, a.' *T. kilianae*; 2b. *T. sanniae*; 3a. *T. orientalis*; 3c. *T. vandenbergae*; 3d. *T. magnus*; 4a. *T. pakistanensis*; 4b. *T. rinae*; 4c. *T. petrusalberti*; 4d, d'. *T. eburneus*; 4e, e'. *T. sparsus*; 4f. *T. philipi*; 4g. *T. nanjingensis*.

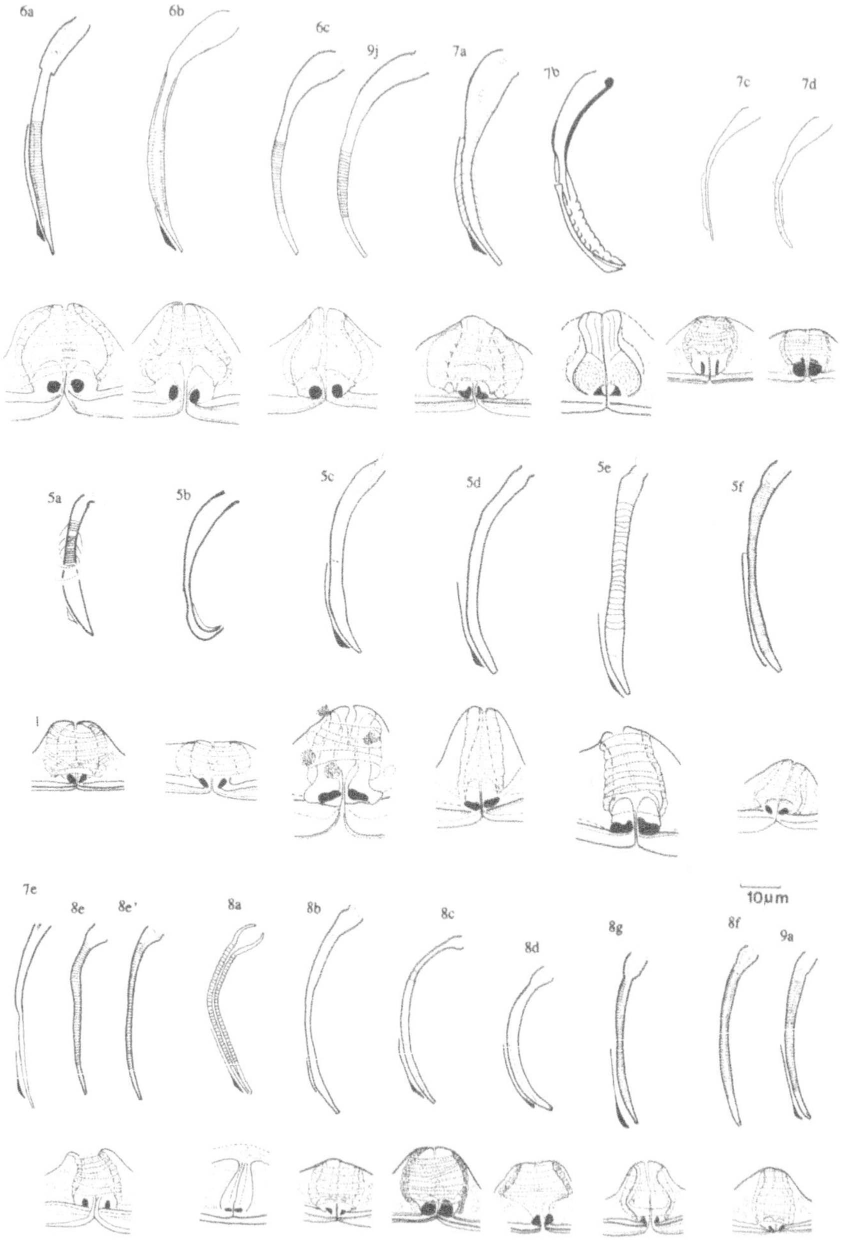

Fig. 3.9. Relation between spicule shape and vaginal sclerotizations in *Trichodorus* (continued). 6a. *T. taylori*; 6b. *T. minzi*; 6c, 9j. *T. persicus*; 7a. *T. dilatatus*, 7b. *T. yokooi*; 7c. *T. primitivus*; 7d. *T. aquitanensis*; 5a. *T. cylindricus*; 5b. *T. carlingi*; 5c. *T. cottieri*; 5d. *T. obscurus*; 5e. *T. elegans*; 5f. *T. borneoensis*; 7e, 8e, 8e'. *T. hooperi*; 8a. *T. complexus*; 8b. *T. coomansi*; 8c. *T. variopapillatus*; 8d. *T. similis*; 8f, 9a. *T. cedarus*; 8g. *T. paracedarus*.

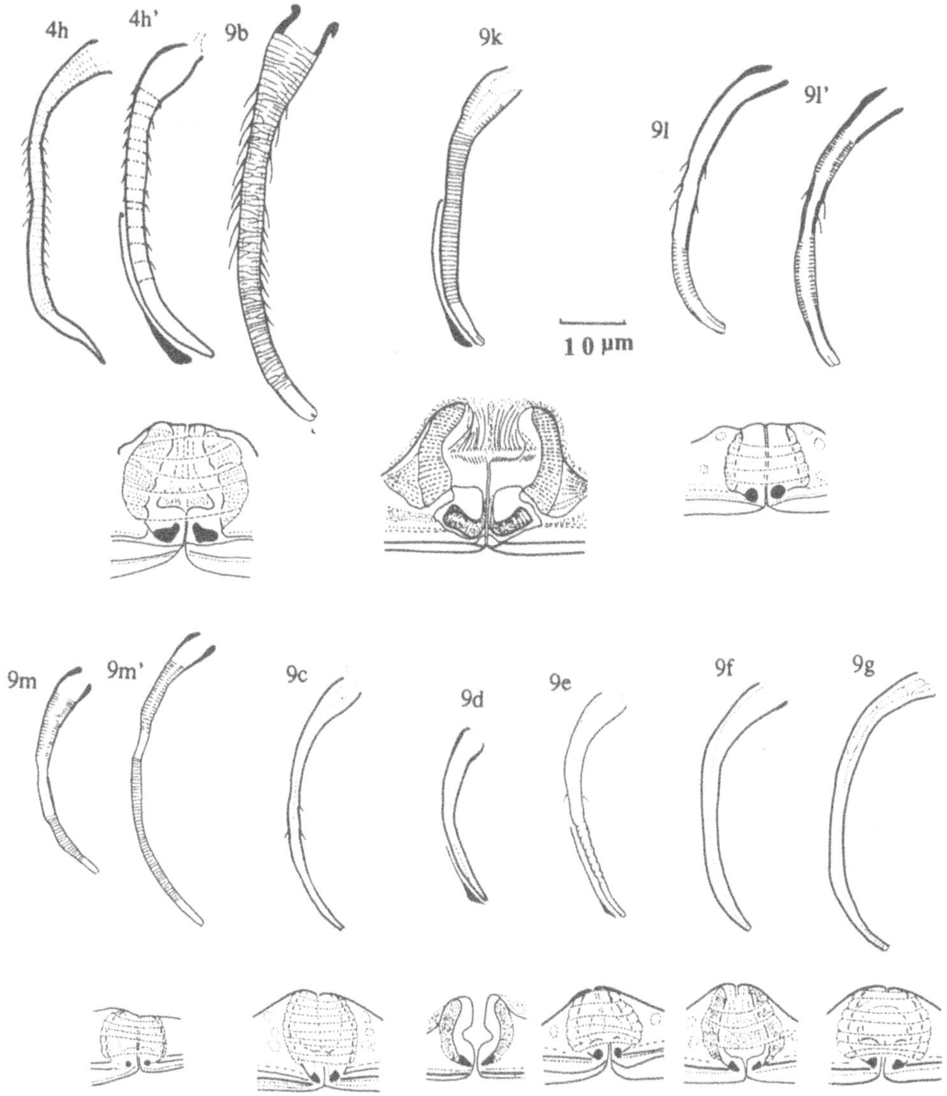

Fig. 3.10. Relation between spicule shape and vaginal sclerotizations in *Trichodorus* (continued). 4h, 4h, 9b. *T. californicus* (4h', 9b paratypes; 4h from Alaska; 9b, 4h after Bernard, 1992a); 9k. *T. intermedius*; 9l, l'. *T. paucisetosus*; 9m, m'. *T. elefjohnsoni*; 9c. *T. giennensis*; 9d. *T. borai*; 9e. *T. tricaulatus*; 9f. *T. aequalis*, 9g. *T. obtusus*.

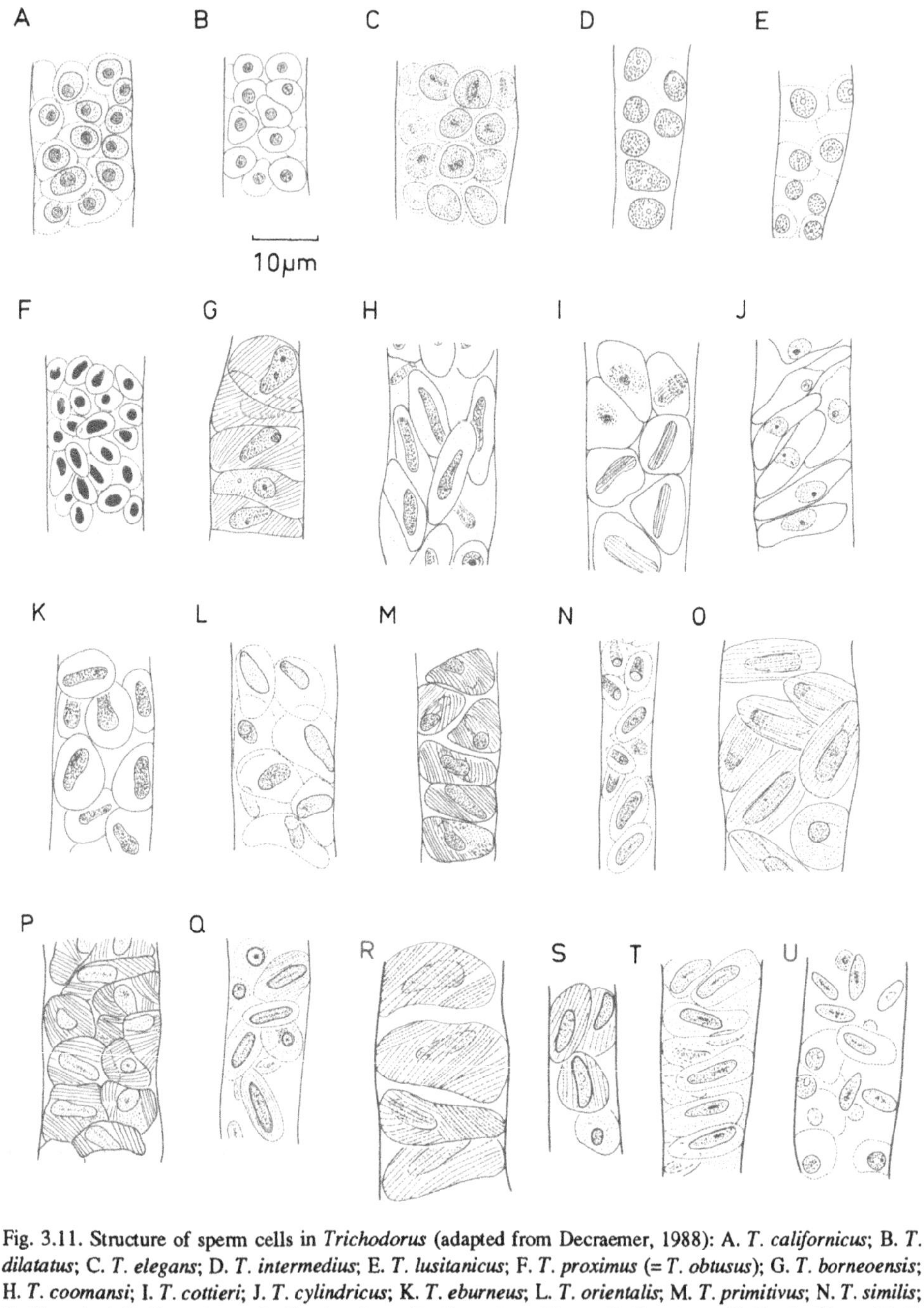

Fig. 3.11. Structure of sperm cells in *Trichodorus* (adapted from Decraemer, 1988): A. *T. californicus*; B. *T. dilatatus*; C. *T. elegans*; D. *T. intermedius*; E. *T. lusitanicus*; F. *T. proximus* (= *T. obtusus*); G. *T. borneoensis*; H. *T. coomansi*; I. *T. cottieri*; J. *T. cylindricus*; K. *T. eburneus*; L. *T. orientalis*; M. *T. primitivus*; N. *T. similis*; O. *T. taylori*; P. *T. cedarus*; Q. *T. tricaulatus*; R. *T. variopapillatus*; S. *T. velatus*; T. *T. viruliferus*; U. *T. pakistanensis* (Courtesy *Bulletin Koninklijk Belgisch Instituut voor Natuurwetenschappen*).

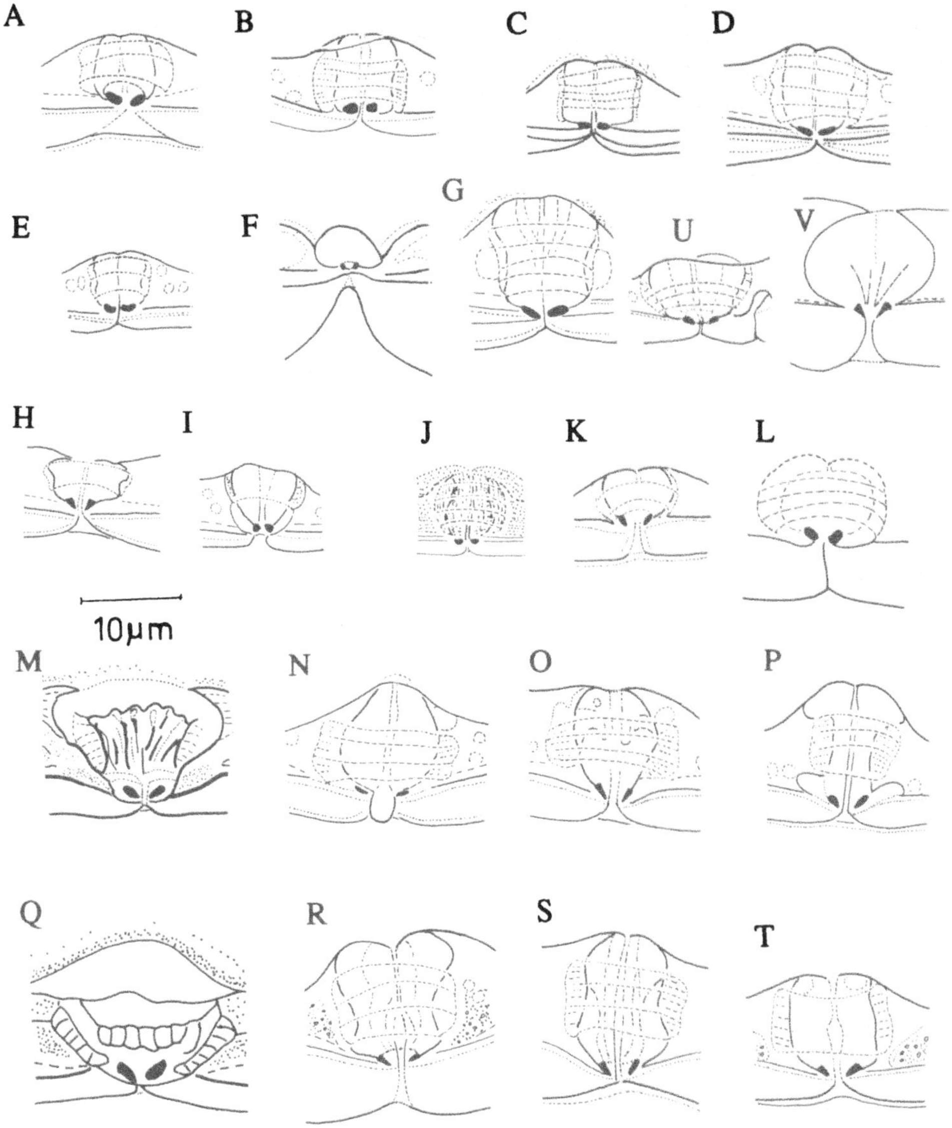

Fig. 3.12. Vaginal region and its variability in *Paratrichodorus* (adapted from Decraemer, 1989b). A-B. *P. nanus*; C-D, G. *P. minor*; E. *P. renifer*; F. *P. psidii* (after Nasira & Maqbool, 1994); H. *P. anthurii*; I: *P. acutus*; J-L. *P. rhodesiensis*; M. *P. grandis*; N-P. *P. anemones*; Q. *P. hispanus*; R-T. *P. pachydermus*; U. *P. porosus*; V. *P. sacchari* (Courtesy *Nematologica*).

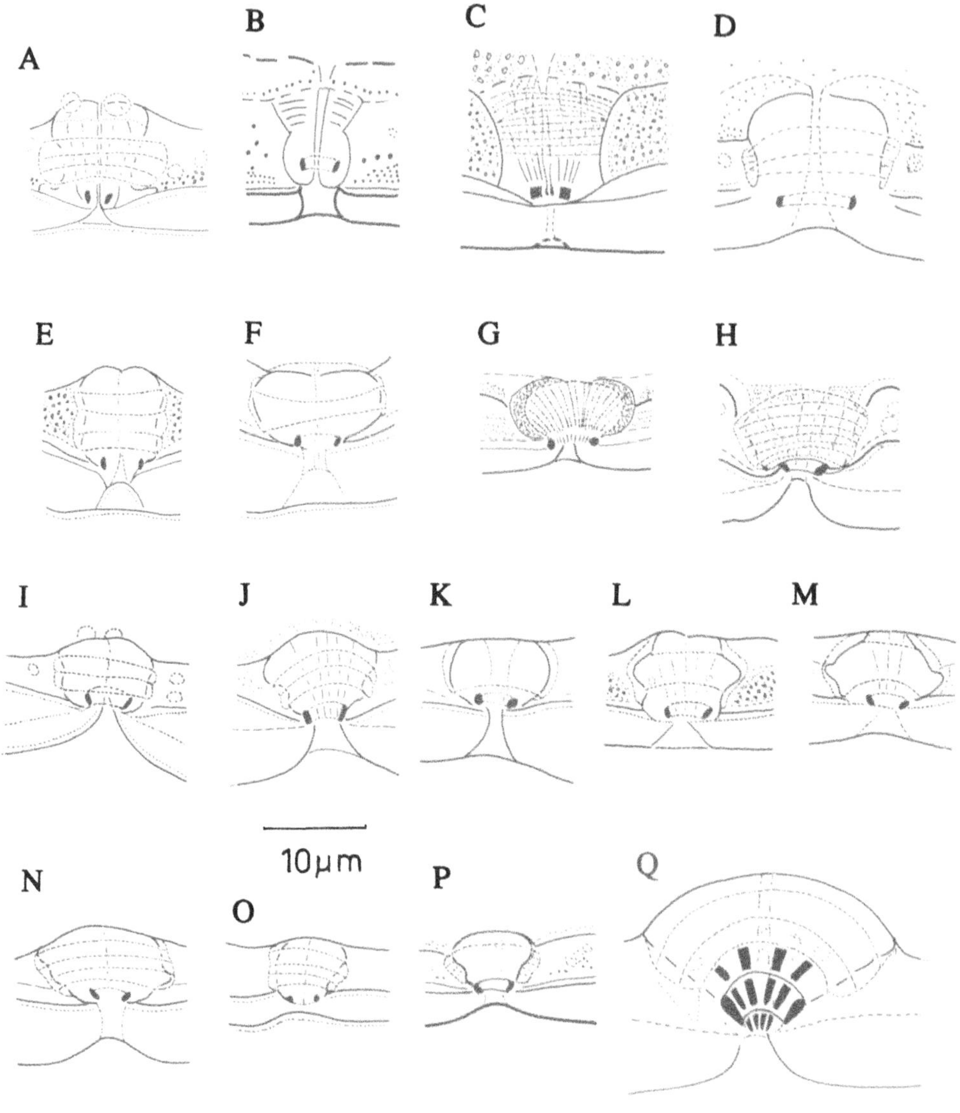

Fig. 3.13. Vaginal region and its variability in *Paratrichodorus* (continued). A. *P. weischeri*; B. *P. tunisiensis*; C. *P. acaudatus*; D. *P. paramirzai* (after Siddiqi, 1991); E-F. *P. lobatus*; G. *P. paraporosus* (after Khan *et al.*, 1989); H. *P. queenslandensis*; I. *P. allius*; J. *P. orrae*; K. *P. catharinae*; L-M. *P. teres*; N. *P. meyeri*; O. *P. mirzai*; P. *P. faisalabadensis* (after Nasira & Maqbool, 1994); Q. *P. macrostylus*.

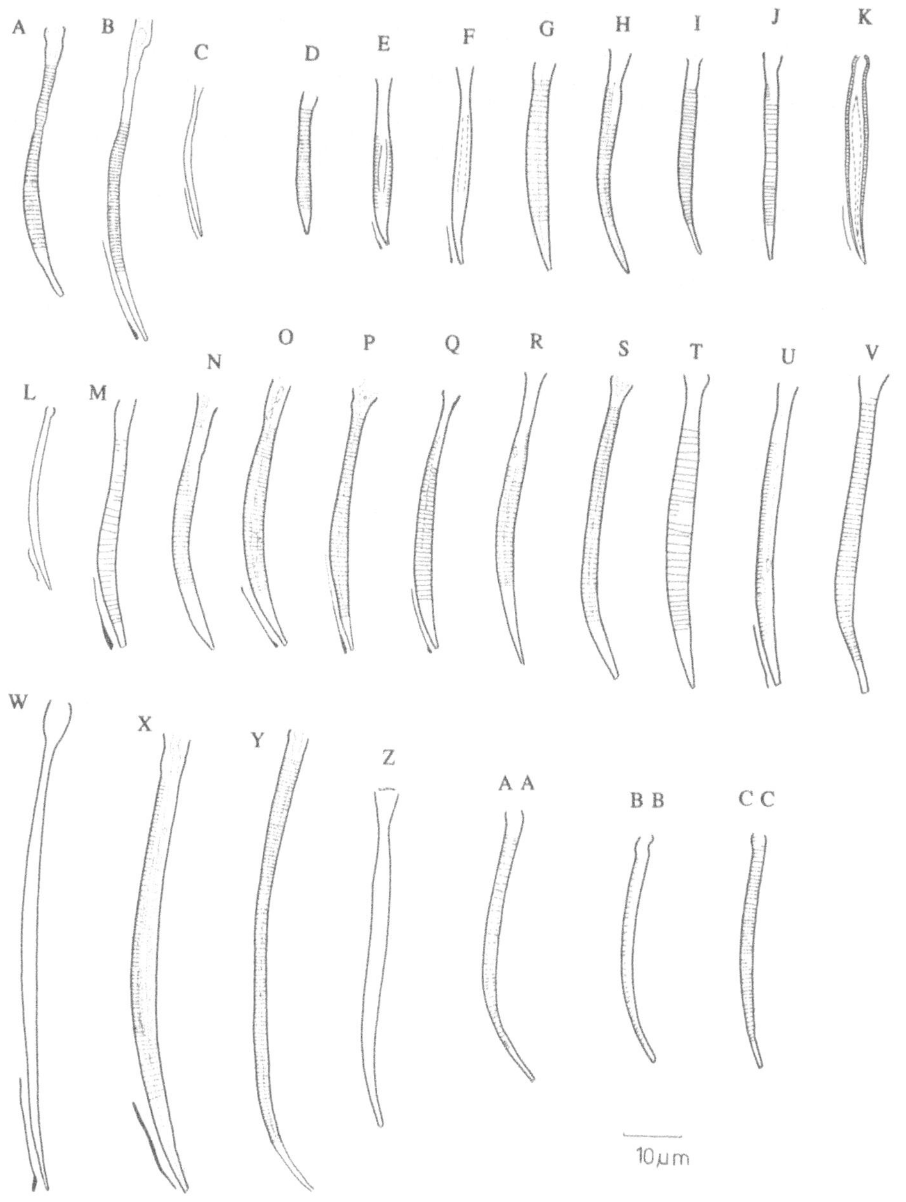

Fig.3.14. Spicule shape in *Paratrichodorus* (adapted from Decraemer, 1989a): A. *P. anemones*; B. *P. hispanus*; C. *P. weischeri*; D. *P. allius*; E. *P. psidii*; F. *P. faisalabadensis* (E-F. after Nasira & Maqbool, 1994); G-H: *P. grandis*; I. *P. mirzai*; J. *P. porosus*; K. *P. paraporosus* (after Khan, Jairajpuri & Ahmad, 1989); L. *P. delhiensis* (after Khan, Saha & Lal, 1993); M. *P. paramirzai* (after Siddiqi, 1991); N. *P. pachydermus*; O. *P. orrae*; P. *P. queenslandensis*; Q. *P. sacchari* (after De Waele & Kilian, 1992); R. *P. teres*; S. *P. lobatus*; T. *P. catharinae*, U. *P. meyeri*; V. *P. acaudatus*; W. *P. atlanticus*; X. *P. macrostylus*; Y. *P. minor*; Z. *P. alleni*; AA *P. tunisiensis*; BB. *P. rhodesiensis*; CC. *P. nanus*.

Fig. 3.15. Male tail and copulatory apparatus in *Paratrichodorus* (adapted from Decraemer, 1989b): A. *P. acaudatus*; B. *P. alleni*; C. *P. allius*; D. *P. anemones*; E. *P. atlanticus*; F. *P. catharinae*; G. *P. delhiensis* (after Khan, Saha & Lal, 1993); H. *P. faisalabadensis* (after Nasira & Maqbool, 1994); I-J. *P. grandis*; K. *P. hispanus*; L. *P. lobatus*; M. *P. macrostylus*; N. *P. minor*; O. *P. meyeri*.

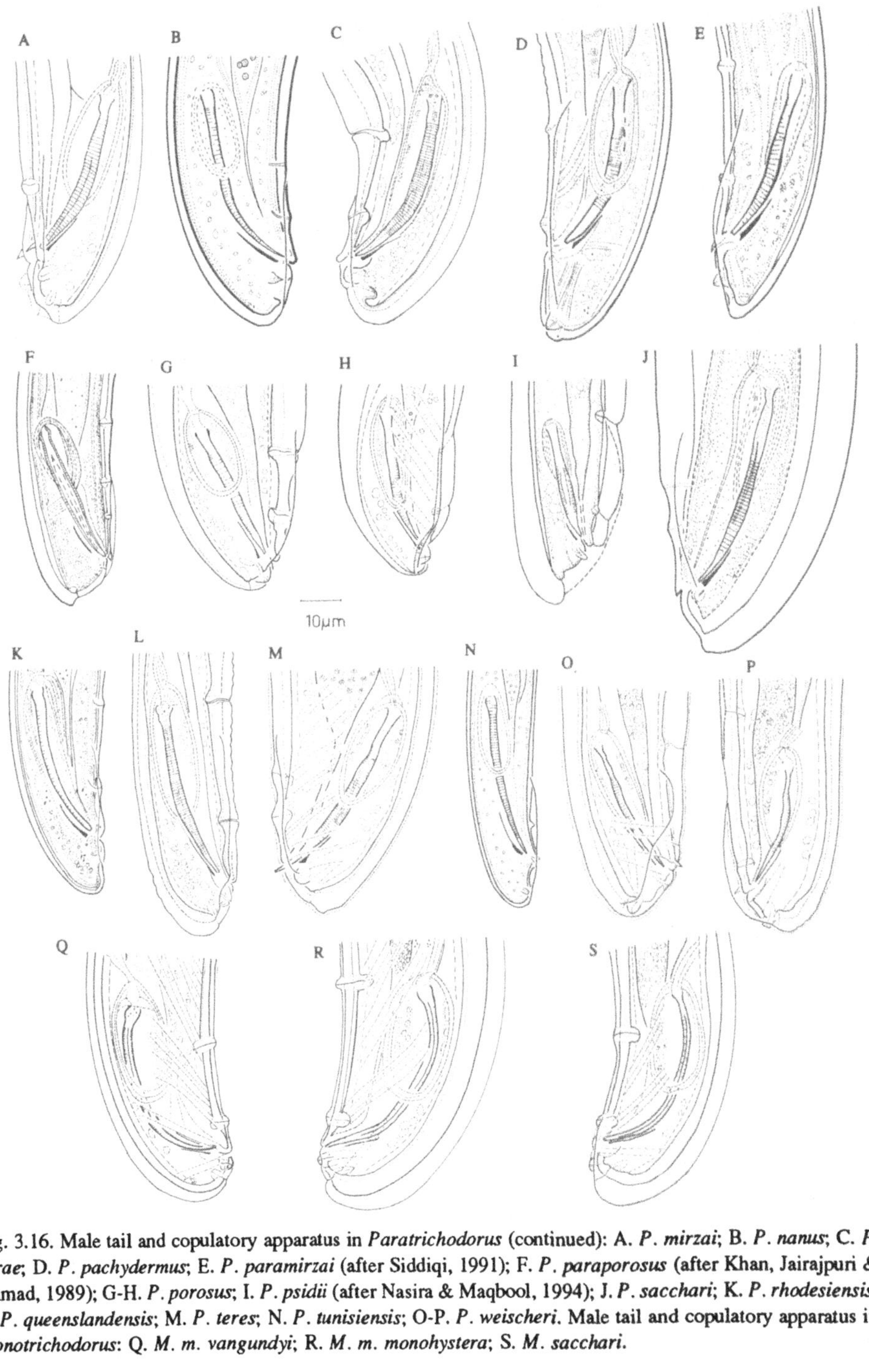

Fig. 3.16. Male tail and copulatory apparatus in *Paratrichodorus* (continued): A. *P. mirzai*; B. *P. nanus*; C. *P. orrae*; D. *P. pachydermus*; E. *P. paramirzai* (after Siddiqi, 1991); F. *P. paraporosus* (after Khan, Jairajpuri & Ahmad, 1989); G-H. *P. porosus*; I. *P. psidii* (after Nasira & Maqbool, 1994); J. *P. sacchari*; K. *P. rhodesiensis*; L. *P. queenslandensis*; M. *P. teres*; N. *P. tunisiensis*; O-P. *P. weischeri*. Male tail and copulatory apparatus in *Monotrichodorus*: Q. *M. m. vangundyi*; R. *M. m. monohystera*; S. *M. sacchari*.

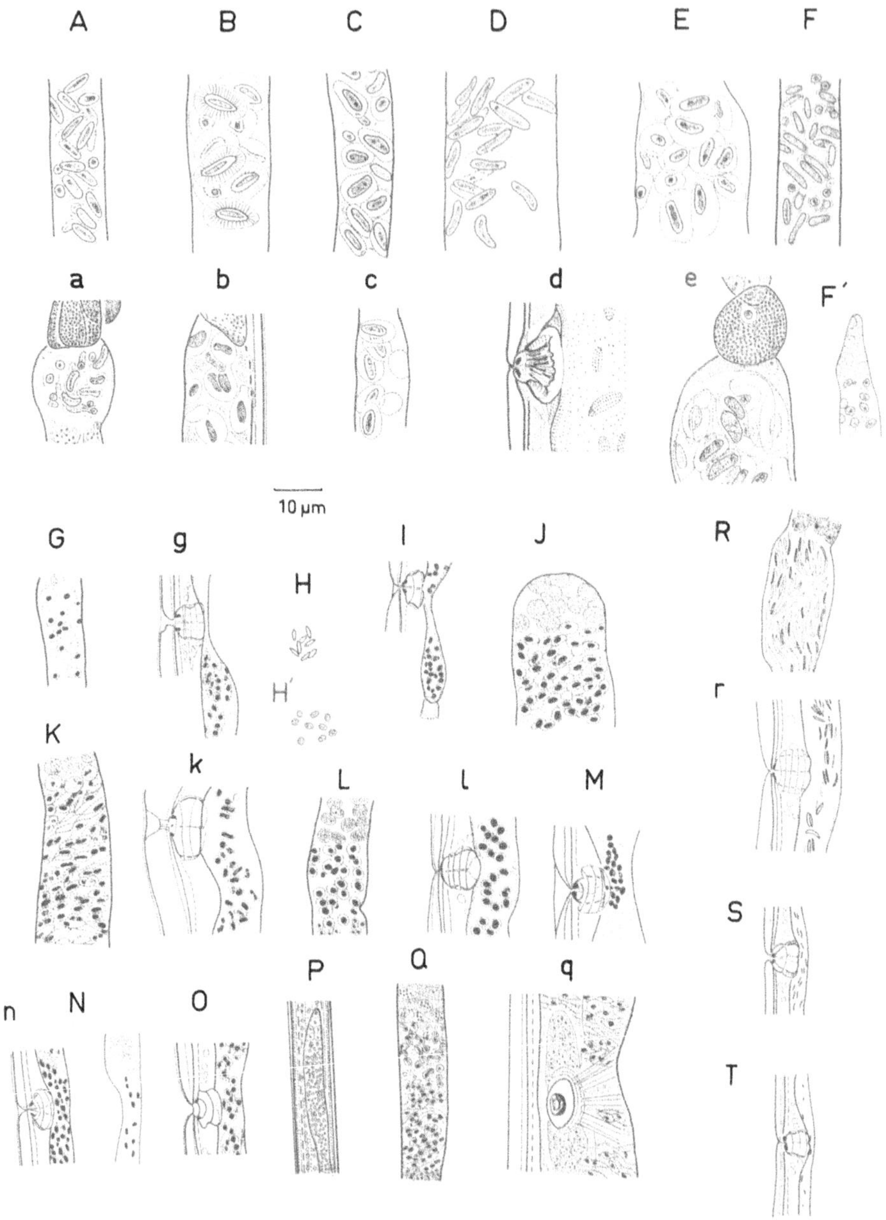

Fig. 3.17. Structure of sperm cells in *Paratrichodorus* (adapted from Decraemer, 1989b): A (♂), a (♀). *P. anemones*; B (♂), b (♀). *P. weischeri*; C (♂), c (♀). *P. pachydermus*; D (♂), d (♀). *P. grandis*; E (♂), e (♀). *P. hispanus*; F. (♀) *P. porosus*; F' (♂). *P. mirzai*; G (♂), g (♀). *P. allius*; H (♂). *P. alleni*; H' (♂). *P. acaudatus*; I (♀). *P. anthurii*; J (♂). *P. catharinae*; K (♂), k (♀). *P. lobatus*; L, R (♂), l, r (♀). *P. minor*; M (♀). *P. nanus*; N (♂), n (♀). *P. rhodesiensis*; O (♀). *P. tansaniensis* (= *P. allius*); P (♂). *P. tunisiensis*; Q (♂), q (♀). *P. teres*; S (♀). *P. acutus*; T (♀). *P. renifer* (Courtesy *Nematologica*).

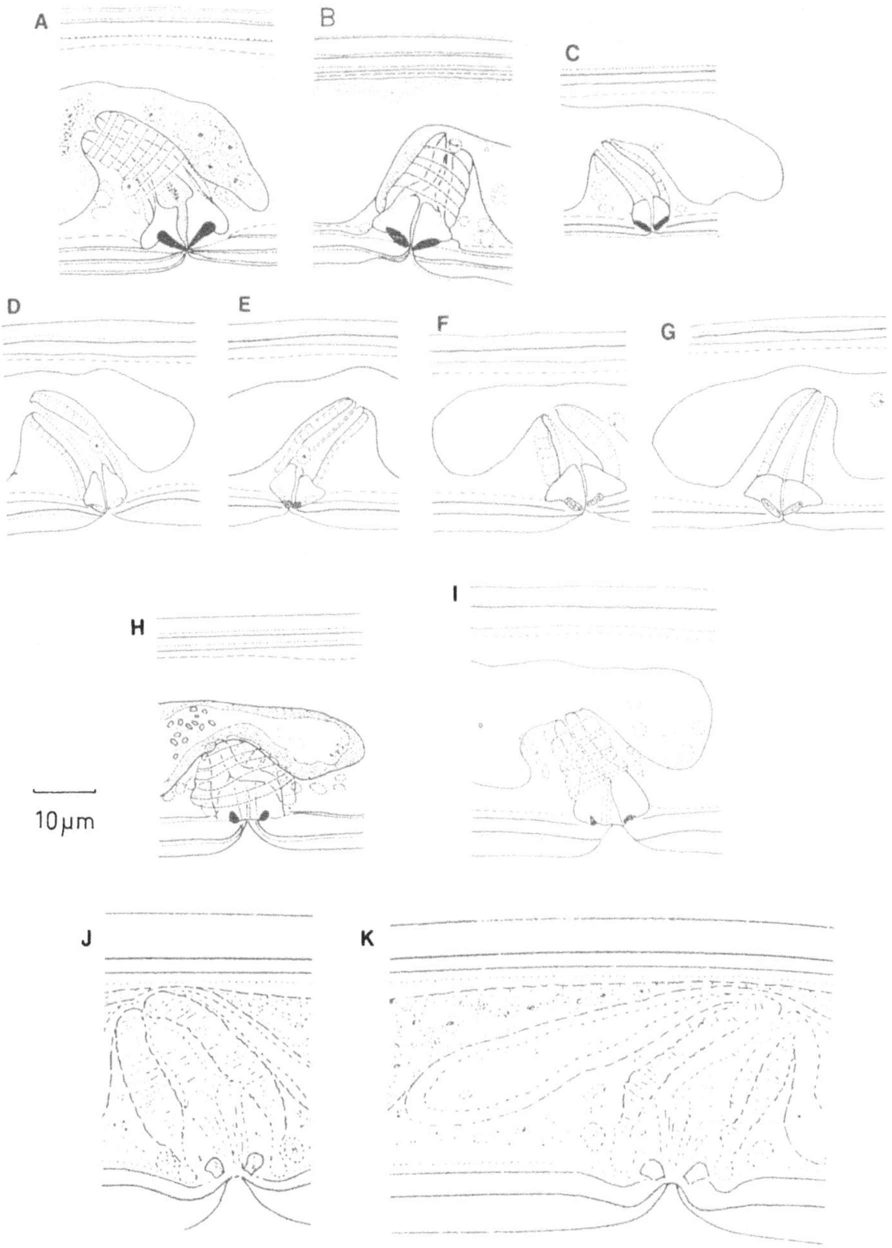

Fig. 3.18. Vaginal region in *Monotrichodorus*: A. *Monotrichodorus monohystera monohystera*; B. *M. m. vangundyi* (paratype); C. *M. m. monohystera* (= *M. parvus*, paratype); D-E. *M. m. monohystera* (= *M. acuparvus*, paratypes); F-G. *M. m. monohystera* (= *M. proporifer*, paratype); H-I. *M. sacchari*, paratypes; J-K. *M. muliebris*, after Andrassy, 1989.

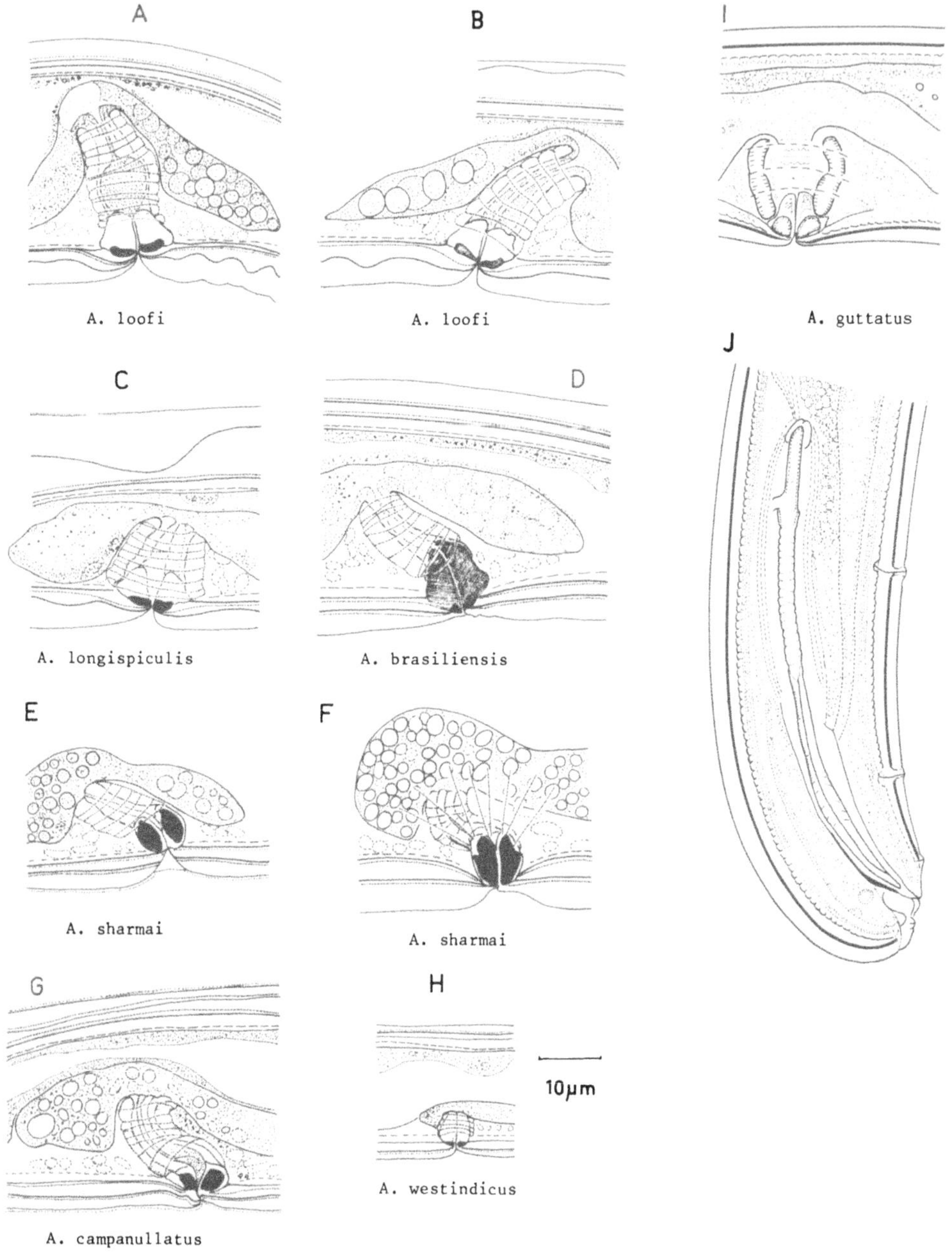

Fig. 3.19. Vaginal region in *Allotrichodorus*: A-B. *A. loofi*; C. *A. longispiculis*; D. *A. brasiliensis*; E-F. *A. sharmai*; G. *A. campanullatus;* H. *A. westindicus;* I. *A. guttatus* . Tail region and copulatory apparatus: J. *A. guttatus* (I, J after Rodriguez-M *et al.*, 1978).

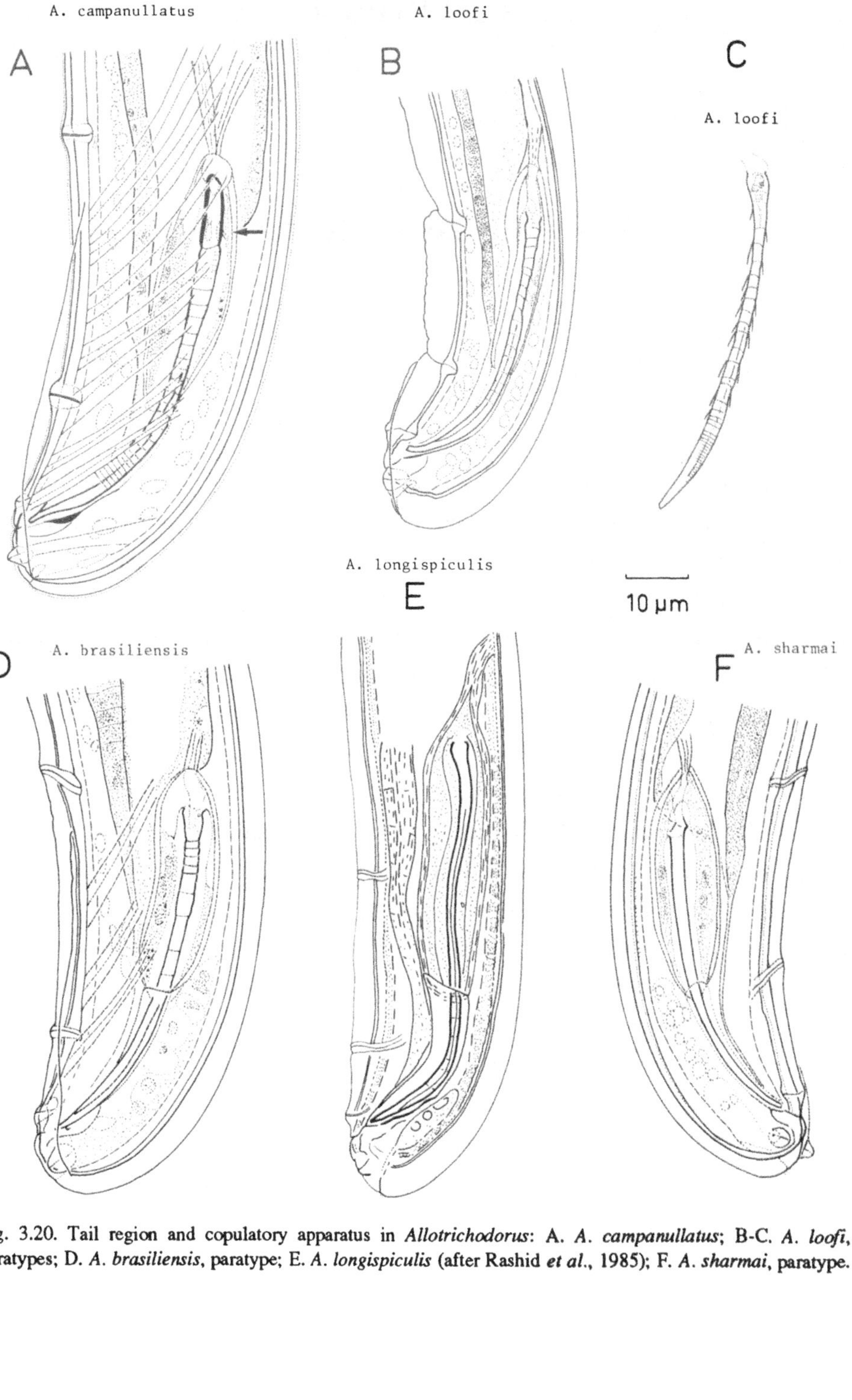

Fig. 3.20. Tail region and copulatory apparatus in *Allotrichodorus*: A. *A. campanullatus*; B-C. *A. loofi*, paratypes; D. *A. brasiliensis*, paratype; E. *A. longispiculis* (after Rashid *et al.*, 1985); F. *A. sharmai*, paratype.

CHAPTER 4

SPECIES DESCRIPTIONS

All valid species of the genera *Trichodorus, Paratrichodorus, Monotrichodorus* and *Allotrichodorus* have included in their descriptions morphometric data (based on populations described in the literature), an elaborate species diagnosis, relationships with other species of the genus, remarks on the variability of diagnostic characters and their distribution and hosts. The elaborate species diagnoses consist of the specific diagnostic features and some additional features in order to allow interspecific comparison and use of the identification keys. The species are illustrated by line drawings and by photographs taken with a Polyvar microscope with interference contrast. Most illustrations are from type specimens.

The morphometric data include mean values from different populations, for body length, onchiostyle length, spicule length and % V. These data are presented in parenthesis, beginning with the type population, and the standard deviation is included when available. The mid-range (indicated by *) is calculated as an approximation of the mean value of each measurement in populations for which only the range of a morphometric character is available.

The relationships of the species are re-examined as the published data are no longer adequate, due to the increase in the number of species described during the last decade. In addition, most of the species diagnoses have been amended accordingly.
In the morphometric data the term EP stands for the secretory-excretory pore (S-E pore).

4.1. The genus *Trichodorus* Cobb, 1913

Trichodorus aequalis Allen, 1957
Figs 3.2.A; 3.4 (9f); 3.6.K; 3.10 (9f); 4.16E

Measurements

Holotype male: L= 810 µm (840 µm), a= 23 (17), b= 5.6, c= 67 (60), T= 69% (60%), onchiostyle= 54 µm (56 µm), spicule length= 42 µm (43 µm), gubernaculum= 23 µm, distance CP1-CP2= 24 µm, distance CP2-EP= 13 µm, pharynx= 145 µm, anterior end to EP= 104 µm (from Loof, 1973, after reexamination of holotype male; original data by Allen, 1957 in parenthesis).

Males: L= 620-891 µm (750 µm*; 676 µm ± 42; 891 µm*), a= 17-25, b= 4.5-6.6, T= 53-70%, onchiostyle= 41.4-79 µm (53µm*; 44.6µm ± 1.5; 79 µm), spicule length= 35-47 µm (42.5 µm *; 40 µm ± 0.7, 47 µm), gubernaculum= 12-23 µm, distance CP1-CP2= 14-25 µm, distance CP2-EP= 4.9-17 µm, pharynx= 117-171 µm, anterior end to EP= 91-112 µm.

Females: L= 601-821 µm (785µm*; 708µm ± 53), a= 17-26, b= 4.0-6.8, V= 49-56% (52.5%*; 53.6% ± 1.5), onchiostyle= 42.3-74 µm (62 µm*; 45 µm ± 1.5) (from Allen, 1957; Shishida, 1979; Bernard, 1992a: ♂ only).

Emended diagnosis

T. aequalis male is distinguished by the two ventromedian cervical papillae anterior to the S-E pore, the position of the lateral cervical pore at the level of the anterior cervical papilla, the position of the precloacal supplements (three supplements, the posterior supplement near the proximal end of retracted spicules, showing a small range of variability), the spicule shape (slightly ventrally curved, shaft about equally wide, slightly tapered posteriorly, smooth, without bristles), the evenly rounded outline of the tail with terminal cuticle not or only very slightly thickened, and by elliptical sperm with elongate nuclei. The female is characterized by the presence of a lateral body pore about four body widths anterior to the vulva and one within one body width posterior to the vulva (on each side), the shape of the vaginal sclerotizations in lateral view (small triangular to oval), the vagina pear-shaped and pore-like vulva. The pharyngeal bulb is offset in males and females.

Relationships

T. aequalis is closest to *T. sparsus* in having two well developed ventromedian cervical papillae, the position of the posterior ventromedian precloacal supplement (near the head of the retracted spicules), the length of spicules and gubernaculum in males, and by the pore-like vulva, the two pairs of lateral body pores (one prevulvar, one postadvulvar), the small (rounded) triangular sclerotized pieces in lateral view in females. They also have comparable onchiostyle lengths. The species differ in the shape of the spicules (slender, provided with bristles in *T. sparsus vs* stouter spicules, without bristles in *T. aequalis*), and by the tail shape (with thickened cuticle at tip in *T. sparsus vs* not thickened in *T. aequalis*). *T. aequalis* is also close to *T. tricaulatus*. Females of both species can only be differentiated by the different type of pharyngo-intestinal junction (bulb offset *vs* ventrally overlapping pharyngeal glands in *T. tricaulatus*). Males have a comparable spicule shape but differ by its ornamentation (shaft smooth *vs* shaft with striae and bristles in *T. tricaulatus*), the number of ventromedian cervical papillae (2 *vs* 3 in *T. tricaulatus*), the position of the precloacal supplements (posterior supplement at level of spicule head *vs* all precloacal supplements anterior to the retracted spicules in *T. tricaulatus*) and by the terminal tail cuticle (non-thickened *vs* thickened in *T. tricaulatus*).

Remarks

1. According to Shishida (1979), juveniles are of similar appearance to the adult female, except for the tail shape which is somewhat acute in several juvenile specimens.
2. One male specimen from a total of six, was observed with four precloacal supplements (Loof, 1973).
3. Loof received specimens (from south-west England) that were regarded as a new species. He considered it to be closest to *T. aequalis* and *T. sparsus*. Loof (1973) made a comparison of type material of *T. aequalis* and *T. sparsus* Szczygiel, 1968, and described the new species *T. hooperi*. He treated these three species as members of a *Trichodorus aequalis*- complex. Loof (1973) considered the species of the complex to be more variable than several other *Trichodorus* species at that time e.g. in onchiostyle length and in number of precloacal supplements. In *T. aequalis* and *T. sparsus*, populations occur with short (under 60 µm) and long onchiostyles (about 70 µm).

Subsequently, the *T. aequalis* complex has become a much larger group of related species: *T. coomansi*, *T. giennensis*, *T. orientalis*, *T. paucisetosus*, *T. parorientalis* and *T. tricaulatus* (Shishida, 1979; Bernard, 1992a and Decraemer *et al.*, 1993). They are characterized in the males by two well developed ventromedian cervical papillae (except in *T. tricaulatus* with three papillae) anterior to the S-E pore (except in *T. parorientalis* which has the S-E pore between the ventromedian cervical papillae), the posterior precloacal supplement lying at the level of the spicule head (except *T. carlingi* and *T. tricaulatus* where it is anterior to the retracted spicules, and in *T. hooperi* and *T. coomansi* where it is clearly in the region of the retracted spicules); and in the females, by small, almost triangular vaginal sclerotizations (except in *T. hooperi* and *T. carlingi* where larger, rounded), and two pairs of lateral body pores (one prevulvar, one postadvulvar) (except in *T. hooperi* which has one pair). Bernard (1992a) described *T. carlingi* with females morphologically very similar to *T. aequalis* and *T. sparsus*, from which they cannot be reliably separated. Males of *T. carlingi*, however, differ from the former species in having a totally different spicule shape.

The main characters identifying the *T. aequalis*-group (2 ventromedian cervical papillae, striated spicules, three precloacal supplements with the posterior one at the head of the retracted spicules and two pairs of lateral body pores in females) are all characters which either individually or combined, are found in about half or more of the species of the genus. So, we can no longer refer to a *T. aequalis*-species group. Initially "similar" spicule shapes when examined closely are different. Even the three species originally composing the *T. aequalis* complex (*T. aequalis, T. sparsus* and *T. hooperi)* can be distinguished by comparing their spicule shape, the thickened or non-thickened tail cuticle at the tip and the position of the posterior precloacal supplement in relation to the retracted spicules. Also, in females by the number of lateral body pores, the shape of the vaginal sclerotizations in lateral view and the shape of the vulva.

Distribution and host

T. aequalis has a limited distribution, restricted to U.S.A., China and Japan. It was originally described from Mt. Diablo (type locality), Orinda, and several other localities in California, from Idaho Springs in Colorado and from Canby and Brooks in Oregon (Allen, 1957). The *T. aequalis* specimens reported from Europe (Poland, Germany, the Netherlands, France, Switzerland, Austria and Turkey) in fact belong to *T. sparsus* (Wyss, 1970; Loof, 1973) and the Italian specimens from northern Italy appear to be *T. sparsus* (see photographs in Mancini *et al.*, 1979) or *T. taylori* (De Waele *et al.*, 1982). Spanish specimens formerly recorded as *T. aequalis* by Pena-Santiago *et al.* (1987) belong to *T. giennensis*. Several of the species of the "*T. aequalis*" group appear to occur sympatrically e.g. *T. aequalis* and *T. tricaulatus* in Japan; *T. sparsus* and *T. hooperi* in Great Britain.

T. aequalis is a polyphagous species with a wide host range: toyon, manzanita (*Arbutus* sp.), *Camellia japonica*, *Erythrina cristagalli* (cockspur coral tree), *Eucalyptus*, cactus, *Ligustrum sinense* (privet), *Rhododendron*, white spruce stand lacking ground cover, *Scale cereale* (rye) (Allen, 1957; Shishida, 1979; McGowan, 1977; Mead, 1988; Miller, 1977; Ridings, 1980; Wang & Wu, 1991; Bernard, 1992a).

Trichodorus aquitanensis Baujard, 1980

Figs 3.1.P, p; 3.4 (7d); 3.5.L; 3.9 (7d); 4.1.A-H; 4.2.A-D

Measurements

Holotype female L= 800 µm, a= 22.5, b= 5.3, V= 54.3%, onchiostyle= 58 µm, anterior end to EP= 85 µm.
Females: L= 710-980 µm (780 µm), a= 22-29.5 (25), b= 4.6-6.5 (5.4), V= 52-59% (56%), onchiostyle= 47-58 µm (53 µm).
Males: L= 680-1000 µm (810 µm), a= 22.5-33.5, b= 4.2-7.9, T= 44.5-66% (59%), onchiostyle= 47-62 µm (53 µm), spicule length= 31.5-41.5 µm (37 µm), gubernaculum= 11-13.5 µm, anterior end to EP= 80-102 µm (from Baujard, 1980).

Emended diagnosis

T. aquitanensis is characterized in the male by the presence of three ventromedian cervical papillae (the anterior two in the onchiostyle region), the position of the lateral cervical pore (in the onchiostyle region, anterior to the second ventromedian cervical papilla), by the shape of the spicules (with wide proximal part, tapering to a fine shaft), the shape and position of the gubernaculum (anteriorly convex, lying largely in between the spicules) and by the position of the precloacal supplements (three supplements, the posterior one within the region of the retracted spicules). In the female by the shape of the vaginal sclerotizations in lateral view (large quadrangular, close to one another), the presence of three lateral body pores on each side (one at the level or slightly prevulvar, two postvulvar), the shape of the vulva (a transverse slit) and by the barrel-shaped vagina. The pharyngeal bulb is offset in males and females.

Relationships

T. aquitanensis is closest to *T. azorensis*, *T. primitivus*, *T. similis* and *T. variopapillatus*, but is distinguished by the three ventromedian cervical papillae in the males (except *T. primitivus*) with the anterior one and the lateral cervical pores in the onchiostyle region. *T. aquitanensis* has a spicule shape comparable with that of *T. primitivus* (proximally wide, tapering distally to a slender shaft) but differs by the wider anterior part. Its spicule shape is clearly different from that of *T. similis* and *T. variopapillatus* (spicules with offset manubrium) and *T. azorensis* (spicules with a clear constriction at mid-shaft). Females of *T. aquitanensis* resemble those of *T. azorensis* and *T. variopapillatus* by the shape of the vaginal sclerotizations (large quadrangular in lateral view) but differ from those of *T. similis* (large, rounded, triangular) and those of *T. primitivus* (elongate, oval). Also, *T. aquitanensis* resembles *T. similis, T. primitivus* and *T. variopapillatus* by the transverse slit-like vulva.

Remarks

1. A study of paratype specimens revealed some variability in the size and shape of the vaginal sclerotizations in females which, in some specimens, may be partly due to the presence of a secretion plug. In lateral view, the sclerotized pieces of the vagina lie close to one another (Fig. 4.2.B,C) compared to the figure in the original description (Fig. 1G in Baujard, 1980) where they are clearly separated. Apparently, this was caused by a secretion plug, a feature also observed in other specimens (Fig. 4.2.D).
2. Among male paratypes, specimens were observed (slide no 15725) with spicules showing at mid-shaft, i.e. at the level of the distal border of the capsule of suspensor muscles, a slight constriction and an irregular outline of the shaft giving the impression of the presence of bristles (Fig. 4.1.E).
3. *T. aquitanensis* has large sperm cells with sausage-shaped nucleus and fibrils.
4. Baujard (1980) noted that the lateral body pores in females could only be observed with the aid of interference contrast.

Distribution and host

T. aquitanensis is only known from its type locality, Pissos, Department Landes, France. It is a polyphagous species with the following hosts: *Pinus pinaster*, *Erica cinerea*, *E. scoparia*, *Calluna vulgaris*, *Ilex europaeus*, *Pteris aquilina*, *Simaethis planifolia* and *Quercus loza* (Baujard, 1980).

Trichodorus azorensis Almeida, De Waele, Santos & Sturhan, 1989
Figs 3.1.O; 3.3 (1c); 3.7.A; 3.8 (1c); 4.2.E, M

Measurements

Holotype female: L= 706 μm, a= 24.3, b= 4.7, V= 58.9%, onchiostyle= 63 μm, anterior end to EP= 93 μm.
Females: L= 601-769 μm (690 μm ± 50), a= 19.1-28.7, b= 4.2-5.8, V= 53.1-61.5% (57% ± 2), onchiostyle= 55-63 μm (59μm ± 2), anterior end to EP= 72-103 μm.
Males: L= 647-807 μm (706 μm ± 50), a= 20.8-28.8, b= 4.4-5.8, T= 57.9-67.2% (61.8% ± 3.2), onchiostyle= 55-66 μm (60 μm ± 3), spicule length= 29-34 μm (31μm ± 1), gubernaculum= 13-18 μm, anterior end to EP= 75-117 μm, distance CP1-CP2= 16-37 μm, distance CP2-CP3= 21-34 μm, distance CP3-EP= 11-22 μm (from Almeida *et al.*, 1989).

Emended diagnosis

T. azorensis female is characterized by the shape of the vaginal sclerotizations in lateral view (large square-shaped pieces, slightly separated), the rhomboid vagina and the presence of one pair of postadvulvar lateral body pores. The male by the presence of three conspicuous ventromedian cervical papillae anterior to the S-E pore (the anterior one usually situated opposite the onchiostyle region), the position of the lateral cervical pore (opposite the posterior half of the onchiostyle region i.e. in between the anterior two ventromedian cervical papillae), the spicule shape (ventrally curved, smooth, with a slight constriction with delicate bristles at mid-shaft or anterior to it, distal end usually not grooved), the spicule length (29-34 μm), the position of the three ventromedian, precloacal supplements (the posterior one lying slightly anterior, opposite or slightly posterior to the head of the retracted spicules) and the non-thickened terminal tail cuticle. The pharyngeal bulb is offset in male and female.

Relationships

T. azorensis resembles *T. beirensis*, *T. lusitanicus* and *T. viruliferus* by the general spicule shape (shaft having a wider proximal part, and a more or less mid-shaft constriction with bristles) but differs in details (degree and location of the spicule shaft constriction). It is distinguished from these species by the characteristic shape of the vaginal sclerotizations in lateral view (two conspicuous square-shaped pieces *vs* two small oval pieces in *T. viruliferus*, two large regular triangular pieces in *T. lusitanicus* and two large rounded triangular to oval pieces in *T. beirensis*) and also in the onchiostyle length (more than 55 μm *vs* usually less than 55 μm in the other three species mentioned). *T. azorensis* females are further distinguished from those of *T. beirensis* in usually having one pair of lateral body pores (no lateral body pores present in *T. beirensis*) and also by the shape of the vagina (rhomboid *vs* cylindrical in *T. beirensis* (Almeida *et al.*, 1989).

Remarks

Exceptionally, males may have two or four ventromedian cervical papillae, all situated anterior to the S-E pore and females one or no lateral body pore (Almeida *et al.*, 1989).

Distribution and host

T. azorensis has a distribution limited to the Azores, with Lagoa das Furnas, Sao Miguel the type locality. It is polyphagous with a diverse host range: *Ilex europaeus*, *Rubus* sp., *Plantago lanceolata*, *Myrica faia*, *Pittosporum* sp., *Senecio* sp., *Pteridium aquilinum*, *Selaginella kraussiana*, Leguminosae, grasses and moss (Almeida *et al.*, 1989).

Trichodorus beirensis Almeida, De Waele, Santos & Sturhan, 1989

Figs 3.1.K,k,k'; 3.3 (1a); 3.7.B; 3.8 (1a); 4.2.F-L

Measurements

Holotype female: L= 750 µm, a= 25, b=4.7, V= 57.3%, onchiostyle= 43 µm, anterior end to EP= 103 µm.
Females: L= 660-1050 µm (787 µm ± 62, 844 µm, 873 µm, 894 µm, 935 µm, 941 µm), a= 20.2-37.9, b= 3.8-7.4, V= 46.9-61.7% (56.4% ± 2.6), onchiostyle= 39-56 µm (43 µm ± 3, 45 µm, 48 µm, 49 µm, 55 µm), anterior end to EP= 76-128 µm (102 µm ± 13).
Males: L= 642-971 µm (767 µm ± 52, 794 µm, 863 µm, 914 µm), a= 22.8-38.3, b= 4.5-6.7, T= 53.1-80.2%, onchiostyle= 32-54 µm (41 µm ± 3, 43 µm, 44 µm, 54 µm), spicule length= 27-35 µm (30 µm ± 2, 31 µm, 33 µm, 35 µm), gubernaculum= 9-16 µm, anterior end to EP= 77-118 µm, distance CP1-CP2= 14-60 µm, distance CP2-CP3= 18-35 µm, distance CP3-EP= 4-40 µm (from Almeida *et al.*, 1989: 6 populations of ♀, 4 of ♂).

Emended diagnosis

T. beirensis female is characterized by the shape of the vaginal sclerotizations in lateral view (large rounded triangular, rarely oval-shaped pieces), the vulva shape (a transverse slit), the cylindrical vagina and the absence of lateral body pores. The male by the presence of three conspicuous ventromedian cervical papillae anterior to the S-E pore (the anterior one usually situated in the onchiostyle region), the position of the lateral cervical pore (slightly anterior or posterior to the onchiostyle base), the spicule shape (smooth, with clear constriction with bristles posterior to mid-shaft, distal end usually grooved), spicule length (27-35 µm), and the position of the three ventromedian precloacal supplements (the posterior supplement slightly anterior, opposite or slightly posterior to the head of the retracted spicules). The pharyngeal bulb is offset in male and female.

Relationships

T. beirensis is distinguished from the related species *T. viruliferus*, *T. lusitanicus* and *T. azorensis* by the characteristic shape of the vaginal sclerotizations in lateral view (two large rounded triangular to almost oval-shaped species *vs* two small oval-shaped pieces in *T. viruliferus*, two large regular triangular pieces in *T. lusitanicus* and two large square-shaped pieces in *T. azorensis*) and by the absence of lateral body pores (one pair of lateral body pores in *T. viruliferus* and *T. azorensis*, one or two pairs in *T. lusitanicus*). Furthermore, *T. beirensis* differs from *T. azorensis* in the onchiostyle length (less than 55 µm *vs* more than 55 µm in *T. azorensis*) (Almeida *et al*, 1989). *T. beirensis* males slightly differ in details of spicule shape (clearly indented and usually with distinct terminal groove *vs* slightly indented usually without groove in *T. azorensis*).

T. beirensis females also resemble those of *T. variopapillatus* by having large rounded triangular vaginal sclerotizations and a transverse slit-like vulva. They differ in females essentially in the shape of the vagina (respectively narrow cylindrical *vs* wider barrel-shaped in *T. variopapillatus*) and by the absence of lateral body pores *vs* one prevulvar and one postvulvar lateral body pore in *T. variopapillatus*. In males, both species differ in the shape of the spicules (spicules withknob-like manubrium in *T. variopapillatus*).

Remarks

The following variations were observed: four out of 30 females had a short dorsal intestinal overlap; 13 out of 30 females were described with a distortion of the sclerotized pieces due to an intrusion of a substance of unknown origin in the vagina (Almeida *et al.*, 1989) and one out of 50 males had four ventromedian cervical papillae anterior to the S-E pore.

Distribution and host

T. beirensis is only recorded from Portugal, with Tapada do Salgeiro, Celorico da Beira the type locality. It is polyphagous with the following host range: rye (*Secale cereale*), rye-grass (*Lolium multiflorum*), hop (*Humulus lupulus*) (Almeida *et al.*, 1989).

Trichodorus borai Rahman, Jairajpuri & Ahmad, 1985
Figs 3.2.B; 3.4 (9d); 3.6.J; 3.10 (9d)

Measurements

Holotype male: L= 610 µm, a= 22, b= 5.0, T= 68%, onchiostyle= 50 µm, spicule= 29 µm, gubernaculum= 14 µm.

Males: L= 490-600 µm (560 µm), a= 21-26 (22), b= 4.1-4.8 (4.5), T= 59-66% (64%), onchiostyle= 48-51 µm (50 µm), spicule length= 27-33 µm (29 µm), gubernaculum= 13-15 µm, anterior end to EP= 68-75 µm, distance CP-EP= 22-29 µm.

Females: L= 540-630 µm (600 µm), a= 21-24 (22), b= 4.1-5.3 (4.8), V= 56-58% (57%), onchiostyle= 46-51 µm (49 µm), anterior end to EP= 68-78 µm (from Rahman *et al.*, 1985).

Emended diagnosis

T. borai male is characterized by a single ventromedian cervical papilla situated in the onchiostyle region, the position of the lateral cervical pore (just behind the base of the onchiostyle), the spicule shape (smooth, ventrally curved, broadened anteriorly, gradually narrowing to a pointed tip), spicule length (27-33 µm), the position of the posterior of the three precloacal supplements (just posterior to the head of the retracted spicules), and the non-thickened terminal tail cuticle. The female by the size, shape and orientation of the vaginal sclerotizations in lateral view (small, roughly ovoid, oblique to the body wall), the shape of the vulva (irregular pore-like), the barrel-shaped vagina. The pharyngeal bulb is offset in male and female.

Relationships

T. borai male resembles 12 other species of the genus in having only one ventromedian cervical papilla but differs from each by having it positioned in the onchiostyle region. It is also characterized by relatively small spicules, as in *T. azorensis*, *T. beirensis*, *T. lusitanicus*, *T. viruliferus* and *T. velatus*, but may be distinguished from all these species by the spicule shape (without constriction of the spicule shaft and without velum).

Remarks

No lateral body were pores observed in female.

Distribution and hosts

T. borai has only been reported from its type locality North Lakhimpur, Assam, India, associated with roots of bamboo (*Bambusa* sp.) (Rahman *et al*., 1985).

Trichodorus borneoensis Hooper, 1962
Figs 3.2.C; 3.3 (5f); 3.6.O; 3.9 (5f); 3.11.G; 4.3.A-D

Measurements

Holotype male: L= 770 µm, a= 20.5, b= 4.8, onchiostyle= 51 µm, spicule length= 51 µm, gubernaculum= 22 µm.

Males: L= 600-880 µm (720 µm), a= 16.6-23.4, b= 4.6-5.6, onchiostyle= 51-60 µm (53 µm), spicule length= 45-52 µm (49 µm).

Females: L= 650-920 µm (740 µm), a= 16.9-25.6, b= 4.9-5.9, V= 54-61% (57%), onchiostyle= 53-57 µm (54 µm) (from Hooper, 1962).

Emended diagnosis

T. borneoensis is characterized in the male by two ventromedian cervical papillae lying close together and immediately anterior to the S-E pore, the shape and ornamentation of the spicules (almost straight spicules, with slightly marked head, shaft slightly wider posteriorly, striated except at both extremities), by the position of the three precloacal supplements with the posterior two within the region of the retracted spicules (the second supplement opposite the spicule head). In the female by the shape of the vaginal sclerotizations in lateral view (small rounded triangular to oval), the pear-shaped vagina, the pore-shaped vulva and a pair of postadvulvar lateral body pores. The pharyngeal bulb is offset in males and females.

Relationships

T. borneoensis resembles *T. dilatatus*, *T. intermedius*, *T. nanjingensis* and *T. obscurus* in having two ventromedian precloacal supplements within the region of the retracted spicules (the second supplement about the level of the spicule head, except in *T. intermedius*), but differs from these species by the shape of the spicules (with shaft slightly widened posteriorly). Apart from *T. nanjingensis*, *T. borneoensis* can also be distinguished from former species by the male possessing two ventromedian cervical papillae (one papilla in *T. dilatatus* and *T. intermedius*, no papillae in *T. obscurus*).

Remarks

The position of the lateral cervical pore was not mentioned in the original description.

Distribution and host

T. borneoensis was originally described from North Borneo, from soil around the roots of Manila Hemp (*Musa textilis*). It has also been recorded from the rhizosphere of coral berry (*Ardisis crenata*) and cabbage palm (*Sabal palmetto*) in Florida (Mead, 1989b; Miller, 1974, both data not checked).

Trichodorus californicus Allen, 1957

Figs 3.1.A, 3.3 (4h, 4h'); 3.4 (9b); 3.6.E; 3.10 (4h, 4h', 9b); 4.16.F

Measurements

Holotype male: L= 1060 µm, a= 21, b= 6.6, T= 71%, onchiostyle= 82 µm, spicule length= 70 µm, gubernaculum= 32 µm.

Males: L= 670- 1060 µm (865 µm*), a= 15- 21, b= 4.2-6.6, T= 60-75%, onchiostyle= 54-82 µm (68 µm*), spicule length= 43-70 µm (56.5 µm*), gubernaculum= 20-32 µm.

Females: L= 790-1250 µm (1020 µm*), a= 15-20, b= 4.4-6.2, V= 57-61%, onchiostyle= 55-80 µm (67.5 µm*) (from Allen, 1957).

Emended diagnosis

T. californicus male can be distinguished by the single ventromedian cervical papilla immediately anterior to the S-E pore, the position of the lateral cervical pore (at the level of the ventromedian cervical papilla), the position of the three precloacal supplements (the posterior supplement near the cloacal opening, the second supplement shortly anterior to the head of retracted spicules), the shape of the spicules (manubrium usually offset, spicules ventrally bent at the extremities, in between shaft largely straight and setose) and terminal tail cuticle not thickened. The female by the shape of the vaginal sclerotizations in lateral view (large, roughly triangular, with inner side concave), the pear-shaped vagina and by one pair of postadvulvar lateral body pores. The pharyngeal bulb is offset.

Relationships

T. californicus resembles *T. dilatatus* in the female by the well developed triangular vaginal sclerotizations with inner side concave. Also in the male by the position of the posterior ventromedian precloacal supplement close to the cloacal opening (the second one at the level of the spicule head in *T. dilatatus*, shortly anterior to it in *T. californicus*) and by the presence of one ventromedian cervical papilla. It differs from *T. dilatatus* in having differently shaped spicules (manubrium more or less marked by the absence of striae and by being slightly wider than the striated shaft which is about equally wide in *T. californicus vs* manubrium wide and long, shaft finely striated and tapered to a fine distal end in *T. dilatatus*).

Remarks

With the SEM, the spicule shaft appeared surrounded by a sheath of small scales and the distal end of the spicule has a ventral longitudinal groove (Rodriguez-M & Bell, 1978). In Alaskan specimens, the spicules are often clearly bent ventrally in their posterior third, then recurved dorsally and the distal region of the spicule tapers noticeably to the tip. The spicular striae appear less dense than in the Californian specimens and are absent in the tapered distal region (Bernard, 1992a).

Distribution and host

T. californicus was originally described from specimens from Moraga Ridge, Conta Costa County (type locality) and from several other places in California, and from Echo Lake, Colorado. It has also been reported from South Dakota, Wisconsin, and many sites in Alaska between the Alaska Range and the Yukon River. Its occurrence in Florida appears anomalous (Bernard, 1992a). Bernard (1992a) determined the historical biogeography of *T. californicus* from the current distribution pattern, post glacial events, and morphology (see chapter on ecology).

T. californicus is a polyphagous nematode which is most abundant and common in taiga and other coniferous rhizospheres. It has also been reported from forest and mixed-range biotopes in the Pacific Coast Range, California, from *Poa pratensis* in South Dakota and from *Vaccinium macrocarpon* (cranberry), *Zea mays* (corn), *Acer* spp. and *Pinus* spp. in Wisconsin. According to Bernard (1992a) its occurrence in Wisconsin may be due to dispersal by human activity e.g. introduction in association with plant material.

Trichodorus carlingi Bernard, 1992
Figs 3.2.BB; 3.3 (5b); 3.7.C; 3.9 (5b); 4.3.E-J

Measurements

Holotype male: L= 813 µm, a= 28, b= 4.8, T= 60%, onchiostyle= 61µm, spicule length= 44 µm, gubernaculum= 17 µm, anterior end to EP= 120 µm.

Males: L= 682-955 µm (828 µm ± 84.6), a= 20-28, b= 4.8-5.4, T= 55-63%, onchiostyle= 57-67 µm (62 µm ± 3.7), spicule length= 44-47 µm (46 µm ± 0.9), gubernaculum= 15-18 µm, anterior end to EP= 106-127 µm, distance CP1-CP2= 11-26 µm, distance CP2-EP= 7-13 µm.

Females: L= 711-886 µm (801 µm ± 53.4), a= 20-24, b= 4.5-5.5, V= 50-57% (54% ± 2.2), onchiostyle= 56-65 µm (59 µm ± 3.2), anterior end to EP= 103-118 µm (from Bernard, 1992a).

Emended diagnosis

T. carlingi male is distinguished from all other species of the genus by their spicule shape (proximally wide, tapering to just posterior mid-shaft, distal part wider, strongly curved ventrally and provided with a velum; shaft smooth). It is further characterized by two ventromedian cervical papillae situated between the nerve ring and the S-E pore, the position of the lateral cervical pore about the level of the anterior ventromedian cervical papilla, ovate striated sperm with about oval nuclei, the position of the three precloacal supplements (all anterior to the retracted spicules) and the bluntly digitate tail with thickened terminal cuticle. The female is characterized by the pore-like vulva, the shape of the vaginal sclerotizations in lateral view (small trapezoidal pieces), vagina very short (about 1/3rd of corresponding body width) barrel-shaped and two lateral body pores on each side (one prevulvar, one posterior to the vulva and advulvar). The pharyngeal bulb is offset in males and females.

Relationships

T. carlingi is morphologically quite similar to *T. aequalis* except for the spicule shape and ovate sperms with striated surfaces and almost oval nuclei in *T. carlingi* (elliptical with elongate nuclei in *T. aequalis).* Females of *T. carlingi* are very similar to *T. aequalis* and to *T. sparsus* and according to Bernard (1992a) cannot be reliably separated from them.

Distribution and host

T. carlingi occurs at several sites in Alaska. It is polyphagous, and has been found in the rhizosphere of black spruce (*Picea mariana*, mixed black spruce-willow (*Salix* sp.), aspen (*Populus tremuloides*), Labrador tea (*Ledum decumbens*) groundcover and mixed paper birch (*Betula payrifera*) -black spruce and mixed grass (Bernard, 1992a).

Trichodorus cedarus Yokoo, 1964
Figs 3.2.D; 3.4 (8f, 9a); 3.6.P; 3.9 (8f, 9a); 4.4.A-I; 6.8.A-J

syn. *T. kurumeensis* Yokoo, 1966
syn. *T. longistylus* Yokoo, 1964

See chapter on economically important species.

Trichodorus complexus Rahman, Jairajpuri & Ahmad, 1985
Figs 3.1.G; 3.4 (8a); 3.6.D; 3.9 (8a)

Measurements

Holotype male: L= 740 µm, a= 24.6, b= 3.8, T= 40%, onchiostyle= 45 µm, spicule length= 43 µm, gubernaculum= 17 µm, anterior end to EP= 90 µm, distance CP-CP2= 14 µm, distance CP2-EP= 6 µm. Females: L= 660-770 µm (700 µm), a= 22-23, b= 4.2-5.5, V= 53-56% (54.5%), onchiostyle= 42-45 µm (43.6 µm), anterior end to EP= 83 µm (from Rahman *et al.*, 1985).

Emended diagnosis

T. complexus male is characterized by two ventromedian cervical papillae just anterior to the S-E pore, the length (43 µm) and shape of the spicules (manubrium knob-like, shaft largely cylindrical, finely striated, ventrally bent), the arrangement of the three precloacal supplements (the posterior supplement at level of the spicule head) and the non-thickened terminal tail cuticle. In the female by the shape of the vaginal sclerotizations (small rounded triangular to oval pieces, parallel to the body wall), the pore-like vulva, vagina about cylindrical and one pair of postadvulvar lateral body pores. Also, by the length of the onchiostyle (42-45 µm) and the offset pharyngeal bulb in males and females.

Relationships

T. complexus females have vaginal sclerotized pieces, in lateral view, parallel to the longitudinal body axis, a character observed only in *T. cottieri* and *T. obscurus*. They differ from both species in having small rounded triangular/oval sclerotizations instead of being large and approximately rectangular. The females of *T. complexus* also differ from those of *T. cottieri* and *T. obscurus* by the shape of the vulva: pore-like *vs* a transverse slit. *T. complexus* males are distinguished from those of the former species by the shape of their spicules (shaft about equally wide and with a knob-like offset manubrium *vs* shaft posteriorly enlarged, without marked manubrium in *T. cottieri* and *T. obscurus*). *T. complexus* is similar to *T. similis* by the shape of the spicules, but differs from it in possessing two ventromedian cervical papillae in the male *vs* three in *T. similis*, and in the female by the shape of the vaginal sclerotizations (larger rounded triangular in *T. similis)* and the shape of the vulva (pore-like *vs* a transverse slit in *T. similis*).

Remarks

The presence of a lateral cervical pore was not mentioned in the original description.

Distribution and host

T. complexus was described from specimens from Jumi, Urunachal Pradesh, India where it was found in the rhizosphere of cane (*Calamus viminalis*)(Rahman *et al.*, 1985).

Trichodorus coomansi De Waele & Carbonell, 1983

Figs 3.2.E,e; 3.4 (8b); 3.6.B; 3.9 (8b); 3.11.H; 4.5. G-I

Measurements

Holotype male: L= 991 µm, a= 21.5, b= 6.1, T= 69.1%, onchiostyle= 65 µm, spicule length= 51 µm, gubernaculum= 26 µm, anterior end to EP= 114 µm, distance CP1-EP= 10 µm, distance EP-CP2= 30 µm.
Males: L= 828-1130 µm (954 µm, 938 µm), a= 21.0-35.3, b= 4.7-6.4, T= 60.7-79.7%, onchiostyle= 50-66 µm (63 µm, 53 µm), spicule length= 40- 54 µm (52 µm, 44 µm), gubernaculum= 18-26 µm, anterior end to EP**= 102-126 µm, distance CP1-EP**= 5-21 µm, distance EP-CP2**= 6-30 µm.
Females: L= 795-1014 µm (913 µm, 906 µm), a= 17.7-32.9, b= 4.7-6.5, V= 51.8-65.2% (55.1%, 56.8%), onchiostyle= 50-65 µm (61 µm, 54 µm), anterior end to EP= 104-159 µm (from De Waele & Carbonell, 1983).
** Males with EP anterior or posterior to CP1 and CP2 omitted.

Emended diagnosis

T. coomansi is distinguished by the male having two ventromedian cervical papillae (one anterior and one posterior to the S-E pore), the position of the lateral cervical pore at the level of the S-E pore, the shape of the spicules (slightly arcuate, manubrium knob-like, shaft striated, slightly narrower at mid-length), the position of the three precloacal supplements with the posterior supplement shortly posterior to the spicule head of retracted spicules and the symmetrical tail with thickened terminal cuticle. The female by the shape of the vaginal sclerotizations in lateral view (small triangular), two lateral body pores on each side (one prevulvar and one post-advulvar), and rhomboid-shaped vagina (when relaxed). The pharyngeal bulb is usually offset in males and females.

Relationships

T. coomansi males resemble *T. parorientalis* and *T. rinae* by having two ventromedian cervical papillae with the S-E pore situated between and the posterior precloacal supplement situated near the head of the retracted spicules. *T. coomansi* females have small, more or less triangular-shaped vaginal sclerotizations unlike those in *T. parorientalis* and *T. rinae*. *T. coomansi* males differ from *T. parorientalis* and *T. rinae* by the shape of the spicules (curved ventrally with marked knob-like manubrium *vs* manubrium not knob-like and with a mid-shaft constriction in *T. parorientalis* and only proximally curved, then straight and without knob-like manubrium in *T. rinae*) and the terminal tail cuticle (thickened in *T. coomansi vs* slightly thickened in *T. rinae* and not thickened in *T. parorientalis*). *T. coomansi* females may be distinguished from *T. parorientalis* by the rhomboid-shaped vagina *vs* a pear-shaped vagina in *T. parorientalis*, and from *T. rinae* which has only one postadvulvar lateral body pore on each side.

Remarks

Exceptionally, both medioventral cervical papillae are posterior (1♂) or anterior (2♂) to the S-E pore. Five out of a population of 18 females had a short dorsal intestinal overlap (De Waele & Carbonell, 1983).

Distribution and host

T. coomansi was originally described from specimens from the north-west slope of Mount Kenya. It is a polyphagous species with the following hosts: *Arundinaria alpina* (Poaceae), various Pteridophyta (*Adianthum poirettii*, *Cheilanthes farinosa*, *Polystichum setiferum, Asplenium monanthes, Pteris catoptera), Erica arborea* (Ericaceae), *Alchemilla argyrophylla* (Rosaceae), *Hypericum keniense* (Clusiaceae), *Lobelia keniensis, Senecio keniodendron* (Lobeliaceae), *Podocarpus milanjianus* and *Lobelia* sp. (De Waele & Carbonell, 1983).

Trichodorus cottieri Clark, 1963
Figs 3.1.E; 3.3 (5c); 3.5.Q; 3.9 (5c); 3.11.I; 4.5.A-F

Measurements

Holotype male: L= 830 µm, a= 16.1, b= 5.01, T= 62%, onchiostyle= 60 µm, spicule length= 68 µm, gubernaculum= 23 µm.

Males: L= 800-1030 µm (880 µm ± 75), a= 16-19, b= 4.3-5.3, T= 53-62%, onchiostyle= 53-64 µm (58.5 µm*), spicule length= 44-53 µm (48.5 µm*).

Females: L= 700-1010 µm (890 µm ± 83), a= 14- 20, b= 3.6-6.6, V= 52-60% (56%*), onchiostyle= 57-68 µm (62.5 µm*) (from Clark, 1963).

Emended diagnosis

T. cottieri male is characterized by the spicule shape (largely straight except for the slightly curved ends, with an enlarged distal third of the shaft and shaft largely striated), the position of the three precloacal supplements with the posterior supplement just posterior to mid-shaft of retracted spicules, terminal tail cuticle not widened, the presence of a single ventromedian cervical papilla just anterior to the S-E pore and the position of the lateral cervical pore slightly posterior to the S-E pore. In the female by the shape of the vagina (almost cylindrical), the shape of the vaginal sclerotizations in lateral view (large, more or less rectangular, parallel to the longitudinal body axis), the transverse slit-like vulval opening and one pair of postadvulvar lateral body pores. The pharyngeal bulb is offset in males and females.

Relationships

T. cottieri is closest to *T. obscurus* with the male having a similar spicule shape with posteriorly enlarged shaft and with the female having a similar vulva (a transverse slit) and the orientation of their vaginal sclerotizations in lateral view (parallel to the longitudinal body axis). Both species differ in males, by the number of ventromedian cervical papillae (one in *T. cottieri*, absent in *T. obscurus*) and the terminal tail cuticle (not thickened in *T. cottieri vs* thickened in *T. obscurus*), and in females, by the slightly different shape and size of the vaginal sclerotized pieces (roughly rectangular in *T. cottieri*, larger and more trapezoid in *T. obscurus*).

Distribution and host

T. cottieri is known from the type population from New Zealand, from soil around the roots of *Carpodetus serratus* and *Weinmannia racemosa* (Clark, 1963). Recently, it has also been recorded from Alachna County, Gainesville, Florida, in association with *Elytratia caroliniensis* (tubiflora) (Esser, 1993) (not checked).

Trichodorus cylindricus Hooper, 1962

Figs 3.1.I, i, i'; 3.3 (5a); 3.5.R; 3.9 (5a); 3.11.J; 4.6A-J

Measurements

Holotype male: L= 680 µm, a= 21.8, b= 5.1, T= 64%, onchiostyle= 38 µm, spicule length= 36 µm, gubernaculum= 17 µm.

Males: L= 500-800 µm (710 µm, 679 µm ± 88), a= 17.6-28.6, b= 4.1-6.2, T= 57-68%, onchiostyle= 37-48 µm (40 µm, 44 µm ± 3.4), spicule length= 28- 49 µm (37 µm, 40 µm*, 33 µm ± 4.0), anterior end to EP= 67-98 µm.

Females: L= 419-770 µm (690 µm, 642 µm ± 87), a= 16.8-26.3, b= 4.2-6.0, V= 53-65% (58%, 56% ± 1.7), onchiostyle= 35-52 µm (40 µm, 45 µm ± 4.0), anterior end to EP= 90-105 µm. (from Hooper, 1962; Brzeski, 1974, ♂ only; Gomez-Barcina & Castillo, 1988).

Emended diagnosis

T. cylindricus male is characterized by the habitus (with straight posterior body region), the presence of caudal alae (although inconspicuous), the spicule shape and ornamentation (manubrium smooth, not marked; anterior part of shaft with striae and bristles, posterior shaft enlarged and smooth), the number and position of the ventromedian cervical papillae (three, all papillae anterior to the S-E pore; anterior papilla in the region of the onchiostyle), the position of the lateral cervical pore just posterior to the base of the onchiostyle, the non-thickened terminal tail cuticle and the three precloacal supplements with the posterior supplement clearly within the region of retracted spicules. In the female by the barrel-shaped vagina, the size and shape of the vaginal sclerotizations in lateral view (well developed, rounded triangular with their tips directed towards the vulva), the short transverse slit-like vulva and a single pair of postadvulvar body pores. The pharyngeal bulb has a short pharyngeal overlap or both a short pharyngeal and a short dorsal intestinal overlap in males and females.

Relationships

T. cylindricus males resemble those of *T. cottieri* and *T. obscurus* in having spicules with a posteriorly enlarged shaft and the posterior precloacal supplement located clearly within the region of retracted spicules. They differ in having smaller spicules of different shape, a larger number of ventromedian cervical papillae (three *vs* respectively one and none) and by the habitus of the posterior body region (straight, provided with a bursa in *T. cylindricus vs* ventrally curved, without bursa in *T. cottieri* and *T. obscurus*). Furthermore, males of *T. cylindricus* agree with those of *T. paracedarus* by the straight posterior body region, presence of caudal alae and less developed copulatory muscles than typical for the genus, but differ mainly by the different spicule shape. *T. cylindricus* females have a transverse slit-like vulva as do those of *T. cottieri* and *T. obscurus* but differ by the orientation and shape of the vaginal sclerotizations in lateral view (medium-sized triangular, with tips toward the vulva).

Remarks

The sclerotized vaginal pieces may show some slight variability in size and shape (rounded triangular to oval).

Distribution and host

T. cylindricus is common in Europe. It was originally described from England, but is also recorded from Poland, Belgium, France, The Netherlands, Germany, Switzerland, Spain and recently reported from Florida. The species usually occurs in sandy soils and appears not as polyphagous as many of the other species of the genus. It is most common in soil from pastures and from grasses along road sides, but has also been found in the rhizosphere of potatoes, lettuce, sugar beet, strawberry, *Cupressus sempervirens* and rarely from deciduous woodland (*Crataegus* sp., *Populus* sp., *Alternanthera ramosissima, Betula nigra*= river birch), cereal crops (maize, barley) or beggar weed and St Augustine grass (Hooper, 1962; Rau, 1975; van Hoof, 1968; Brzeski & Szczygiel, 1974; Heathcote, 1973; Gügerli, 1977; Gomez-Barcina & Castillo, 1988; Anonymous, 1988b; Mead, 1979, 1990).

Trichodorus dilatatus Rodriguez-M & Bell, 1978
Figs 3.1.B; 3.4 (7a); 3.5.K; 3.9 (7a); 4.7.A-I

Measurements

Holotype male: L= 900 µm, a= 18.0, b= 4.9, T= 69.0%, onchiostyle= 65 µm, spicule length= 49 µm, gubernaculum= 19 µm, anterior end to EP= 132 µm, distance CP1-EP= 10 µm.

Males: L= 720-1030 µm (870 µm ± 100), a= 14-19, b= 4.5-6.1, T= 50- 70%, onchiostyle= 63-69 µm (63 µm ± 2.3), spicule length= 48-55 µm (53 µm ± 2.3), gubernaculum= 11-14 µm.

Females: L= 720-1030 µm (870 µm ± 100), a= 13-17.7, b= 5.6-6.1, V= 54-61.7% (58% ± 2.3), onchiostyle= 60.5-69 µm (65 µm ± 2.9) (from Rodriguez-M & Bell, 1978).

Emended diagnosis

T. dilatatus male can be distinguished by the spicule shape (proximal part wide, uniformly tapered to a fine distal end, shaft striated, provided with bristles), by the location of the three precloacal supplements (posterior two supplements within the region of retracted spicules, the middle one at about the spicule head, with a small range of variation), large rounded sperm with rounded nucleus, the presence of a single ventromedian cervical papilla just anterior to the S-E pore, the position of the lateral cervical pore anterior to the ventromedian cervical papilla and the non-thickened terminal tail cuticle. In the female, by the size and shape of the vaginal sclerotizations (medium-sized, roughly rounded triangular with inner side concave), the transverse vulva slit, the pear-shaped vagina and one pair of postadvulvar lateral body pores. The pharyngeal bulb is offset in males and females.

Relationships

T. dilatatus resembles *T. minzi*, *T. taylori* and *T. yokooi* in possessing spicules with an enlarged and long proximal end, but differs from *T. minzi* and *T. taylori* in the non-offset manubrium. *T. dilatatus* females resemble those of *T. minzi* by the pear-shaped vagina but differs from them in other characters as shown in table 4.1.

Distribution and host

T. dilatatus is known only from the original record in California, with cottonwood (*Populus* sp.) and manzanita (*Arctostaphylos* sp.) as hosts (Rodriguez-M & Bell, 1978).

Table 4.1.

	T. dilatatus	*T. minzi*	*T. taylori*	*T. yokooi*
n° cp	1	2	2	3
n° SP in region of spicules	2	1	1	0
shape of vaginal sclerotizations	large triangular, inner side concave	large rounded to oval	large rounded	medium-sized, triangular, tips toward vulva
vulva shape	transverse slit	?	?	rounded

Trichodorus eburneus De Waele & Carbonell, 1983

Figs 3.2.F, f, f'; 3.3 (4d, d'); 3.5.F; 3.8 (4d, d'); 3.11.K; 4.8.A-E

syn. *Trichodorus proximus* apud Baujard, 1983

Measurements

Holotype male: L= 750 µm, a= 27.0, b= 4.3, T= 78%, onchiostyle= 48 µm, spicule length= 50 µm, gubernaculum= 23 µm, anterior end to EP= 109 µm, distance CP1-EP= 13 µm.

Males: L= 610-1100 µm (787µm; 693 µm; 740 µm; 920 µm ± 82; 710 µm ± 63; 650 µm ± 89; 750 µm ± 63), a= 15-32, b= 4.3-7.6, T= 59.2-65%, onchiostyle= 40-67.5 µm (52 µm, 44 µm; 55 µm; 57 µm ± 1.8; 55 µm ± 1.5; 55 µm ± 8.0; 53 µm ± 6.8), spicule length= 42-55 µm (48 µm; 47 µm; 50 µm; 52 µm ± 2.1; 51 µm ± 2.5; 46 µm ± 2.8; 50 µm ± 1.6), gubernaculum= 17-26 µm, anterior end to EP= 82.5-140 µm, distance CP1-EP= 12-27 µm.

Females: L= 550-1210 µm (903 µm*; 679 µm; 670 µm ± 88; 930 µm ± 86; 720 µm; 690 µm ± 80), a= 16-39, b=4.0-7.3, V= 51-60% (56%; 55.2%; 56.5%; 55% ± 1.6; 57%; 56% ± 2.5), onchiostyle= 42- 60 µm (49.5 µm*; 47 µm; 56 µm; 57 µm ± 1.9; 57 µm; 49 µm ± 7.0), anterior end to EP= 89- 150 µm (from De Waele & Carbonell, 1983: 2 populations; Baujard, 1983: 4 populations of ♀, 5 of ♂).

Emended diagnosis

T. eburneus male is distinguished by a combination of the following characters : the shape of the spicules (usually slightly cephalated; proximally ventrally bent, shaft straight except at distal tip, striated, without bristles), three precloacal supplements with the posterior supplement at the level of the spicule head, the possession of a single ventromedian cervical papilla in front of the S-E pore, the position of the lateral cervical pore slightly anterior to the ventromedian cervical papilla and the slightly thickened terminal tail cuticle. In the female by the size and shape of the vaginal sclerotizations in lateral view (small triangular to rounded), the barrel-shaped vagina, one pair of lateral body pores well anterior to the vulva and one pair of subventral (instead of lateral) body pores within one body width posterior to the vulva. The pharyngeal bulb is offset in males and females.

Relationships

T. eburneus has one ventromedian cervical papilla as in thirteen other species of the genus but among them only *T. philipi* and some specimens of *T. obtusus* have similar spicules i.e. with the proximal end ventrally curved and shaft largely straight and with the posterior precloacal supplement at the level of the head of the retracted spicules; in *T. obtusus* no supplements within the region of retracted spicules. *T. eburneus* females differ from the former species, by the size and shape of their vaginal sclerotizations in lateral view (small (rounded) triangular in *T. eburneus vs* slightly larger oval or triangular in *T. obtusus* and larger, rounded in *T. philipi*). Also, *T. eburneus* may be distinguished from *T. philipi* by its longer onchiostyle (40-67.5 µm *vs* 26- 30.5 µm) and longer gubernaculum (17-26 µm *vs* 10.5-15 µm).

Remarks

A dorsal overlap of the pharyngeal bulb by the intestine was rarely observed in the type population (one male out of nine). However, in the several populations described by Baujard (1983) 75% of the specimens showed a dorsal overlap of the pharynx by the intestine. The author remarked that this was not due to fixation. A few abnormalities were observed among the populations from West Africa, e.g. a male without ventromedian cervical papillae and lateral cervical pores, a male with an additional precloacal supplement between the second and third supplement and a female with an additional lateral pore, anterior to the vulva.

Distribution and host

T. eburneus was described from specimens from the Ivory Coast. Subsequently, Baujard (1983) reported the presence of *T. proximus* (= *T. obtusus*) in Senegal, Upper Volta and the Ivory Coast but De Waele (1986) identified these as *T. eburneus*.

T. eburneus is a polyphagous species with the following hosts: *Diospyros chevalieri* and *Uapaca guineensis* from primary forest, grasses (*Loudetia simplex* and

Hyparrhenia sp.) from the Guinea savanna, *Arachis hypogea* (peanut), *Icacina senegalensis, Saccharum officinarum* (sugar-cane), *Oryza sativa* (rice) and unidentified grass in the National Park of Niokolokoba in Senegal.

Trichodorus elefjohnsoni Bernard, 1992

Figs 3.2.U; 3.4 (9m, m'); 3.7.D; 3.10 (9m, m'); 4.9.D-F

Measurements

Holotype male: L= 594 µm, a= 18, b= 4.9, T= 66%, onchiostyle= 52 µm, spicule length= 41 µm, gubernaculum= 17 µm, anterior end to EP= 91 µm, distance CP1-CP2= 25 µm, distance CP2-EP= 9 µm.

Males: L= 521-731 µm (608 µm ± 57.5), a= 17-24, b= 4.1-5.7, T= 56-68%, onchiostyle= 45-60 µm (53 µm ± 3.5), spicule length= 33-50 µm (40 µm ± 4.8), gubernaculum= 14-24 µm, anterior end to EP= 73-97 µm, distance CP1-CP2= 13-30 µm, distance CP2-EP= 4-9 µm.

Females: L= 516-731 µm (626 µm ± 59.4), a= 17-24, b= 3.9-5.6, V= 53-60% (56% ± 1.9), onchiostyle= 45-60 µm (52 µm ± 4.5), anterior end to EP= 73-102 (from Bernard, 1992b).

Emended diagnosis

T. elefjohnsoni male is distinguished by the following combination of characteristics: two ventromedian cervical papillae anterior to the S-E pore (anterior papilla near onchiostyle base), the position of the lateral cervical pore just posterior to the anteriormost ventromedian cervical papilla, slender spicules (with wider, smooth head, shaft tapering posteriorly to a fine bifid tip, without bristles, but with numerous fine striations in two groups separated by a slightly narrower smooth part), three precloacal supplements with the posterior supplement varying in position from the proximal end of the spicules on specimens with short spicules to clearly within the region of the retracted spicules on specimens with longer spicules, the tail shape (with rounded to bluntly digitate terminus with slightly thickened cuticle). In the female by the vaginal sclerotizations in lateral view (small rounded to ovate), the very short (about 1/3rd of the corresponding body width) barrel-shaped vagina, the pore-like vulva and one pair of postadvulvar lateral body pores.

Relationships

T. elefjohnsoni males most closely resemble *T. orientalis* De Waele & Hashim, 1984 in having slender, medially narrower spicules (Bernard, 1992b), but differ by differences in the striation of the spicule shaft (fine striae only in distal part of shaft in *T. orientalis*) and sharp constriction of the shaft in *T. orientalis vs* a longer narrower shaft zone in *T. elefjohnsoni*). Females of *T. orientalis* have minute, triangular vaginal sclerotizations, whereas in *T. elefjohnsoni* they are larger rounded to ovate pieces.

T. elefjohnsoni also is somewhat similar to *T. taylori* De Waele, Mancini, Roca & Lamberti, 1982 and to *T. persicus* De Waele & Sturhan, 1987. Males of *T. taylori* are larger, with longer, stouter spicules, which are weakly curved and striated only on the calamus compared to those of *T. elefjohnsoni*. Males in the Iranian species *T. persicus* also have stouter spicules, striated only on their distal halves, and the females, have two pairs of lateral body pores near the vulva, whereas there is just one pair in *T. elefjohnsoni* (Bernard, 1992b).

Remarks

The posterior body region is curved in fixed specimens, the copulatory muscles are well developed and a rudimentary bursa may be observed which is obscure, being more visible in living active specimens than in fixed specimens. Bernard (1992b) observed that during movements of the males, the alae tended to flatten, then reappear. He considered that the alae in *Trichodorus* are formed by temporary muscular activity and are not fixed cuticular expansions as in many Tylenchida. A rudimentary bursa has also been observed in mounted specimens of several other *Trichodorus* species (see chapter 2).

The length of the spicules in *T. elefjohnsoni* (SD= 4.8, CV= 12.0) is much more variable than usually reported for *Trichodorus* species. Bernard (1992b) suggested that the variability of the spicule length between populations may be due to severely restricted gene flow because of the numerous gorges and fast-flowing streams that dissect the Great Smoky Mountains National Park, which probably tend to produce isolated populations affected by different host and micro-environmental conditions, or simply to random variation inherent in small sample sizes. The five populations were compared on the basis of 2 to 6 specimens. The maximum range in spicule length within a population is 7 µm with the range of the calculated mean values (from 32.5 to 46.5 µm) between the populations being twice as large.

Distribution and host

T. elefjohnsoni has recently been described from several localities in the Great Smoky Mountains National Park, Alaska at altitudes between 800 and 1144 m. It is a polyphagous species, found in the rhizosphere of tuliptree (*Liriodendron tulipifera*), mountain apple (*Acer pensylvanicum*), blackberry (*Rubus* sp.). It has also been recorded from mixed forest either of silverbell (*Halesia carolina*), buckeye (*Aesculus glabra*) and hemlock (*Tsuga canadensis*), or of mixed forest dominated by yellow birch (*Betula lutea*), buckeye, and red oak (*Quercus rubra*) or mixed forest dominated by cucumbertree (*Magnolia acuminata*), buckeye, and tuliptree (Bernard, 1992b).

Trichodorus elegans Allen, 1957

Figs 3.1.D; 3.3 (5e); 3.6.M; 3.9 (5e); 3.11.C; 4.8.F-H, 4.9.A-C

Measurements

Holotype male: L= 1100 µm, a= 15, b= 4.4, T= 63%, onchiostyle= 140 µm, spicule length= 62 µm, gubernaculum= 22 µm.

Males: L= 950-1170 µm (1060 µm*), a= 14-16, b= 4.0-4.7, T= 55-63%, onchiostyle= 140-155 µm (147.5 µm*), spicule length= 55-62 µm (58.5 µm*), gubernaculum= 20-22 µm.

Females: L= 860-1230 µm (1045 µm*), a= 14-16, b= 4.0-4.8, V= 56-65% (60.5%*), onchiostyle= 140-150 µm (147.5 µm*) (from Allen, 1957).

Emended diagnosis

T. elegans is characterized by its long onchiostyle (140-155 µm), and long body length (860-1230 µm). In the male, the species is distinguished by a single ventromedian cervical papilla anterior to the S-E pore, the position of lateral cervical pores at the same level as the ventromedian cervical papilla, three precloacal supplements with the posterior supplement opposite the posterior half of the retracted spicules, the shape of the spicules (shaft about straight and striated), the non-thickened terminal tail cuticle and in the female by the size and shape of the vaginal sclerotizations (large, roughly rectangular with indented inner side, and largely parallel to the body axis), the barrel-shaped vagina and one pair of postadvulvar lateral body pores. The pharyngeal bulb is offset in males and females.

Relationships

T. elegans is most closely related to *T. cottieri* in having one ventromedian cervical papilla and the posterior precloacal supplement just posterior to mid-spicule when spicules are retracted in the male, and by the orientation of the vaginal sclerotizations, parallel to the longitudinal body axis in the female. It differs from *T. cottieri* in the shape of the spicules (straight, shaft about equally wide *vs* ventrally curved at both ends and shaft clearly widened posteriorly in *T. cottieri)* in the male, and by the shape of the vaginal sclerotizations (roughly rectangular with indented inner side *vs* rectangular in *T. cottieri*) in the female. *T. elegans* may also be distinguished by the very long onchiostyle (140-155 μm), the longest in the genus.

Distribution and host

T. elegans is known by the type population from Kaniksu National Forest, Idaho, found in the rhizosphere of pine (Allen, 1957).

Trichodorus giennensis Decraemer, Roca, Castillo, Pena-Santiago & Gomez-Barcina, 1993

Figs 3.2.G.g; 3.4. (9c); 3.7.E.; 3.10 (9c); 4.10.A-F

Measurements

Holotype male: L= 805 μm, a= 24.5, b= 5.2, T= 64.1%, onchiostyle= 58 μm, spicule length= 44 μm, gubernaculum= 22 μm, anterior end to EP= 112 μm, distance CP1-CP2= 24 μm, distance CP2-EP= 16 μm.

Males: L= 550-1005 μm (734 μm ± 29.5; 861 μm ± 63; 781 μm ± 82.2; 787.5 μm ± 3.4; 887.5 μm ± 96.8), a= 16.0-31.0, b= 3.5-7.5, T= 50.9-72.5%, onchiostyle= 44.5-60.0 μm (57.4 μm ± 1.8; 52.5 μm ± 3.0; 54.0 μm ± 2.7; 55.5 μm ± 2.1; 53.5 μm ± 2.0), spicule length= 39.5-50.0 μm (43.0 μm ± 1.9; 46.0 μm ± 2.2; 44.5 μm ± 2.1; 43.5 μm ± 2.4; 46.0 μm ± 2.8), gubernaculum= 11.5-20 μm, anterior end to EP= 80.5-135.5 μm, distance CP1-CP2= 14.0-36.0 μm, distance CP2-EP= 6.0-22.0 μm.

Females: L= 700-1177 μm (762.0 μm ± 46.2; 875.5 μm ± 58.5; 927.5 μm ± 163.0; 797.5 μm ± 20.3; 861.0 μm ± 70.0), a= 16.5-27.5, b= 4.0-7.5, V= 49.0-62.0% (58.1% ± 1.9; 56.0% ± 1.4; 55.0% ± 3.0; 55.5% ± 0.9; 57.5% ± 3.4), onchiostyle= 48-62.5 μm (57.3 μm ± 3.5; 54.0 μm ± 3.5; 53.0 μm ± 3.0; 53.0 μm ± 3.3; 54.5 μm ± 1.4), anterior end to EP= 82.5-135.5 μm (from Decraemer *et al.*, 1993: 5 populations).

Diagnosis

T. giennensis is characterized by an onchiostyle of average length, between 52.5 and 57.5 μm (several populations), no overlap of the pharyngeal glands but usually with a dorsal intestinal overlap. It is distinguished in the male: by two ventromedian cervical papillae between onchiostyle base and S-E pore, the position of the lateral cervical pore near the nerve ring (i.e. near the anterior ventromedian cervical papilla or in between both cervical papillae), the shape of the spicules (a widened manubrium, a slender shaft distally tapered, at mid-level with a slight narrowing and usually provided with a few bristles), by the arrangement of the three precloacal supplements (the posterior one within the region of the retracted spicules near the manubrium), large sperm cells with a strong fibrillar appearance and by the tail shape (asymmetrically rounded to slightly digitate) and usually slightly thickened terminal tail cuticle. In the female, by a barrel-shaped vagina (when relaxed), small triangular to irregularly rounded vaginal sclerotized pieces, clearly separated in lateral optical section, and by the presence of a single pair of postadvulvar lateral body pores, rarely with a prevulvar lateral body pore on each side at about five body widths anterior to the vulva.

Relationships

T. giennensis closely resembles the species formerly considered as the "*T. aequalis* group" (see higher). It is closest to *T. parorientalis* and *T. sparsus* having similar morphometric data, number of ventromedian cervical papillae (2), a comparable general spicule shape, but a more slender shaft and usually with a minor narrowing and only a few bristles in *T. giennensis vs* shaft equally wide and with numerous bristles (when observable) in *T. sparsus*, and shaft with a more pronounced constriction in *T. parorientalis*.

T. giennensis differs from *T. parorientalis* and *T. sparsus* in the female by a single pair of postadvulvar body pores in the type population and most other specimens, rarely two pores i.e. a lateral postadvulvar and a prevulvar lateral body pore on each side vs two lateral body pores in the other species.

Remarks

The following rarities were observed: two out of 67 males had only one ventromedian cervical papilla; two males had four precloacal supplements and one with five precloacal supplements; some specimens had an obviously swollen end cuticle (swelling influenced by fixation). In females the vagina is barrel-shaped when relaxed but otherwise up to long cylindrical and some variability in size and shape of the small vaginal sclerotizations was observed (see diagnosis), also, a secretion plug may be present at the vagina (observed in one third of the females in two populations). Occasionally, the pharynx shows a small overlap of the pharyngeal glands (e.g. of 9 µm) and exceptionally the anterior dorsal intestinal overlap is lacking.

Third-stage juveniles were identified based upon the body length (568-641 µm), the onchiostyle length (45-52 µm with a 16-18 µm long replacement onchium) and the reproductive system (31-34 µm) with a clear spicular primordium in the male juvenile.

Distribution and host

T. giennensis was described from natural vegetation in a mountainous area in southeastern Spain. It was found in the rhizosphere of *Paeonoi-Quercetum rotundifoliae* (holm-oak forest), *Olea europea* (olive), *Asphodelus* sp., *Alnus glutinosa* L., *Fraxinus angustifolia* and grass (Decraemer *et al.*, 1993).

Trichodorus hooperi Loof, 1973

Figs 3.2.S; 3.4 (7e, 8e,e'); 3.5.N; 3.9 (7e, 8e,e'); 4.11.A-G; 4.12.A-D

Measurements

Holotype male: L= 900 µm, a= 22, b= 5.1, T= 64%, onchiostyle= 60 µm, spicule length= 51 µm.

Males: L= 790-1050 µm (920 µm*), a= 19-25, b= 4.1-5.4, T= 54-68%, onchiostyle= 54-62 µm (58 µm*), spicule length= 44-53 µm (48.5 µm*), gubernaculum= 19 µm, distance CP1-CP2 = 16-30 µm, distance CP2-EP= 5-12 µm.

Females: L= 710-1100 µm (905 µm*), a= 18-24, b= 3.4-6.2, V= 53-59% (56%*), onchiostyle= 53-60 µm (56.5 µm*) (from Loof, 1973).

Emended diagnosis

T. hooperi male is characterized by the shape of the spicules (manubrium wider, short knob-like or anterior part longer, gradually tapering to an almost straight fine shaft, often with a minor narrowing at mid-corpus, finely striated, without bristles), the asymmetric rounded to digitate tail with thickened terminal cuticle, two ventromedian cervical papillae anterior to the S-E pore, the position of the pair of lateral cervical pores posterior to the nerve ring, i.e. between the ventromedian cervical papillae, three precloacal supplements with the posterior supplement usually just posterior to the head of retracted spicules. In the female by a transverse slit-like vulva, the presence of only one pair of posterior advulvar body pores, the shape of the vagina being almost pear-shaped (conoid in original description) and the shape of the vaginal sclerotized pieces in lateral view (medium-sized, rounded). The pharyngeal bulb is offset in males and females.

Relationships

T. hooperi male resembles *T. coomansi*, *T. giennensis* and *T. sparsus* by the two ventromedian cervical papillae, the position of the posterior supplement within the region of retracted spicules, and the thickened terminal tail cuticle. *T. hooperi* male differs from these species by the different spicule shape (shaft very fine and almost straight, without bristles *vs* shaft ventrally bent, with a few bristles in *T. giennensis*, without bristles in *T. coomansi* and with or without bristles in *T. sparsus*). *T. hooperi* female differs from *T. sparsus* by the shape of the vulva (a transverse slit *vs* a pore in *T. sparsus*), and from the three species by the shape of the vagina (pear-shaped vs barrel-shaped in *T. coomansi*, *T. giennensis* and *T. sparsus*), the shape of the vaginal sclerotizations in lateral view (small rounded *vs* small oblique triangular or rounded triangular in *T. coomansi*, *T. giennensis* and *T. sparsus*).

Remarks

Data on the presence of lateral cervical pores in male were not given in the original description and, *T. hooperi* was characterized by smooth spicules i.e. without striae. Examination of several slides with paratype specimens revealed that they all had striated spicules with some having the manubrium widened and appearing offset from the striated shaft.

The following variations were found among the type material: one male with the S-E pore between the two ventromedian cervical papillae (Loof, 1973), two males with four ventromedian precloacal supplements.

Distribution and host

T. hooperi was described from specimens from a sandy loam soil in mixed woodland with herbaceous undergrowth in England (Loof, 1973). It was also reported from Florida (Mead, 1986b) and from China by Wang & Wu (1991) (both data not checked).

Trichodorus intermedius Rodriguez-M & Bell, 1978

Figs 3.1.C; 3.4 (9k); 3.6.N; 3.10 (9k); 3.11.D; 4.12.E-H

Measurements

Holotype male: L= 720 μm, a= 20.0, b= 4.6, T= 63%, onchiostyle 62 μm, spicule length= 42 μm,

gubernaculum= 19 µm, anterior end to EP= 122 µm, distance CP1-EP= 22µm.
Males: L= 720-950 µm (830 µm ± 70), a= 18-24, b= 3.9-5.9, T= 63-72.8%, onchiostyle= 62-71.5 µm (69.0 µm ± 2.5), spicule length= 42-50 µm (45.0 µm ± 3.0), gubernaculum= 18-22 µm.
Females: L= 680-1130 µm (850 µm ± 100), a= 17.2-22, b= 4.3-6.4, V= 53-64% (57.0% ± 2.7), onchiostyle= 62-75.2 µm (69.0 µm ± 3.4), anterior end to EP= 78.4-138 µm (from Rodriguez-M & Bell,1978).

Emended diagnosis

T. intermedius male is characterized by the single ventromedian cervical papilla situated anterior to the S-E pore, the shape of the spicules (slightly cephalated, shaft about equally wide, straight, slightly tapered posteriorly to a curved end and provided with striae and bristles), three precloacal supplements with two within the region of the retracted spicule (the second supplement opposite mid-shaft), the rounded tail with terminal cuticle slightly thickened. The female by the size and shape of the vaginal sclerotizations (large, roughly triangular with tips bent toward interior of vagina, inner side concave), the rhomboid-shaped vagina, the transverse slit-like vulva and one pair of postadvulvar lateral body pores. The pharyngeal bulb is offset in males and females.

Relationships

T. intermedius resembles *T. borneoensis*, *T. dilatatus*, and *T. obscurus* in having two ventromedian precloacal supplements within the region of the retracted spicules, but differs from these species by the characters in Table 4.2.
T. intermedius has most characters in common with *T. dilatatus* and *T. obscurus*.

Remarks

No data available on the lateral cervical pore in the male.
Spicule shaft largely provided with bristles, visible in extruded spicules. SEM reveals large membranous scales on the spicule shaft and the distal end of the spicule with a small invagination (Rodriguez-M & Bell, 1978).

Distribution and host

T. intermedius is described from specimens from Oak Grove Campground, San Diego, California. It is polyphagous with the following hosts: live oak (*Quercus virginiana* Mill), grass and willow (*Salix* sp.), *Chenopodium* sp. (goosefoot), *Ambrosia* sp., *Pinus* sp., *Salvia leurophylla* (gray sage), and redwood (Rodriguez-M & Bell, 1978).

T. intermedius was sometimes found mixed with *Paratrichodorus minor*, *P.*

allius or *Trichodorus aequalis* (Rodriguez-M & Bell, 1978).

Table 4.2

	Trichodorus intermedius	*Trichodorus dilatatus*	*Trichodorus borneoensis*	*Trichodorus obscurus*
position 2nd precloacal supplements	mid-shaft	at spicule head	at spicule head	just anterior to spicule head
shape spicule	shaft equally wide	long, enlarged proximal part	shaft equally wide	shaft equally wide
n° CP	1	1	2	0
shape vaginal sclerotization	roughly triangular, tips bent toward interior vagina	roughly triangular, inner side concave	small oval to rounded triangular	trapezoidal, parallel to longitudinal body axis
shape vulva	transverse slit	transverse slit	pore	transverse slit
length onchiostyle	62-75.2 µm	60.5-69 µm	51-60 µm	60-70 µm

Trichodorus kilianae Decraemer & Marais, 1993
Figs 3.2.H; 3.3 (2a, a'); 3.7.F; 3.8 (2a, a'); 4.14.C-E

Measurements

Holotype male: L= 993 µm, a= 33.1, b= 7.5, T= 58.5%, onchiostyle= 46 µm, spicule length= 46 µm, gubernaculum= 20 µm, anterior end to EP= 106 µm, distance CP1-EP= 10 µm.

Males: L= 830-1048 µm (926 µm ± 76.4), a= 23.5-34.7, b= 5.5-7.3, T= 50.6-62.6%, onchiostyle= 45-49 µm (47 µm ± 1.4), spicule length= 43-46.5 µm (45 µm ± 1.2), gubernaculum= 16-20 µm, anterior end to EP= 88.5-109 µm.

Females: L= 878-962 µm (927 µm ± 32.7), a= 23.8-32.3, b= 5.7-8.3, V= 47.1-56.9% (51.4% ± 3.2), onchiostyle= 46-49 µm (47.7 µm ± 1.1), anterior end to EP= 81-117 (from Decraemer & Marais, 1993).

Diagnosis

T. kilianae is characterized by a rather long (830-1048 µm) and slender body, a relatively short onchiostyle (45-49 µm) and by the pharyngeal bulb being offset or with a short dorsal intestinal overlap (exceptionally in female). The male is further distinguished by the length (43-46.5 µm) and shape of the spicules (with marked manubrium, a slender shaft: anterior half convex dorsally, with septum visible in front of a distinct indentation, posterior half straight, except for distal end; one or two bristles may be present at indentation level), three precloacal supplements with the posterior supplement at the level of the head of retracted spicules, the presence of a single ventromedian cervical papilla, usually situated just anterior to the S-E pore opposite anterior part pharynx, and a non-thickened terminal tail cuticle. The female is further characterized by minute triangular with tip pointed to vulva (rarely oval) vaginal sclerotizations in lateral optical view, a pore-like vulva, the barrel-shaped vagina (42-54% of the corresponding body width long) and the presence of one pair of postadvulvar sublateral body pores.

Relationships

T. kilianae closely resembles *T. vandenbergae* in the male, by the possession of a single ventromedian cervical papilla, a more or less comparable spicule shape (proximally curved, shaft straight, with indentation) and the non-thickened terminal tail cuticle; in the female, by the pore-like vulva. It differs from *T. vandenbergae* by a somewhat longer body (830-1048 µm *vs* 681-801 µm) and shorter onchiostyle (45-49 µm *vs* 50-58 µm). In the male, by details of the spicule shape (finer with a longer, irregular constriction, with or without a few bristles and a septum just anterior to it, absent in *T. vandenbergae*), by the more anterior position of posterior precloacal supplement (opposite the manubrium of the retracted spicules *vs* about mid-spicule), by the smaller distance between the ventromedian cervical papilla and S-E pore (3.8-16.4 µm *vs* 13.4-28.6 µm). In the female by minute triangular vaginal sclerotizations *vs* minute rounded to rounded triangular pieces in *T. vandenbergae*, by one pair of sublateral postadvulvar body pores *vs* one pair of subventral postadvulvar pores and one pair of lateral prevulvar pores (1.7-4.5 body widths anterior to vulva).

Remarks

The following peculiarities were observed: one male with the ventromedian cervical papilla posterior to the S-E pore; one male with four precloacal supplements; one female with a dorsal intestinal overlap, all the others with an offset pharyngeal bulb and one female with on the right side one postadvulvar body pore, but two pores on the left body side.

Third stage juveniles were observed. They resemble females in most respects. The body length was 590-640 µm, the onchiostyle 40-44 µm with a 15.5 µm long replacement onchium; the genital system was 32-40 µm long.

Distribution and host

T. kilianae was described from specimens from the Ndubai plantage of Sappi Forest, eastern Transvaal, South Africa, from loamy soil in the rhizosphere of *Eucalyptus nitens* (Decraemer & Marais, 1993).

Trichodorus lusitanicus Siddiqi, 1974

Figs 3.1.L, l; 3.3 (1b); 3.5.E; 3.8 (1b); 3.11.E; 4.13A-D

Measurements

Holotype male: L= 760 µm, a= 27, b= 5.5, T= 60%, onchiostyle= 40 µm, spicule length= 32.5 µm, gubernaculum= 15 µm.

Males: L= 690-1020 µm (730 µm; 850 µm ± 63; 765 µm; 892 µm; 870 µm; 958 µm; 929 µm; 823 µm), a= 19.2-38.6, b= 3.8-8.8, T= 42.3-74.4%, onchiostyle= 33-59 µm (41 µm; 53 µm ± 3; 52 µm; 52µm; 50 µm; 50 µm; 54 µm; 51 µm), spicule length= 29-38 µm (32 µm; 36 µm ± 3; 31 µm; 34 µm; 33 µm; 34 µm; 33 µm; 35 µm), gubernaculum= 9-18 µm, anterior end to EP= 76-135 µm, distance CP1-CP2= 14-53 µm, distance CP2-CP3= 16-34 µm, distance CP3-EP** = 4-32 µm (from Siddiqi, 1974; Almeida *et al.*, 1989: 7 populations).

Females: L= 668-998 µm (720 µm; 804 µm ± 76; 771 µm; 866 µm; 859 µm; 896 µm; 805 µm; 928 µm; 885 µm), a= 20.9-38, b= 3.5-6.9, V= 50.5-64.5% (57%; 56.6% ± 2; 56.9%; 57%; 58.1%; 56.2%; 57%; 55.2%; 56.2%), onchiostyle= 35-58 µm (42.5 µm; 52 µm ± 2; 53 µm; 52 µm; 49 µm; 52 µm; 52 µm; 52 µm; 53 µm), anterior end to EP= 70-126 µm (from Siddiqi, 1974; Almeida *et al.*, 1989: 8 populations).

** males with only two cervical papillae: distance CP2-EP.

Emended diagnosis

T. lusitanicus is characterized in the male: by the shape and ornamentation of the spicules (ventrally curved, not cephalated, anteriorly wide, mid-shaft with irregular constriction zone, provided with bristles, posteriorly shaft tapered to distal end), two or three ventromedian cervical papillae (respectively one or two in the onchiostyle region), lateral cervical pore on both sides near the base of the onchiostyle (slightly anterior or posterior), three precloacal supplements all anterior to the region of retracted spicules, and asymmetrical tail with terminal cuticle not thickened. In the female by the size and shape of the vaginal sclerotizations in lateral view (large, regular to roughly triangular pieces with obtuse angles), the pore-like vulva, the rhomboid-shaped vagina, and one lateral postadvulvar body pore or with two lateral body pores (one postadvulvar, one prevulvar) on each side. The pharyngeal bulb is usually offset.

Relationships

T. lusitanicus is closest to *T. azorensis*, *T. beirensis*, *T. velatus* and *T. viruliferus* in general spicule shape (with marked constriction provided with bristles at about mid-shaft), the possession of two or three ventromedian cervical papillae with at least one in the region of the onchiostyle. It differs from these species in details of the spicule shape (difference in the position, length and degree of the indentation or the absence of a velum as in *T. velatus*), by the position of the posterior precloacal supplement (anterior to the retracted spicules in *T. lusitanicus vs* at the head of the retracted spicules in *T. azorensis*, *T. beirensis* and *T. velatus*). The female of *T. lusitanicus* differs from the four species (except *T. viruliferus*) in the pore-shaped vulva *vs* a transverse slit in the other species, and by the size and shape of the vaginal sclerotizations being large, rounded triangular pieces but smaller than in *T. beirensis vs* large quadrangular pieces in *T. azorensis*, long oval, parallel to the vaginal lumen in *T. viruliferus* and smaller, roughly triangular in *T. velatus*.

Remarks

Males in the type population have only two ventromedian cervical papillae whereas populations from Portugal have either two or three with no variation within populations. Also, in the Portuguese populations with three ventromedian cervical papillae eight out of 30 males have CP2 just posterior to the base of the onchiostyle.

Usually, there is no pharyngeal overlap, occasionally a minute one.

Some females showed a distortion of the vaginal sclerotization due to the presence of a secretion plug (="intrusion of a substance of unknown origin" in Almeida *et al.*, 1989).

Twenty out of 30 females from the Portuguese populations had two lateral body pores and the other specimens only one postadvulvar pore on each side.

Distribution and host

T. lusitanicus has been reported only from Portugal. It is polyphagous, occurring in soil around the roots of tomato plants, of *Olea europaea* (common olive), and of cork trees (*Quercus suber*) (Siddiqi, 1974; Almeida *et al.*, 1989).

Trichodorus magnus Decraemer & Marais, 1993
Figs 3.2.I; 3.3 (3d); 3.7.G; 3.8 (3d); 4.14A-B

Measurements

Holotype male: L= 1202 µm, a= 18.4, b=6.9, T= 47.4%, onchiostyle= 68.5 µm, spicule length= 60 µm, gubernaculum= 26 µm, anterior end to EP= 129 µm, distance CP1-EP= 4.5 µm.

Males: L= 1059-1232 µm (1184 µm ± 54.2), a= 18.4-28.6, b= 5.1-7.2, T= 47.4-71.8%, onchiostyle= 62.5-71 µm (66.5 µm ± 3.9), spicule length= 52-57 µm (55.5 µm ± 1.6), gubernaculum= 23.5-26 µm, anterior end to EP= 127-151 µm.

Females: L= 970-1295 µm (1128 µm ± 118.7), a= 19.1-22.8, b= 5.7-7.7, V= 51.1-57.6% (55.5% ± 2.5), onchiostyle= 68-70 µm (69 µm ± 0.8), anterior end to EP= 124-136.5 (from Decraemer & Marais, 1993).

Diagnosis

Trichodorus magnus is characterized by a long body (970-1295 µm) and a relatively long onchiostyle (62.5-71 µm). The male can be identified by the possession of one medioventral cervical papilla (just anterior or posterior to the S-E pore), the shape of the ventrally curved spicules (manubrium slightly widened, shaft stout with an indentation at 41-45 % of the spicule length from the anterior end and with a narrower distal end with septum, a pair of bristles may be present at indentation), three ventromedian precloacal supplements with the posterior one usually just anterior to the spicule head when spicules retracted and a short broadly rounded tail with non-thickened cuticle. The female is characterized by small rounded triangular to oval vaginal sclerotizations in lateral view, a pore-like vulva, the barrel-shaped vagina and the presence of a pair of postadvulvar sublateral body pores. The pharyngeal bulb is offset.

Relationships

T. magnus most closely resembles *T. obtusus* in body, onchiostyle and spicule lengths, presence of one ventromedian cervical papilla close to the S-E pore, a non-thickened terminal tail cuticle, the position of the posterior precloacal supplement (just anterior to or at level of manubrium) and the pore-like vulva. In both sexes, *T. magnus* differs most distinctly from *T. obtusus* by the presence of large sperm cells with a long sausage-shaped nucleus instead of small sperm cells with small nucleus in *T. obtusus*. *T. magnus* males differ from those of *T. obtusus* by the position of the ventromedian cervical papilla which is just anterior or just posterior to the S-E pore *vs* always anterior to the S-E pore in *T. obtusus*, by the marked indentation of the spicule shaft and by the narrower distal end provided with a septum *vs* shaft about equally wide in *T. obtusus*. Females can be distinguished from those of *T. obtusus* by their having only one pair of postadvulvar sublateral body pores *vs* one pair of postadvulvar lateral body pores and by one pair of prevulvar lateral body pores in *T. obtusus*.

Remarks

Position of the lateral cervical pore on both sides variable (up to 15 µm apart), either of them can be situated anterior, at the level, or posterior to the ventromedian cervical papilla or the S-E pore in the male. Postadvulvar sublateral body pores in the female not always at the same level on both sides of the body, up to 13 µm apart.

Distribution and host

T. magnus was described from specimens from Jessievale Staatsbos, eastern Transvaal, South Africa, from sandy-loam soil from the rhizosphere of grass (natural vegetation) (Decraemer & Marais, 1993).

Trichodorus minzi De Waele & Cohn, 1992

Figs 3.1.S, s; 3.4 (6b); 3.7.H; 3.9 (6b); 4.13.E-I

Measurements

Holotype female: L= 1053 µm, a= 19.9, b= 6.4, V= 51.9%, onchiostyle= 60 µm, anterior end to EP= 129 µm.
Females: L= 822-973 µm (937 µm ± 60), a= 14.4-19.2, b= 4.1-6.1, V= 49.6-55.3% (52.3% ± 1.6), onchiostyle= 53-64 µm (58 µm ± 3), anterior end to EP= 89-113 µm.
Males: L= 783-1072 µm (919 µm ± 90), a= 14.5-23.2, b= 4.5-6.7, T= 49.3-84.6%, onchiostyle= 53-62 µm (57 µm ± 2), spicule length= 63-71 µm (66 µm ± 2), gubernaculum= 30-37 µm, anterior end to EP= 101-141 µm, distance CP1-CP2= 16-36 µm, distance CP2-EP= 5-45 µm (from De Waele & Cohn, 1992).

Emended diagnosis

T. minzi male is characterized by two ventromedian cervical papillae anterior to the S-E pore, a pair of lateral cervical pores slightly anterior to the anterior ventromedian cervical papilla or between both cervical papillae, the shape of the spicules (manubrium long, wide, marked from the shaft with an anterior slight indentation zone, smooth, then shaft wider and striated, tapered at distal end; no bristles) and the spicule length (63-71 µm), the position of the three precloacal supplements with the posterior supplement clearly within the region of the retracted spicules and opposite the indentation zone, the usually small dorsal overlap of the pharyngeal bulb by the intestine, the asymmetrical tail with terminal cuticle mostly slightly thickened. The female is distinguished by the shape of the vaginal sclerotizations in lateral view (large round to about oval), the pear-shaped vagina and two lateral body pores (one prevulvar and one posterior advulvar) on each side.

Relationships

T. minzi resembles *T. dilatatus*, *T. taylori*, *T. persicus* and *T. yokooi* in general spicule shape i.e. with a long, wide, proximal part or manubrium, but differs in the manubrium being less sharply offset by a constriction than in *T. taylori* but more marked than in *T. dilatatus*, *T. persicus* and *T. yokooi*. *T. minzi* also differs from these species (except *T. taylori*, *T. persicus*) in having two ventromedian cervical papillae in males and a dorsal intestinal overlap *vs* three ventromedian cervical papillae in *T. yokooi* and one in *T. dilatatus*, both species also having an offset pharyngeal bulb. *T. minzi* resembles *T. petrusalberti* in having long spicules, but may be distinguished from it by their shape. *T. minzi* differs from all other species of the genus by the very long gubernaculum (30-37 µm).

T. minzi female has similar vaginal sclerotizations in lateral view as *T. taylori* and *T. persicus* (large rounded) but different from those in *T. yokooi* (medium-sized triangular) and *T. dilatatus* (large roughly triangular with inner side concave).

Remarks

1. The following peculiarities were observed: one out of fourteen females had no prevulvar lateral body pores; a male had only one ventromedian cervical papilla anterior to the S-E pore.
2. Eight out of fourteen females had a short (5-23 µm) dorsal overlap of the pharyngeal bulb by the intestine, and 28 out of 33 males.
3. According to the authors of the species, objects of unknown origin were observed in the uterus. From their drawing (Fig. 1 F) these structures appear similar to the inclusions observed in several specimens of other *Trichodorus* species such as *T. variopapillatus* (Geraert *et al., 1980*).

Distribution and host

T. minzi was described from specimens from central Israel, where it was found in wheat fields with stunted plants (De Waele & Cohn, 1992).

Trichodorus nanjingensis Liu & Cheng, 1990

Figs 3.2.J, j; 3.3 (4g); 3.7.I; 3.8 (4g); 4.15.A-G

Measurements

Holotype male: L= 917 µm, a= 19, b= 6.5, T= 63%, onchiostyle= 44 µm, spicule length= 47 µm, gubernaculum= 21.2 µm.

Males: L= 777-1133 µm (970 µm ± 86; 948 µm ± 64; 938 µm ± 86; 1026 µm± 32), a= 18-28.8, b= 3.7-7.6, T= 53-76.3%, onchiostyle= 45-52 µm (47 µm ± 1.3; 51 µm ± 0.7; 49 µm ± 1.5; 50.7 µm ± 0.8), spicule length= 43-53 µm (48 µm ± 2.8; 50.5 µm ± 2.0; 48 ± 1.2; 48 µm ± 2.2), gubernaculum= 18-24 µm, anterior end to EP= 99-135 µm, distance CP1-CP2= 8- 30 µm, distance CP2-EP= 5-21.5 µm.

Females: L= 884-1315 µm (1014 µm ± 77; 1062 µm ± 106; 1059 µm ±67; 979 µm ± 57), a= 20-28; b= 5.2-8.4, V= 52-63% (55% ± 1.7; 56% ± 3.2; 56% ± 1.4; 54% ± 0.8), onchiostyle= 41-53 µm (47 µm ± 2.1; 49.3 µm ± 2.4; 49 µm ± 1.4; 50.8 µm ± 1.2), anterior end to EP= 109-147 µm (from Liu & Cheng, 1990; Decraemer & Cheng, 1994, 2 populations).

Emended diagnosis

T. nanjingensis male is characterized by two ventromedian cervical papillae anterior to the S-E pore and nerve ring, the shape of the spicules (largely straight except for a slight bend proximally and at distal tip; manubrium short, slightly marked; shaft striated, bristles rarely observed) and the spicule length (43-53 µm), the position of the three precloacal supplements with the posterior supplement clearly within the region of the retracted spicules. The female is distinguished by a pore-like vulva, the size and shape of the vaginal sclerotizations in lateral view (small to almost medium-sized triangular or rounded), the pear-shaped vagina, the absence of spermathecae and sperm equally spread throughout the uteri and one pair of postadvulvar lateral body pores. The pharyngeal bulb is offset.

Relationships

T. nanjingensis males share a typical spicule shape (largely straight but proximal part ventrally curved) with six other species of the genus. They most closely resemble *T.*

petrusalberti and *T. sparsus* in having also two ventromedian cervical papillae anterior to the S-E pore and the posterior precloacal supplement within the region of the retracted spicules as in *T. petrusalberti* and a thickened terminal tail cuticle as in *T. sparsus*, but differ from *T. petrusalberti* by shorter spicules (43-53 µm *vs* 62-74 µm) and thickened end cuticle, and from *T. sparsus* by details of the spicule shape and ornamentation (usually without bristles and manubrium marked *vs* numerous bristles and manubrium not marked). Females of *T. nanjingensis* have a pore-like vulva and small triangular vaginal sclerotizations as in *T. petrusalberti* and *T. sparsus* but differ by the number of lateral body pores (one postadvulvar pair *vs* one preadvulvar and one postadvulvar pair), the shape of the vagina (pear-shaped *vs* rhomboid in *T. petrusalberti* and barrel-shaped in *T. sparsus*) and the presence of sperm throughout the uterus *vs* sperm in spermathecae in *T. petrusalberti* and *T. sparsus*.

Remarks

For the first time, the occurrence of an inner onchium in adult specimens of a trichodorid species appears as an incidental maintenance of a juvenile character. Decraemer & Cheng (1994) observed in a population from Suzhou two males with an inner onchium but two other males and eight females from the same sample lacked the inner onchium.

Variability was observed in the following diagnostic features: In males: - the number of ventromedian cervical papillae usually 2, may vary between 0 and 2; - although rarely, males with 4 precloacal supplements (SP) were found, the majority has 3 SP, the typical number for the genus (Table 4.3); - the second supplement (SP2) varies in position from near the spicule head to further anterior; - the striated ornamentation of the spicules varied in appearance from weakly developed to strongly marked; - the bristles are usually not observed or are obscure when the scales are not detached from the spicule corpus; - the position of the S-E pore varies in relation to the posterior ventromedian cervical papilla (CP2) (range between 5 and 21.5 µm) and to the pharyngo-intestinal junction (ratio anterior end to S-E pore / pharynx length (%)= 59.4-90.1%, range for mean values from 71.4 to 80.2%); - the terminal tail cuticle shows various degrees of thickening: from the rarely observed thin cuticle, to a more usual moderately swollen cuticle or an obviously swollen terminal cuticle; the subventral body wall may show a slight bursa-like protrusion in some specimens; both tail features seem influenced by fixation.

In females, some minor variability was observed in the size (1.5 - 2.5 µm) and shape of the vaginal sclerotized pieces (from triangular to almost rounded), and in the shape and length of the vagina in longitudinal optical section (from a distinct pear-shape to an elongate cylindrical shape and 16-18 µm long in Linbao specimens, 16-21 µm in specimens from Suzhou). The pair of postadvulvar lateral body pores are located at 12-28 µm or 7-10 µm respectively in specimens from Suzhou and from Linbao.

Table 4.3. Variation in number of ventromedian cervical papillae (CP) and precloacal supplements (SP) in males of two populations of *T. nanjingensis*.

	0 CP	1 CP	2 CP	3 SP	4SP
Pop. Suzhou (n=34)	1	3	30	33	1
Pop. Linbao (n=73)	0	8	65	70	3

Distribution and host

T. nanjingensis was originally described from specimens from Nanjing, Jiangsu Province, eastern China, with peach roots as host and was further recorded from Suzhou and Linbao, Jiangsu, respectively from the rhizosphere of plum tree and apple tree (Liu & Cheng, 1990; Decraemer & Cheng, 1994).

Trichodorus obscurus Allen, 1957

Figs 3.1.F, f; 3.3 (5d); 3.5.S; 3.9 (5d); 4.16.A-D

syn. *T. primitivus* according to Thorne (1939) and Goodey (1951)

Measurements

Holotype male: L= 1050 µm, a= 21, b= 5.5, T= 62%, onchiostyle= 60 µm, spicule length= 58 µm, gubernaculum= 21 µm.

Males: L= 690-1100 µm (865 µm*; 1100 µm), a= 14-22, b= 4.2-5.7, T= 61-70%, onchiostyle= 60-77 µm (68.5 µm*;-), spicule length= 50-62 µm (56 µm*; 56 µm), gubernaculum= 20-22.5 µm.

Females: L= 720-1170 µm (945 µm*), a= 16-24, b= 4.5-6.3, V= 54-58% (56%*), onchiostyle= 60-72 µm (66 µm*) (from Allen, 1957; Thorne, 1939).

Emended diagnosis

T. obscurus is characterized in the male by the absence of ventromedian cervical papillae, a lateral cervical pore just anterior to the S-E pore, the shape of the spicules (manubrium slightly marked, shaft about equally wide with a minor narrowing at level of distal end of capsule of suspensor muscles, spicules striated except at the extremities) and by the arrangement of the three ventromedian precloacal supplements (the posterior one near the cloacal opening; the second supplement just anterior to the manubrium when spicule retracted). In the female by the shape of the vaginal sclerotizations in lateral view (roughly trapezoidal, largely parallel to the longitudinal body axis), the shape of the vulva (a transverse slit) and the presence of a lateral body pore at the level of the vulva on each side. The pharyngeal bulb is offset in males and females.

Relationships

Allen (1957) considered *T. obscurus* to be close to *T. californicus*. He could not separate females of both species, but males could be differentiated by the absence of ventromedian cervical papillae in *T. obscurus vs* one papilla in *T. californicus*, and by the position of the ventromedian precloacal supplements (a character, which appears similar in both species).

Based on a study of type specimens, *T. obscurus* and *T. californicus* could be more clearly differentiated. The females of *T. obscurus* and *T. californicus* differ in the shape of the vaginal sclerotizations in lateral view (roughly triangular with inner side concave in *T. californicus vs* roughly trapezoidal in *T. obscurus*). Males can be distinguished by the shape and ornamentation of the spicules (manubrium slightly marked, spicule ornamented with striae except at the smooth ends in *T. obscurus vs* a clearly offset manubrium, shaft provided with striae and bristles, except at the smooth extremities in *T. californicus*).

Remarks

The terminal tail cuticle in males appears thickened in the original figure of a type specimen, but non-thickened in the specimens described by Thorne (1939) and spicules are more curved, shaft wider and manubrium not offset in Thorne (1939).

Distribution and hosts

T. obscurus was originally described as *T. primitivus* from Arlington, Virginia and from cultivated and virgin soil in Utah, Colorado and California (Thorne, 1939). It was also found in Oregon and in soil associated with roots of grass sod and of Manzanite (Allen, 1957). It has been recorded from China by Wang & Wu (1991) (specimens not checked).

Trichodorus obtusus Cobb, 1913

Figs 3.2.K. 3.4. (9g, 9h); 3.6.G, L; 3.10 (9g); 3.11.F; 4.20.A-E

syn. of *T. primitivus* in Micoletzky (1922), Thorne (1939), Goodey (1951) and Allen (1957)

species inquirenda in Hunt (1993)

syn. *T. proximus* Allen, 1957, n.syn.

For discussion on the taxonomic status of *T. obtusus* see chapter 3.

Measurements

Male type: L= 1100 µm, length of stoma= 12.1 µm, anterior end to nerve ring= 110 µm, pharynx= 187 µm, anterior end to cloacal opening= 1083 µm, a= 22.7, b= 5.9 (calculated from formula in Cobb, 1913)

Female type: L= 1000 µm, length of stoma= 14 µm, anterior end to nerve ring= 110 µm, pharynx= 220 µm, V= 54%, anterior end to anus= 996 µm, a= 21.3, b= 4.5 (calculated from formula in Cobb, 1913)

Males: L= 850-1320 µm (1120 µm*; 1100 µm; 1140 µm), a= 20.4-33.7, b= 6.6-7.9, c= 60-89.2, onchiostyle= 48-65 µm (56.5 µm*; 60 µm; 57 µm), spicule length= 43-65 µm (56.5 µm*; -; 58 µm), gubernaculum= 17-26 µm

Females: L= 1100-1500 µm (1310 µm*; 1100 µm; 1240 µm), a= 19-30.9, b= 6.7-9.3, V= 49-60% (52%*; 49%; 51%), onchiostyle= 49-70 µm (57.5 µm*; 60 µm; 60 µm)(from Allen, 1957; Thorne, 1974; Norton *et al.*, 1982).

Emended diagnosis

T. obtusus is characterized in the male by the shape of the spicules (usually ventrally curved, proximal end slightly wider than shaft which is about equally wide, slightly tapered distally, smooth), the arrangement of the three ventromedian precloacal papillae (the posterior one usually just anterior to the head of the retracted spicules), one ventromedian cervical papilla anterior to the S-E pore, one pair of lateral cervical pores at the level of the ventromedian cervical papilla, tail with non-thickened terminal cuticle. In the female by a pore-like vulva, one or two postadvulvar lateral body pores on each side, the size and shape of the vaginal sclerotizations in lateral view (small triangular to oval) and the barrel-shaped vagina (relaxed condition). The pharynx is offset in males and females.

Relationships

T. obtusus males are closest to *T. eburneus* by the presence of one ventromedian cervical papilla, the position of the posterior precloacal supplement (usually just anterior to the retracted spicules as in many specimens of *T. eburneus*) and by the shape of the spicules (slightly ventrally curved or only proximal part clearly ventrally bent and shaft mainly straight). They differ by having smaller sperm and by the curvature of the spicules (usually totally curved *vs* only proximal part ventrally bent). *T. obtusus* females differ from those of *T. eburneus* in the slightly different shape of the vaginal sclerotizations in lateral view (larger triangular, rarely oval *vs* smaller (rounded) triangular pieces or rounded in *T. eburneus)* and by the position of the postadvulvar body pore(s) (sublateral instead of subventral in *T. eburneus).* Both species also differ in geographical distribution with *T. obtusus* limited to the U.S.A. and *T. eburneus* to Africa.

Remarks

Some males have spicules which are largely straight except for the ventrally curved proximal end (Decraemer, 1991: Fig. 5d).

The prevulvar anterior body pore may be absent on both sides (Norton *et al.*, 1982).

Specimens described by Baujard (1983) as *T. proximus* were subsequently identified as *T. eburneus* (De Waele, 1986).

Distribution and host

T. obtusus is restricted to the U.S.A. It was originally described from soil associated with roots of grasses, from Arlington, Virginia, U.S.A. (Cobb, 1913) and later from native turf, South Dakota (Thorne, 1974). Since *T. proximus* is here considered as a new synonym of *T. obtusus*, *T. obtusus* also occurs in Florida, Iowa, Kansas, Michigan, New York (Allen, 1957; Thorne, 1974; Norton *et al.*, 1982; Noffsinger, 1984; McGowan, 1988b; Esser, 1991b). The only record outside the U.S.A. is from the south of France (Scotto la Massese, 1985), but it appeared to be *P. pachydermus* (pers. commun., Scotto la Massese). Known hosts are tomato, St Augustine grass, pear, live oak, *Opuntia* sp., *Borrichia frutescens*, *Rhus copallina*, *Scutellaria multiglandulosa*, *Cichorium intybus* (Baldwin, 1977; Harrison & Smart, 1975; Rhoades, 1965; McGowan, 1987, 1988b; Esser,

1991b; Lehman, 1990; Mead, 1984; Anonymous, 1988c). Other plant associations are big bluestem, side oats grama, cabage palmetto, potato, Bermuda grass, Kentucky bluegrass, small leaved linden, *Eucalyptus* sp.,*Rhododendron* sp., *Magnolia virginiana* (Noffsinger, 1984).

Trichodorus orientalis De Waele & Hashim, 1984

Figs 3.2.L; 3.3 (3a); 3.5.D; 3.8 (3a); 3.11.L; 4.17.A-E

Measurements

Holotype female: L= 679 µm, a= 22.6, b= 4.9, V= 56.8%, onchiostyle= 48 µm, anterior end to EP= 87 µm.
Females: L= 651-941 µm (686 µm; 722 µm; 735 µm; 820 µm ± 109; 754 µm ± 25), a= 17.9-28.3, b= 4.4-6.3, V= 52.7-59.4% (55.3%; 54.3%; 57.6%; 56% ± 2; 55.4% ± 1.5), onchiostyle= 47-58 µm (54 µm; 48 µm; 52 µm; 53 µm ± 2; 52 µm ± 2), anterior end to EP= 74-117 µm.
Males: L= 609-888 µm (699 µm*; 665 µm; 772 µm; 866 µm; 756 µm ± 78), a= 20.2-28.6, b= 4.4-7.5, T= 61.5-75.5%, onchiostyle= 47-54 µm (53.5 µm*; 47 µm; 49 µm; 52 µm; 51 µm ± 1), spicule length= 34-43 µm (38.5 µm*; 34 µm; 39 µm; 41 µm; 40 µm ± 1), gubernaculum= 4-7 µm, anterior end to EP= 82-115 µm, distance CP1-CP2= 16-35 µm, distance CP2-EP= 5-15 µm (from De Waele & Hashim, 1984: 3 populations; De Waele & Cohn, 1992: 2 populations).

Diagnosis

T. orientalis is characterized in the male by a short anterior dorsal intestinal overlap, by the shape of the spicules (with widened head, a slender shaft with constriction at mid-length and fine striae distally), two ventromedian cervical papillae anterior to the S-E pore (anterior one in the posterior onchiostyle region or just posterior to it), a pair of lateral cervical pores between both ventromedian cervical papillae (i.e. near the nerve ring), three precloacal supplements with the posterior supplement near the head of the retracted spicules, and the asymmetrical tail with terminal cuticle thickened. In the female by the shape of the vaginal sclerotizations in lateral view (minute triangular), the barrel-shaped vagina, and one pair of lateral postadvulvar body pores.

Relationships

T. orientalis resembles *T. kilianae*, *T. sanniae* and *T. vandenbergae* by the shape of the spicules in the male (manubrium wider, shaft almost straight except at both ends and with a constriction at mid-shaft). It differs from those species by having two ventromedian cervical papillae, the anterior one at the base of the onchiostyle region *vs* one ventromedian cervical papilla in the other species, by the thickened terminal tail cuticle, thin end cuticle in the other species. *T. orientalis* females have vaginal sclerotizations similar to those in *T. kilianae* and *T. vandenbergae* (minute triangular) but they differ from the dot-like sclerotizations in *T. sanniae*. Also, *T. orientalis* may be distinguished from *T. vandenbergae* by the presence of a dorsal intestinal overlap (also in *T. sanniae* and several specimens of *T. kilianae)* vs an offset bulb in *T. vandenbergae*.

Remarks

One male was observed with only one ventromedian cervical papilla. The pharynx rarely has an offset bulb.

In *T. orientalis* the body appears slightly flattened ventrally at the level of the copulatory apparatus, giving the impression of the beginning of a bursa.

Distribution and host

T. orientalis has been described from Jordan, Iran and Israel. It is polyphagous and was found in soil around the roots of tomato (*Lycopersicon esculentum*), garden bean *(Phaseolus vulgaris)*, grapevine (*Vitis* sp.) (on moist loamy soil), avocado and natural vegetation (De Waele & Hashim, 1984; De Waele & Cohn, 1992).

Trichodorus pakistanensis Siddiqi, 1962

Figs 3.2.V; 3.3 (4a); 3.5.G; 3.8 (4a); 3.11.U; 4.18.A-K

syn. *Trichodorus litchi* Edward & Misra, 1970 by Siddiqi (1974)

Measurements

Holotype male: L= 920 µm, a= 30, b= 6.3, T= 59%, onchiostyle= 40 µm.

Males: L= 640-1200 µm (970 µm; 755 µm*; 837 µm ± 99.0; 850 µm), a= 13-34, b= 4.4-7.4, T= 50-70%, onchiostyle= 39-48 µm (42 µm; 42 µm ± 1.2; 42 µm), spicule length=46-58 µm (52.6 µm; 47 µm*; 52 µm ± 0.8; 51 µm), gubernaculum= 11-25 µm, anterior end to EP = 99-116 µm, anterior end to CP1= 79-102 µm, distance CP1-CP2= 8-16.5 µm, distance CP2-CP3= 8-21.5 µm.

Females: L= 690-1220 µm (980 µm; 805 µm*; 873 µm ± 68.3; 984 µm), a= 15-31, b= 4.3-6.8, V= 44-58.5% (54%; 49.5%*; 55% ± 1.6; 56%), onchiostyle= 38-51 µm (41.5 µm; 44.5 µm*; 42 µm ± 1.1; 41 µm) (from Siddiqi, 1962; Edward & Misra, 1970; Xu & Decraemer, 1995: 2 populations).

Emended diagnosis

T. pakistanensis is distinguished by the relatively long body and the length of the onchiostyle (38-51 µm). In the male, the species is characterized by having three ventromedian cervical papillae (two anterior and one posterior to the S-E pore), a pair of lateral cervical pores near the level of the middle ventromedian cervical papilla, the position of the three ventromedian precloacal supplements (posterior supplement clearly within the region of the retracted spicules, shortly posterior to the manubrium) and the shape and ornamentation of the spicules (cephalated, shaft with proximal part dorsally convex, distal half about straight, with obvious striae, except for a short, smooth mid-part), tail obtusely rounded, terminal cuticle not thickened. In the female by the shape of the vaginal sclerotized pieces in lateral view (small, round to oval), shape of the vulva (a transverse slit), the short barrel-shaped vagina (when relaxed) and a pair of postadvulvar lateral body pores. The pharyngeal bulb is offset in males and females.

Relationships

In contrast to nearly all species of the genus, *T. pakistanensis* has the S-E pore situated between the ventromedian cervical papillae, a character it shares only with *T. parorientalis* and *T. rinae*. It differs from these two species in having three ventromedian cervical papillae instead of two. *T. pakistanensis* males have spicules with the proximal part ventrally bent and a largely straight shaft as in seven other species of the genus, among them *T. rinae*. It differs from *T. rinae* by having stouter spicules, with an obvious striation interrupted by a smooth part *vs* slender spicules, with very fine striae in *T. rinae*. The females of *T. pakistanensis* have small rounded sclerotized pieces in lateral view similar to those in *T. parorientalis*, but differently orientated (Figs 3.2.V, N). The females differ from those of *T. parorientalis* by the shape of the vulva (a transverse slit *vs* a pore in *T. parorientalis*; unknown in *T. rinae*) and by having only one pair of lateral body pores as in *T. rinae* vs two pairs in *T. parorientalis*.

Remarks

Vagina variable is size from one third to half the corresponding body width long.

The position of the lateral cervical pore in relation to the ventromedian cervical papilla varies slightly; in many specimens it is at the level of the S-E pore. In the Chinese populations the S-E pore is situated between the CP1 and CP2 in about one third of the specimens examined.

In the Chinese specimens the spicules are provided with a pair of bristles just posterior to the smooth narrower region. The gubernaculum in the Chinese specimens is longer and differently shaped (22-25 µm and proximally hook-shaped *vs* 11-16 µm and straight).

Siddiqi (1974) synonymized *T. litchi* with *T. pakistanensis* without providing any comment. Thus the range of morphometric data such as the body length, the a and V-values were much increased with the *T. litchi* specimens being shorter, broader and with a far lower range for the V value. These differences may be influenced by fixation which would provide, for example, a low a value. Edward & Misra (1970) described the vulva as pore-shaped in *T. litchi* instead of a transverse slit as in the type specimens of *T. pakistanensis*. They also observed in the male and females a slight overlap of the pharyngeal glands *vs* bulb offset in the type specimens of *T. pakistanensis*.

The unidentified male specimen described by Khera & Chaturvedi (1977) from a tea plantation at Lakhanwala, India belongs to *T. pakistanensis*. The authors considered it filled an intermediate position between *T. pakistanensis* and *T. kurumeensis* (=*T. cedarus*), but did not provide any explanation for their opinion. The Indian male specimen largely agrees with the original description of *T. pakistanensis*, although the body length (620 µm) and the length of the gubernaculum (8 µm) are slightly smaller.

Distribution and host

T. pakistanensis has been reported from West Pakistan and India and more recently was found in China (Wang & Wu,1991; Xu & Decraemer, 1995). *T. pakistanensis* has been found in the rhizosphere of Mulberry trees (*Morus* sp.), lychee (*Litchi chinensis*), tea and ramie (*Boehmeria nivea)* (Siddiqi, 1962; Edward & Misra, 1970; Khera & Chaturvedi, 1977; Xu & Decraemer, 1995).

Trichodorus paracedarus Xu & Decraemer, 1995

Figs 3.2.M; 3.4 (8g); 3.7.M; 3.9 (8g); 4.10.G-L

Measurements

Holotype male: L= 680 µm, a= 19, b= 5.1, T= 67%, onchiostyle= 47µm, spicule length= 43 µm, gubernaculum= 20 µm, anterior end to EP= 102 µm, distance CP1-CP2= 8 µm, distance CP2-CP3= 8 µm, distance CP3-EP= 10 µm.

Males: L= 512-733 µm (597 µm ± 48.4; 623 µm ± 59.8), a= 16-22, b= 3.6-6.2,T= 53-72%, onchiostyle= 39-45 µm (42 µm ± 1.5; 41 µm ± 1.4), spicule length= 38-47 µm (41 µm ± 1.1; 42 µm ± 2.2), gubernaculum= 17-21 µm, anterior end to EP= 81-114 µm, distance CP1-CP2= 6-13 µm, distance CP2-CP3= 6-10 µm, distance CP3-EP= 3-21.5 µm.

Females: L= 506-751 µm (591 µm ± 44.2; 646 µm ± 58.0), a= 15-22, b= 3.5-6.3, V= 52-60% (56% ± 1.8; 57% ± 1.5), onchiostyle= 39-45 µm (42 µm ± 1.3; 41 µm ± 1.0), anterior end to EP= 83-112 µm (from Xu & Decraemer, 1995).

Diagnosis

T. paracedarus is characterized in the male by the habitus with straight posterior body region, three ventromedian cervical papillae between the onchiostyle base and the S-E pore (the anterior one at the level of mid-isthmus), the position of the lateral cervical pore (at level of the anterior cervical papilla), by the shape of spicules (slightly ventrally curved to almost straight with distal part slightly bent, marked manubrium, shaft finely striated, mid-shaft may show an irregular outline, four pairs of bristles just posterior to mid-spicule), by three ventromedian precloacal papillae (the posterior one well developed and opposite the distal third of retracted spicules, the anterior two reduced to pore-like structures, positioned well out of reach of retracted spicules), by thickened terminal tail cuticle and the presence of caudal alae. In the female by the transverse slit-like vulva, the pear-shaped vagina, the size and shape of the vaginal sclerotizations in lateral view (small triangular or rarely drop-like), well developed spermathecae and a pair of advulvar lateral body pores posterior to the vulva. The pharyngeal bulb is offset.

Relationships

T. paracedarus is most closely related to *T. cedarus*, by virtue of its general measurements, number and position of ventromedian cervical papillae, shape and ornamentation of spicules and relative position of the anterior two precloacal supplements in male and shape of vaginal sclerotized pieces in female. *T. paracedarus* differs consistently from the various populations of *T. cedarus* reported in a number of characters in male posterior end: posterior end habitus (straight with little developed copulatory muscles *vs* ventrally curved with well developed copulatory muscles in *T. cedarus*; bursa development (present *vs* absent); tail terminus shape (terminal cuticle conspicuously thickened with a ventral indentation *vs* slightly thickened without any ventral indentation); position of the posterior precloacal supplement (opposite the distal third of retracted spicules *vs* opposite the middle of retracted spicules) and structure of anterior two supplements (strongly reduced to pore-like *vs* normally developed). Furthermore in the female by the position of advulvar body pores (situated laterally at more than half body width posterior to the vulva *vs* subventrally at less than half body width posterior to the vulva in *T. cedarus*). Further *T. paracedarus* males agree with those of *T. cylindricus* in their habitus (straight posterior body region), presence of caudal alae and little developed copulatory muscles, all features aberrant for the genus *Trichodorus* but typical of *Paratrichodorus*.

Remarks

Three males out of 34 with only two ventromedian cervical papillae.

The straight posterior body region, associated with weakly developed copulatory muscles and the presence of a bursa in males of *T. paracedarus* are characters reminiscent of *Paratrichodorus* species. *T. paracedarus* is retained in the genus *Trichodorus*: in males, because of the number of ventromedian cervical papillae, the well developed capsule of suspensor muscles, the large sperm cells, an offset pharyngeal bulb and in females, because of the well developed vagina (about half the corresponding diameter long), the well developed vaginal constrictor muscles, the presence of spermathecae, the presence of postadvulvar lateral body pores, well developed vaginal sclerotizations and an offset pharyngeal bulb. Apparently *T. paracedarus* represents an intermediate form between the genera *Trichodorus* and *Paratrichodorus*.

Distribution and host

T. paracedarus is known from two localities in Jiangsu Province, China: the type locality Nanjing and in Lianyungang respectively from clay soil around roots of tomato (*Lycopersicon esculentum*) and from sandy soil around roots of Yoshino cherry (*Prunus yedoensis*).

Trichodorus parorientalis Decraemer & Kilian, 1992
Figs 3.2.N; 3.3 (1f, 3b); 3.7.K; 3.8 (1f, 3b); 4.17.G-I

Measurements

Holotype male: L= 720 µm, a= 18.4, b= 4.7, T= 65.6%, onchiostyle= 52.5 µm, spicule length= 42 µm, gubernaculum= 17.5 µm, anterior end EP= 112 µm, distance CP1-CP2= 24 µm, distance EP-CP2= 7 µm.

Males: L= 722-806 µm (763 µm ± 25.7), a= 18.4-25.9, b= 4.7-5.5, T= 54.9-69.9%, onchiostyle= 51-54 µm (52.2 µm ± 1.3), spicule length= 37.5-44 µm (41.7 µm ± 2.2), gubernaculum= 16.5-21 µm, anterior end to EP= 112-128 µm, anterior end to CP1= 95-115 µm, anterior end to CP2= 119-148 µm.
Females: L= 634-768 µm (703 µm ± 46.7), a= 15.7-25.5, b= 3.7-4.9, V= 50.8-56.6% (54.2% ± 2.9), onchiostyle= 50.5-55 µm (53 µm ± 1.8), anterior end to EP= 94-122 µm (from Decraemer & Kilian, 1992).

Diagnosis

T. parorientalis is characterized by its onchiostyle length (50.5-55 µm) and large sperm cells each with a long, sausage-shaped nucleus. Males of the species can be identified by two well developed medioventral cervical papillae posterior to the onchiostyle base and with the S-E pore situated between, the shape of the spicules (shaft slender, ventrally bent, with a slight constriction at mid-shaft, with or without bristles), the arrangement of the three precloacal supplements with the posterior one at the level of the spicule head when spicule retracted. The female is characterized by the vaginal sclerotizations in lateral view (small oval, close and oblique), the pore-like vulva, the pear-shaped vagina (when relaxed) and the presence of one prevulvar lateral body pore and one postadvulvar lateral body pore on each side. The pharynx has an offset bulb.

Relationships

T. parorientalis closely resembles *T. orientalis* in morphometric data, in having two conspicuous ventromedian cervical papillae, a similar arrangement of the precloacal supplements, spicules with a mid-shaft constriction and sausage-shaped sperm nucleus in males, and small rounded triangular to oval vaginal sclerotizations in females. The males differ from *T. orientalis* in the position of their S-E pore (between the cervical papillae instead of posterior to them as in *T. orientalis*); the non-thickened terminal cuticle (a thickened cuticle in *T. orientalis*) and in spicule shape (constriction less marked and a wider posterior shaft than in *T. orientalis*). Moreover, in *T. orientalis* the body appears slightly flattened ventrally at the level of the copulatory apparatus giving the impression of a minute bursa ventrosublaterally. The females of *T. parorientalis* differ from *T. orientalis* by having slightly larger, oval vaginal sclerotized pieces, closely located in lateral view, a pear-shaped vagina *vs* a barrel-shaped vagina and the presence of a pair of prevulvar lateral pores, absent in *T. orientalis*.

Remarks

A minute intestinal overlap was observed in a male specimen. The lateral cervical pores usually near to the S-E pore, are exceptionally anterior to CP1 or posterior to CP2.

Third stage juveniles have been observed: L= 565-595 µm, onchiostyle= 44-48 µm, with a 19 µm long replacement onchium, and reproductive system 34 µm (juv. female), 42 µm (juv. male) long, consisting of several cells and with a clear spicular primordium in the juvenile male.

Distribution and host

T. parorientalis was described from specimens from Eastern Transvaal, South Africa, from around the roots of an unidentified grass growing between rocks (Decraemer & Kilian, 1992).

Trichodorus paucisetosus Bernard, 1992

Figs 3.2.T; 3.4 (91, 1'); 3.7.L; 3.10 (91, 1'). 4.17.F

Measurements

Holotype male: L= 696 µm, a= 18, b=4.2, T= 59%, onchiostyle= 71 µm, spicule length=46 µm, gubernaculum= 18 µm, anterior end to EP= 110 µm, distance CP1-CP2= 19 µm, distance CP2-EP= 6 µm.

Males: L= 638-911 µm (771 µm ± 76.5), a= 17-27, b= 4.2-5.7, T= 56-64%, onchiostyle= 57-71 µm (67 µm ± 4.7), spicule length= 41-50 µm (46 µm ± 3.1), gubernaculum= 17-22 µm, anterior end to EP= 100-124 µm, distance CP1-CP2= 14-26 µm, distance CP2-EP= 5-11 µm.

Females: L= 706-882 µm (805 µm ± 64.6), a= 16-23, B= 4.4-5.8, V= 52-58% (54% ± 2.3), onchiostyle= 58-72 µm (67 µm ± 5.1), anterior end to EP= 119-131 µm (from Bernard, 1992a).

Emended diagnosis

T. paucisetosus male is characterized by the shape and ornamentation of the spicules (non-cephalated, about equally wide, shaft slightly narrower at about mid-corpus, tapered distally to a bifid tip, finely striated except at both ends and at narrower part provided with a few strong spines), the two ventromedian cervical papillae anterior to the S-E pore, one pair of lateral cervical pores usually between the ventromedian cervical papillae, three ventromedian precloacal supplements with the posterior one at proximal end of the retracted spicules, tail rounded to bluntly protuberant and with tail tip cuticle slightly thickened. The female by the size and shape of the vaginal sclerotizations in lateral view (medium-sized, spherical or broadly oval), the two lateral body pores (one prevulvar, one postadvulvar) on each side, the short barrel-shaped vagina and the pore-like vulva. The pharyngeal bulb is offset in males and females.

Relationships

Bernard (1992a) considered *T. paucisetosus* as a member of the *T. aequalis* complex, being morphologically most similar to *T. sparsus* and *T. nanjingensis*. It differs from *T. sparsus* in spicules (manubrium not widened, shaft with a few prominent bristles, restricted to the middle part of the spicules *vs* manubrium wider, shaft fine, usually with fine bristles over most of the spicule length in *T. sparsus*); and oval or round vaginal sclerotizations (triangular in *T. sparsus*). *T. paucisetosus* can be separated from *T. nanjingensis* by having a longer onchiostyle, different spicules (manubrium slightly marked, shaft slender, striated, bristles rarely present in *T. nanjingensis*), by the oval or round vaginal sclerotizations in lateral view (triangular in *T. nanjingensis*) and by possessing spermathecae (sperm distributed throughout the uteri in *T. nanjingensis)*.

Distribution and host

T. paucisetosus is known only from Alaska. It was found in the rhizosphere of white spruce (*Picea glauca*) with rose (*Rosa acicularis*), spruce-Labrador tea, spruce, mixed aspen-spruce stand with rose groundcover and spruce-birch wood with *Equisetum* sp. groundcover (Bernard, 1992a).

Trichodorus persicus De Waele & Sturhan, 1987

Figs 3.1.R; 3.4 (6c, 9j); 3.5.J; 3.9 (6c, 9j); 4.19.A-E

Measurements

Holotype female: L= 898 µm, a= 23.6, b=5.0, V= 55.4%, onchiostyle= 71 µm, anterior end to EP= 113 µm.

Females: L= 726-949 µm (828 µm ± 71.7), a= 18.6-24.3, b= 4.4-5.7, V= 51.0-58.4% (56.0% ± 1.9), onchiostyle= 63-71 µm (68 µm ± 2.1), anterior end to EP= 94-128 µm.

Males: L= 693-953 µm (833 µm ± 65.3), a= 19.3-27.2, b= 4.3-5.5, T= 59-72%, onchiostyle= 65-72 µm (68 µm ± 2.2), spicule length= 51-61 µm (56 µm ± 3.2), gubernaculum= 6-7 µm, anterior end to EP= 100-137 µm, distance CP1-CP2= 15-39 µm, distance CP2-EP= 4-22 µm (from De Waele & Sturhan, 1987).

Emended diagnosis

T. persicus male is characterized by the spicule shape (slightly ventrally curved, manubrium long, wide and more or less marked, shaft narrow anteriorly and smooth, from mid-shaft on slightly wider, then tapered to a fine distal tip, posterior shaft provided with striae), three ventromedian precloacal supplements with the posterior supplement at the level of the spicule head when spicule retracted, two ventromedian cervical papillae anterior to the S-E pore, one pair of lateral cervical pores somewhat posterior to the level of the nerve ring and tail asymmetrical with middle cuticular layer thickened on dorsal side. The female by having large rounded vaginal sclerotizations in lateral view, the pear-shaped vagina and two pairs of lateral body pores (one prevulvar and one postadvulvar). A dorsal intestinal overlap is usually present in males and females.

Relationships

T. persicus resembles *T. minzi*, *T. paucisetosus* and *T. taylori* by the well developed round vaginal sclerotizations in lateral view in females. It is further similar to *T. taylori* and *T. minzi* by the pear-shaped vagina (about barrel-shaped in *T. paucisetosus*) in females, and by the presence of a dorsal intestinal overlap in males and females (pharyngeal bulb offset in *T. paucisetosus*). It also agrees with *T. taylori* by the short gubernaculum (6-7 µm *vs* 30-37 µm in *T. minzi* and 17-22 µm in *T. paucisetosus*). *T. persicus* differs from these species in the male, by the different shape of the spicules and in the female by two lateral body pores on each side, instead of one pore in *T. taylori*, and larger sclerotized pieces than those in *T. minzi* and *T. paucisetosus*.

Remarks

Rarely (2 out of 12 females, 1 out of 20 males) was the pharyngeal bulb offset. Exceptionally, males occur with one or three ventromedian cervical papilla(e). One male specimen has been observed with the S-E pore between the two ventromedian cervical papillae. Distance between S-E pore and CP2 usually much less than distance CP1-CP2, except in two specimens where the distance S-E pore-CP2 and CP1-CP2 is almost the same and in one specimen where the distance between S-E pore-CP2 is larger than CP1-CP2. One female had only the postadvulvar pair of lateral body pores.

Distribution and host

T. persicus was described from Iran where it was found in moist soil around the roots of grasses (*Oplismenus* sp.), various herbs and box-tree and from soil from an apple and pear orchard (De Waele & Sturhan, 1987).

Trichodorus petrusalberti De Waele, 1988
Figs 3.2.O.; 3.3 (4c); 3.6.I.; 3.8 (4c); 4.14.F-H; 4.15.J

Measurements

Holotype male: L= 873 µm, a= 14.3, b= 5.9, T= 73.2%, onchiostyle= 71 µm, spicule length= 70 µm, gubernaculum= 27 µm, anterior end to EP= 119 µm, distance CP1-CP2= 31 µm, distance CP2-EP= 7 µm.
Males: L= 707-967 µm (884 µm ± 92.4), a= 11.9-17.1, b= 4.7-6.1, T= 58.9-73.2%, onchiostyle= 67-73 µm (70 µm ± 2.1), spicule length= 62-74 µm (67 µm ± 3.3), gubernaculum= 23-29 µm, anterior end to EP= 104-157 µm, distance CP1-CP2= 21-33 µm, distance CP2-EP= 7-12 µm.
Females: L= 809-957 µm (892 µm ± 55.9), a= 13.5-18.8, b= 5.0-6.0, V= 51.5-56.1% (54.1% ± 1.8), onchiostyle= 65-74 µm (71 µm ± 1.4; 69.5 µm), anterior end to EP= 125-143 µm (from De Waele, 1988; Decraemer & Marais, 1993).

Emended diagnosis

T. petrusalberti is characterized by having a stout, usually very corpulent habitus, with the body cuticle as swollen as in species of the genus *Paratrichodorus*, a relatively long onchiostyle (65-74 µm), the presence of an inner stylet 23-27.5 µm long and a short dorsal intestinal overlap. Males can be distinguished by the shape of the spicules (proximal part ventrally curved, manubrium slightly wider than the slender striated shaft without bristles), three precloacal supplements with the posterior supplement clearly within the region of retracted spicules, the two ventromedian cervical papillae anterior to the S-E pore, one pair of lateral cervical pores between the ventromedian cervical papillae and terminal tail cuticle not thickened. Females are characterized by the shape of the vaginal sclerotizations in lateral view (small triangular or tear-drop-like), two pairs of lateral body pores (one prevulvar, one postadvulvar), the rhomboid vagina and a pore-like vulva.

Relationships

T. petrusalberti males are most similar to *T. rinae* in the shape of the spicules (slender, proximally ventrally bent, shaft mainly straight), the presence of two ventromedian cervical papillae and the position of the posterior precloacal supplement (within the region of retracted spicules). Females by the small triangular vaginal sclerotizations in lateral view. Also by the presence of a dorsal intestinal overlap which is present only in about half of the specimens of *T. rinae*. *T. petrusalberti* differs from *T. rinae* by having longer spicules (62-74 µm *vs* 40-55 µm) and a longer onchiostyle (65-74 µm *vs* 41-64 µm).

Remarks

The spicules were described originally as being smooth but upon reexamination appeared to be striated, a feature which is clearly visible in protruded spicules. Spicules may be completely ventrally curved instead of the proximal part only. The gubernaculum is prominent, well sclerotized and striated, especially marked in the thickened distal keel. The subventral body wall has a slight bursa-like protrusion, flanking the cloacal region (Decraemer & Marais, 1993).

The following peculiarities were observed among the type population: one male specimen with the S-E pore between CP1 and CP2; one female with an offset pharyngeal bulb, and two females with an additional third pair of lateral body pores, located prevulvar (De Waele, 1988).

The lined oblong objects in the uterus mentioned in the original description, appear to be swollen (deteriorated?) uterus cells filled with fungi (Decraemer & Marais, 1993).

Distribution and host

T. petrusalberti is known only from the type population from Northern Natal, South Africa, from wet soil around the roots of rice (*Oryza sativa)* De Waele, 1988).

Trichodorus philipi De Waele, Meyer & Van Mieghem, 1990

Figs 3.2.W.; 3.3 (4f); 3.7.J.; 3.8 (4f); 4.10.F

Measurements

Holotype male: L= 534 µm, a= 20.9, b=5.4, T= 72.4%, onchiostyle= 28.5 µm, spicule length= 42.5 µm, gubernaculum= 14.5 µm, anterior end to EP= 93 µm, distance CP1-EP= 16 µm.

Males: L= 534-666 µm (610 µm ± 37.9), a= 19.2-29.4, b= 5.2-6.4, T = 64.8-72.4%, onchiostyle= 27.5-30.5 µm (28.5 µm ± 0.9), spicule length= 39-45 µm (42.5 µm ± 2.1), gubernaculum= 10.5-15.0; anterior end to EP= 89-

106 µm, distance CP1-EP= 14-23 µm.

Females: L= 504-643 µm (572 µm ± 49.5), a= 18.2-23.0, b= 4.5-6.3, V= 50.0-56.7% (52.9% ± 2.0), onchiostyle= 26-28.5 µm (27.0 µm ± 0.9), anterior end to EP= 89-108 µm (from De Waele *et al.*, 1990).

Emended diagnosis

T. philipi male is characterized by the shape of the spicules (slightly cephalated, anteriorly curved, shaft finely striated and mainly straight), one ventromedian cervical papilla anterior to the S-E pore, one pair of lateral cervical pores at the level of ventromedian cervical papilla, three ventromedian precloacal supplements with the posterior one within the region of retracted spicules, near their head and tail terminal cuticle slightly thickened. The female by the shape of the vaginal sclerotizations in lateral view (oval to rounded), the rhomboid vagina and the two pairs of lateral body pores (one prevulvar, one postadvulvar). Males and females have an offset pharyngeal bulb.

Relationships

T. philipi males resemble those of *T. petrusalberti*, *T. rinae* and *T. sparsus* by the shape of the spicules (slender, proximal part ventrally bent, shaft mainly straight), the posterior precloacal supplement in the region of the retracted spicules, shortly posterior the manubrium. Males differ from these species in having only one ventromedian cervical papilla instead of two, and from *T. petrusalberti* and *T. sparsus* by the slightly thickened terminal tail cuticle *vs* tail cuticle thickened in *T. sparsus*, and equally wide in *T. petrusalberti*. *T. philipi* further agrees with *T. sparsus* by possessing an offset pharyngeal bulb instead of a dorsal intestinal overlap as in *T. petrusalberti* and *T. rinae* (partim). *T. philipi* females differ from these species by the size and shape of the vaginal sclerotizations (small, oval to rounded *vs* small, triangular in T. *rinae* and *T. sparsus* and minute triangular in *T. petrusalberti).*

Remarks

Tail terminus cuticle slightly thickened in male specimens without swollen cuticle, more thickened in specimens with swollen cuticle (De Waele *et al.*, 1990).

The vaginal sclerotizations in lateral view were described originally as large, rounded, square-shaped but in the original illustration they appear rather small and oval to rounded in shape.

Distribution and host

T. philipi is known only from the type locality Hottentots-Holland Nature Reserve, Republic of South Africa where it was extracted from wet soil around the roots of native plants of Cape Fynbos on the summit of the Landdroskloof, 1,350 m altitude (De Waele *et al.*, 1990).

Trichodorus primitivus (de Man, 1880) Micoletzky, 1922
Figs 3.2.Z; 3.4 (7c); 3.5.M; 3.9 (7c); 3.11.M; 4.19.F-J, 6.5.A-J

syn. *Dorylaimus primitivus* de Man, 1880
syn. *Trichodorus castellanensis* Arias Delgado, Jimenez Millan & Lopez Pedregal, 1965
syn. *Trichodorus mirabilis* Ivanova, 1977, n.syn.

See chapter on economically important species

Trichodorus rinae Vermeulen & Heyns, 1984
Figs 3.2.P; 3.3 (4b); 3.5.I; 3.8 (4b)

Measurements

Holotype male: L= 924 µm, a= 20.1, b= 6.2, onchiostyle= 60 µm, spicule length= 55 µm, gubernaculum= 25 µm.

Males: L= 582-1034 µm (794 µm), a= 12.9-26.5, b= 3.5-8.7, onchiostyle= 41-64 µm (53 µm), spicule length= 40-55 µm (50 µm), gubernaculum= 20-26.4 µm, anterior end to EP= 64-102 µm, distance between CP1-EP= 8-16 µm, distance between EP-CP2= 5-20 µm.

Females: L= 624-1083 µm (818 µm), a= 15.6-25.5, b= 4.7-8.4, V= 50-61% (54%), onchiostyle= 41-62 µm (51 µm) (from Vermeulen & Heyns, 1984; De Waele, 1988).

Emended diagnosis

T. rinae male can be distinguished by two ventromedian cervical papillae with the S-E pore situated between, the shape of the spicules (cephalated, strongly ventrally curved at proximal end and at distal tip, in between straight, shaft finely striated), three precloacal supplements with the posterior supplement within the region of retracted spicules, at the level of the manubrium, and the terminal slightly thickened tail cuticle. The female by the size and shape of the vaginal sclerotizations in lateral view (minute, triangular), the more or less barrel-shaped vagina and the presence of one pair of postadvulvar lateral body pores. Pharyngeal bulb offset or with an anterior overlap of the intestine.

Relationships

T. rinae males are similar to *T. pakistanensis* and *T. parorientalis* by the position of the S-E pore between the ventromedian cervical papillae. *T. rinae* males further agree with those of *T. parorientalis* in having two ventromedian cervical papillae *vs* three in males of *T. pakistanensis*, and, with *T. pakistanensis* by the shape of the spicules (proximally ventrally bent, shaft mainly straight, striated finely in *T. rinae*, but strongly in *T. pakistanensis* instead of ventrally bent spicules with mid-shaft constriction in *T. parorientalis*). It differs from both former species by the slightly thickened terminal tail cuticle which is not thickened in the other species and by the dorsal intestinal overlap in 50% of the population *vs* an offset pharyngeal bulb in *T. pakistanensis* and *T. parorientalis*. Females of *T. rinae* are similar to *T. pakistanensis* and *T. parorientalis* by the small vaginal sclerotizations showing some differences in shape (triangular in *T. rinae vs* rounded to oval in *T. pakistanensis* and oval in *T. parorientalis*, lateral view). *T. rinae* females further differ from those of *T. parorientalis* by possessing only one pair of lateral body pores, as in *T. pakistanensis*.

Remarks

Half of the type specimens have an offset pharyngeal bulb, the other half a distinct anterior overlap of the intestine (De Waele, 1988). One male has the S-E pore anterior to the ventromedian cervical papillae. The posterior ventromedian cervical papilla exceptionally located at level of the anterior intestine. The lateral cervical pores, with variable position, are from the level of the anterior ventromedian cervical papilla to the level of the S-E pore, on both sides.

Vermeulen & Heyns (1984) considered *T. rinae* close to *T. aequalis*, both species with two ventromedian cervical papillae, one precloacal supplement within the region of retracted spicules in males and similar vaginal sclerotizations in females. They considered as a major difference between females of *T. rinae* and *T. aequalis* : the presence of spermathecae in *T. rinae* and the absence in *T. aequalis*. However, both species have spermathecae.

Distribution and host

T. rinae has been recorded only from Natal, South Africa, from the rhizosphere of unidentified indigenous plants and grasses, *Rubus ludwigii* and *Thalicrum* sp. (Vermeulen & Heyns, 1984).

Trichodorus sanniae Vermeulen & Heyns, 1984

Figs 3.2.X; 3.3 (2b); 3.5.B; 3.8 (2b)

Measurements

Male holotype: L= 767 µm, a= 23.2, b= 5.2, onchiostyle= 41 µm, spicule length= 50 µm, gubernaculum= 16.4 µm, distance between EP -CP1= 5 µm.

Males: L= 600-767 µm (654 µm), a= 16.2-23.2, b= 5.2-7.8, onchiostyle= 41-45 µm (43 µm), spicule length= 40-52 µm (47.3 µm), gubernaculum= 16-16.4, anterior end to EP= 93-109 µm, distance tween EP-CP1= 5-11 µm.

Females: L= 695-706 µm, a= 18.6-20.4, b= 7.8-8.2, V=52-54%, onchiostyle= 43 µm, anterior end to EP= 99-104 µm (from Vermeulen & Heyns, 1984; De Waele, 1988).

Emended diagnosis

T. sanniae male can be distinguished by the shape of the spicules (proximally and at distal end clearly curved ventrad, shaft slender, straight, at mid-length with a constriction preceded by a septum), one ventromedian cervical papilla anterior to the S-E pore, one pair of lateral cervical pores at the base of the pharynx, three ventromedian precloacal supplements with the posterior one opposite the head of retracted spicules and the terminal tail cuticle not thickened. The female by the shape of the vaginal sclerotizations in lateral view (minute dot-like), one pair of postadvulvar lateral body pores, the shape of the vagina (short, about barrel-shaped) and by the shape of the vulva (a longitudinal slit). This species is also characterized by the far posterior position of the S-E pore which is situated at the base of the pharynx and by the presence of a short anterio-dorsal intestinal overlap in both males and females.

Relationships

T. sanniae males resemble *T. kilianae* and *T. vandenbergae* in having one ventromedian cervical papilla anterior to the S-E pore, the shape of the spicules (proximally and distally ventrally bent,in between straight with mid-shaft constriction), the non-thickened terminal tail cuticle. It further resembles *T. kilianae* by the position of the posterior precloacal supplement at the level of the head of retracted spicules in males, by one pair of lateral body pores in females, and by the presence of an intestinal dorsal overlap in both sexes. It differs from *T. kilianae* and *T. vandenbergae* by the posterior position of the S-E pore (at the pharyngo-intestinal junction *vs* more anteriorly), by the shape of the vulva (a longitudinal slit *vs* pore-like in the other two species) and the shape of the minute vaginal sclerotizations in lateral view (dot-like *vs* triangular in *T. kilianae* and rounded triangular in *T. vandenbergae*) in females. *T. sanniae* can be further distinguished from *T. vandenbergae* by the position of the posterior precloacal supplement in males (at spicule head *vs* shortly posterior to the spicule head) and by one pair of lateral body pores *vs* two pairs in *T. vandenbergae*.

Remarks

The pharyngeal bulb is not offset as originally described but always with a short dorsal overlap by the intestine (De Waele, 1988). The interpretation of the presence or absence of a swelling of the tail tip cuticle, when the latter is not pronounced, may be rather subjective and shows a discrepancy between the description and the illustration.

Distribution and host

T. sanniae has been described from Transvaal, Republic of South Africa, from a soil sample collected in a field with natural vegetation (Vermeulen & Heyns, 1984).

Trichodorus similis Seinhorst, 1963

Figs 3.1.H; 3.4 (8d); 3.6.F; 3.9 (8d); 3.11.N; 4.20.F-I, 6.6.A-K

See chapter on economically important species

Trichodorus sparsus Szcygiel, 1968

Figs 3.2.Q, q; 3.3 (4e, e'); 3.4 (9i); 3.5.H; 3.8 (4e, e'); 4.21.A-F

Measurements

Holotype female: L= 910 µm, a= 25, b=5.5, V= 55.4%, onchiostyle= 50 µm.

Females: L= 530-1090 µm (920 µm; 930 µm*; 975 µm*; 950 µm*; 800 µm*; 745 µm*; 750 µm*; 700 µm*; 1010 µm*; 950 µm*; 795 µm*; 795 µm*; 719 µm; 866 µm ± 60.9; 887 µm ± 98.1; 940 µm), a= 16.1-31, b= 4.2-8.0, V= 52-59.3% (54.6%; 54.4% ± 2.3; 55.0% + 1.5; 56%; 57%; 55%), onchiostyle= 45-67 µm (48.3 µm; 50.5 µm*; 55.5 µm*; 54.5 µm*; 54.5 µm*; 54 µm*; 48.5 µm*; 51.5 µm*; 54.5 µm*; 56 µm*; 66.5 µm*; 63 µm*; 59 µm; 55.8 µm ± 2.2; 50.0 µm ± 2.8; 67 µm), anterior end to EP= 96-124 µm.
Males: L= 560-1150 µm (850 µm; 935 µm*; 1035 µm*; 935 µm*; 880 µm*; 745 µm*; 920 µm*; 750 µm*; 715 µm*; 685 µm*; 1070 µm*; 990 µm*; 810 µm*; 825 µm*; 785 µm; 940 µm; 868 µm ± 65.3; 929 µm ± 74.6; 873 µm; 880 µm), a= 15.9-29, b= 4.3-10.3, T= 53-71.6%, onchiostyle= 44-73 µm (48.6 µm; 52.5 µm*; 57 µm*; 53.5 µm*; 55.5 µm*; 58.5 µm*; 55 µm*; 50.5 µm*; 50 µm*; 52 µm*; 57 µm*; 57 µm*; 67.5 µm*; 62.5 µm*; 58 µm; 55.2 µm ± 2.2; 49.8 µm ± 2.1; 51.4 µm; 66 µm), spicule length= 43-65 µm (49.5 µm; 52 µm*; 55.5 µm*; 55 µm*; 55.5 µm*; 53 µm*; 55 µm*; 43 µm; 48.5 µm*; 46.5 µm*; 46.5 µm*; 62 µm*; 54 µm; 54 µm*; 48 µm*; 55 µm; 52.3 µm ± 2.4; 50.6 µm ± 1.5; 50.6 µm; 53 µm), gubernaculum= 18-25 µm, anterior end to EP= 105-130 µm, distance CP1-CP2= 9-42 µm, distance CP2-EP= 3-18 µm (from Szczygiel, 1968; Loof, 1973: 11 populations ♀, 13 ♂; Roca & Lamberti, 1984; Vinciguerra & Giannetto, 1987, ♂ only; Peneva, 1988: 2 populations; new observations, ♂ only; Mancini *et al.*, 1979).

Emended diagnosis

T. sparsus males are characterized by two ventromedian cervical papillae anterior to the S-E pore, on each side a lateral cervical pore between the anterior ventromedian cervical papilla and the S-E pore, the shape of the spicules (anterior end slightly wider, spicule slightly ventrally curved or proximally curved then shaft straight except at distal tip, shaft about equally wide, tapered distally, median part striated and provided with bristles), the tail shape (terminal tail cuticle strongly thickened with the terminus often bulging out drop-like) and three ventromedian precloacal papillae with the posterior one at the level of the head of the retracted spicules or just anterior to it. Females by the pore-like vulva, the shape of the vaginal sclerotizations in lateral view (small triangular to rounded triangular), the barrel-shaped vagina and by two lateral body pores (one prevulvar, one postadvulvar) on each side. The pharyngeal bulb is usually offset in males and females.

Relationships

T. sparsus male is similar to *T. aequalis* in having two well developed ventromedian cervical papillae, and the position of the posterior precloacal supplement (near the head of retracted spicules), and the female in having a pore-like vulva, two pairs of lateral body pores and small (rounded) triangular vaginal sclerotized pieces in lateral view. Both species also have a comparable onchiostyle length. *T. sparsus* can easily be distinguished from *T. aequalis* in males mainly by the spicule shape (shaft about equally wide, tapered distally, median part usually striated and with bristles *vs* spicules ventrally bent, shaft about equally wide, without bristles in *T. aequalis*) and the tail shape (not evenly rounded and usually with a strongly thickened terminal cuticle *vs* non-thickened end cuticle in *T. aequalis*). Females of both species can hardly be distinguished i.e. only upon some differences in the shape of the vagina.

Remarks

The following peculiarities were observed. In the type population: one male specimen with only one ventromedian cervical papilla, and two males with two precloacal supplements (Loof, 1973). In Bulgarian populations: one male with the S-E pore situated between the two ventromedian cervical papillae, two males with a single ventromedian cervical papilla and one female with three lateral body pores (two prevulvar, one postadvulvar on each side) (Peneva, 1988). A slight ventral overlap of the pharynx has also been observed (Roca & Lamberti, 1984). The male tail is often ventrally concave suggesting the presence of lateral alae (Loof, 1973).

Loof (1973) discussed the intraspecific variability of some morphometric data e.g. he noted some variation in body length, but considered that very small values may partly be due to poor fixation. Large variations were observed in the length of the onchiostyle and spicules e.g. specimens from two German populations were conspicuous by their long onchiostyles, up to 73 μm.

Among male specimens of *T. sparsus* from Belgian populations, variability was observed in the size of the thickening of the end cuticle on the tail and one specimen had an equally wide terminal cuticle (= non-thickened). In females, the shape of the vagina may vary (Loof, 1973; Decraemer, 1988) and to a lesser extent also the size and degree of sclerotization of the vaginal sclerotizations (Decraemer, 1988); exceptionally, weak rounded sclerotized pieces have been observed. Moreover, the spicule shape may vary (Decraemer, 1988) e.g. in the degree of the curvature of the spicule shaft, influenced by contraction of the posterior body end upon fixation.

Distribution and host

T. sparsus is widespread in Europe: Austria, Belgium, Bulgaria, France, Great-Britain, Hungary, northern Italy, The Netherlands, Poland, Roumania, Sweden, Switzerland, Turkey (European part), western Germany, although in most countries (except Bulgaria and northern Italy) it is usually rare or occurs in low populations. Until now, *T. sparsus* has not been recorded from the Iberian peninsula. It is a polyphagous species, found in soil around strawberry roots, potato, sugar beet, onions, Eastern white pine (*Pinus strobus*), grapevine, cherry trees, tobacco, olive, poplar, alder, beech, pear, chestnut, *Fagus* sp., pastures, deciduous woodland, *Picea abies*, *Pinus sylvestris*, *Larix decidua* and *Pseudotsuga taxifolia* (Szczygiel, 1968; Rössner, 1969; Loof, 1973; Andrassy, 1973; Brzeski & Szczygiel, 1974; Eriksson, 1974; Rau, 1975; Mancini *et al.*, 1979; Popovici, 1980; Roca & Lamberti, 1984; Peneva, 1988; Choleva, 1988, De Waele & Coomans, 1991)

Trichodorus taylori De Waele, Mancini, Roca & Lamberti, 1982
Figs 3.1.Q,q; 3.4 (6a); 3.6.A; 3.9 (6a); 3.11.O; 4.21.G-I

Measurements

Holotype female: L= 960 µm, a= 17.4, b= 6.8, V= 58.5%, onchiostyle= 70 µm, anterior end to EP= 126 µm. Females: L= 668-1061 µm (939 µm ± 72.8; 747 µm ± 59.7), a= 17.2-21.9, b= 4.1-6.9, V= 53.4-60.8% (57.2% ± 1.9; 58.7% ± 1.3), onchiostyle= 60-72 µm (70 µm ± 2.8; 64 µm ± 2.2), anterior end to EP= 111-132 µm. Males: L= 682-1019 µm (881 µm ± 70.3; 761 µm ± 74.0), a= 16.3-25.3, b= 4.5-6.1, onchiostyle= 61-70 µm (68 µm ± 1.3; 65 µm ± 2.3), spicule length= 57-65 µm (60 µm ± 2.7; 61 µm ± 2.5), gubernaculum= 6-8 µm, anterior end to EP= 111-134 µm, distance CP1-CP2= 11-22 µm, distance CP2-EP= 4-18 µm (from De Waele *et al.*, 1982).

Emended diagnosis

T. taylori male is characterized by the shape of the spicules (manubrium long, heavily built and clearly offset, shaft with middle portion slightly wider and striated, no bristles), two ventromedian cervical papillae anterior to the S-E pore, a pair of lateral cervical pores at about the level of the nerve ring (i.e. anterior to the ventromedian cervical papillae), the three precloacal supplements with the posterior one at the level of the posterior end of the manubrium of retracted spicules and a thickened terminal tail cuticle. In female by the shape of the vaginal sclerotizations in lateral view (rounded), the pear-shaped vagina and one pair of postadvulvar lateral body pores. Usually, the intestine overlaps the pharyngeal bulb dorsally in males and females.

Relationships

T. taylori is similar to *T. minzi* in spicule shape (with long, offset manubrium) but differs by the manubrium being more sharply offset then in *T. minzi*. It further resembles *T. minzi* by the presence of two ventromedian cervical papillae in males, large rounded vaginal sclerotizations in lateral view in females and by the dorsal intestinal overlap. *T. taylori* differs from *T. minzi* by having shorter spicules (57-65 µm *vs* 63-71 µm), a longer onchiostyle (60-72 µm *vs* 53-64 µm), a much shorter gubernaculum (6-8 µm *vs* 30-37 µm) and a thickened terminal tail cuticle *vs* only slightly thickened in *T. minzi*.

Remarks

Some specimens have an offset pharyngeal bulb.

Distribution and host

T. taylori has been reported only from Northern Italy where it was collected from the rhizosphere of Eastern white pine (*Pinus strobus*), olive trees (*Olea europaea*) and grapevine (*Vitis* sp.) (De Waele *et al.*, 1982).

Trichodorus tricaulatus Shishida, 1979

Figs 3.2.Y; 3.4 (9e); 3.5.P; 3.10 (9e); 3.11.Q; 4.22.A-D

Measurements

Holotype male: L= 757 µm, a= 25, b= 5.4, T= 67%, onchiostyle= 43.6 µm, spicule length= 43 µm, gubernaculum= 15 µm.

Males: L= 583-827 µm (727 µm ± 55), a= 19-26, b= 4.2-5.8, T= 55-70%, onchiostyle= 38.6-51 µm (42.6 µm ± 2.3), spicule length= 35-44 µm (39 µm ± 2.6), gubernaculum= 11-19 µm, anterior end to EP= 81-115 µm, distance CP1-CP2= 3.9-10.5 µm, distance CP2-CP3= 8.1-14 µm.

Females: L= 602-785 µm (709 µm ± 53), a= 18-26, b= 4.6-6.2, V= 52-56% (54.1% ± 1.2), onchiostyle= 38.6-44 µm (42.2 µm ± 1.7), anterior end to EP= 77-95 µm (from Shishida, 1979).

Emended diagnosis

T. tricaulatus males are characterized by the shape of the spicules (about straight, shaft with striae except for expanded proximal end and distal tip, mid-shaft with a few bristles), three ventromedian cervical papillae anterior to the S-E pore, a pair of lateral cervical pores at about the level of the nerve ring, three precloacal supplements all anterior to the region of the retracted spicules and the terminal tail cuticle thickened. Females by the shape of the vaginal sclerotizations in lateral view (small, round or rounded triangular), the pore-like vulva, the shape of the vagina (pear-shaped when relaxed) and two pairs of lateral body pores (one pair prevulvar, one pair postadvulvar). Small pharyngeal and intestinal overlaps are present in males and females.

Relationships

T. tricaulatus appears most close to *T. aequalis*, especially in females. Females of both species can only be distinguished by a different type of pharyngo-intestinal junction (usually with ventral overlap by pharyngeal glands *vs* an offset bulb in *T. aequalis*) (see also relationships of *T. aequalis*).

Remarks

The following peculiarities were found: two of 26 males had only one ventromedian cervical papilla and four specimens had two papillae. The vaginal sclerotizations may exceptionally be rounded in females.

T. tricaulatus was described as having the pharynx ventrally overlapping the intestine. However, the original figure (1 A) of a male also shows a short anterio-dorsal overlap of the intestine together with a minute pharyngeal overlap.

Shishida (1979) gave morphometric data of 16 juveniles, probably forming a mixture of fourth and third stage juveniles: L= 463-662 µm (526 µm), a= 14-24, b= 3.8-5, onchiostyle= 35-42 µm (38.6 µm), gonad length= 24-56 µm (40 µm)

Distribution and host

This species is considered native to Meiji Shrine Forest in Japan as it was found associated with natural vegetation (Shishida, 1979). It was collected around the roots of *Quercus acutissima*, *Q. serrata*, *Neolitsea sericea*, *Ilex crenata* (Japanese holly), *Cleyera japonica*, *Styrax japonica* and *Sambucus racemosa*, all of natural growth.

Trichodorus vandenbergae De Waele & Kilian, 1992

Figs 3.2.R, r; 3.3 (3c); 3.7.N; 3.8 (3c); 4.23.A-K

Measurements

Holotype female: L= 750 µm, a= 23.1, b= 4.6, V= 55.2%, onchiostyle= 53 µm, anterior end to EP= 107 µm. Females: L= 681-801 µm (753 µm ± 32), a= 21.4-25.5, b= 4.5-5.5, V= 52.6-61.3% (54.6% ± 2.1), onchiostyle=

50-58 µm (54 µm ± 3), anterior end to EP= 97-115 µm.
Males: L= 720-809 µm (765 µm), a= 21.3-26.4, b= 4.2-5.5, T= 56.4-68.4%, onchiostyle= 55-57 µm (56 µm), spicule length= 43-49 µm (46 µm), gubernaculum= 19-22 µm, anterior end to EP= 103-120 µm, distance CP1-EP= 13-26 (from De Waele & Kilian, 1992).

Emended diagnosis

T. vandenbergae males can be distinguished by one ventromedian cervical papilla anterior to the S-E pore, one pair of lateral cervical pores about at the level of the ventromedian cervical papilla, three ventromedian precloacal papillae with the posterior one within the region of retracted spicules (just posterior to the spicule head), the shape of the smooth spicules (cephalated, proximal half dorsally convex, distal half straight except at tip, slight mid-shaft constriction) and the terminal tail cuticle not thickened. Females by the shape of the vaginal sclerotizations in lateral view (minute, rounded triangular or exc. rounded), a pore-like vulva, the pear-shaped vagina (elongated upon contraction) and one pair of prevulvar lateral body pores and one pair of subventral postadvulvar body pores. The pharyngeal bulb is usually offset in males and females.

Relationships

T. vandenbergae has one ventromedian cervical papilla as in 12 other species of the genus, among them *T. kilianae*, *T. magnus* and *T. velatus* which further agree by having spicules with a mid-shaft constriction. *T. vandenbergae* differs from these three species by one or more characters presented in table 4.4. *T. vandenbergae* also resembles *T. eburneus* but differs in spicule shape (with indentation) and smaller vulvar sclerotizations. *T. vandenbergae* most closely resembles *T. kilianae*.

Remarks

The following exceptions have been described: one female with three lateral body pores (two prevulvar, one postadvulvar) on both sides; one male (out of four) and two females (out of fourteen) with a slight dorsal intestinal overlap (De Waele & Kilian, 1992).

Distribution and host

T. vandenbergae was described from specimens collected from Cape Province, Republic of South Africa, associated with soil around the roots of "witpeer" (*Apodytes dimitiata*) from indigenous forest (from De Waele & Kilian, 1992).

Table 4.4.

	T. vandenbergae	*T. kilianae*	*T. magnus*	*T. velatus*
spicule shape	shaft fine, short constriction	shaft fine, short constriction, preceded by a septum	shaft stout, constriction sharp	spicule stout, shaft with velum, constriction short
spicule length	43-49 µm	43-46.5 µm	52-57 µm	31-37 µm
position SP1	posterior to manubrium	at level manubrium	at level manubrium	at level manubrium
shape vulva	pore	pore	pore	a transverse slit
shape vaginal sclerotizations	small rounded triangular	small triangular	small, rounded triangular	small triangular
n° lateral body pores	2 pair	1 pair	1 pair	1 pair
L	681-809 µm	830-1048 µm	970-1295 µm	616-864 µm

Trichodorus variopapillatus Hooper, 1972
Figs 3.1.N; 3.4 (8c); 3.6.C; 3.9 (8c); 3.11.R; 4.22.E-G

Measurements

Holotype male: L= 826 µm, a= 25, b=5.6, T= 66%, onchiostyle= 52 µm, spicule length= 45 µm, gubernaculum= 19 µm.

Males: L= 774-915 µm (816 µm ± 31; 850 µm), a= 22-28.8, b= 4.7-10.5, T= 59-71%, onchiostyle= 50-56 µm (53 µm ± 2.0; 53 µm), spicule length= 40-53 µm (43 µm ± 1.2; 49 µm), gubernaculum= 15-24 µm, anterior end to EP= 83-85 µm.

Females: L= 694-1006 µm (803 µm ± 54; 880 µm), onchiostyle= 50-56 µm (52 µm; 53µm), a= 19-29.5, b= 4.6-12, V= 50-61% (56% ± 1.0; 55%) (from Hooper, 1972; Roca & Lamberti, 1984).

Emended diagnosis

T. variopapillatus male is characterized by the shape of the spicules (ventrally curved, manubrium knob-like, offset from a fine smooth shaft with slightly bifid tip), two or three ventromedian cervical papillae anterior to the S-E pore with the anterior one in the region of the onchiostyle, the position of the lateral cervical pore (about level with the base of the onchiostyle), tail rounded with non-thickened terminal cuticle and the three ventromedian precloacal supplements with the posterior one opposite or just anterior to the head of retracted spicules. In the female by two lateral body pores (one prevulvar, one postadvulvar), the shape of the vulva (a transverse slit), the more or less rhomboid-shaped vagina and by the shape of the vulvar sclerotizations in lateral view (conspicuous, rounded quadrangular). The pharyngeal bulb is usually offset, rarely with a slight subventral or sublateral overlap of the pharyngeal glands in males and females.

Relationships

T. variopapillatus most closely resembles *T. similis* in males by the shape of the spicules (ventrally bent, manubrium knob-like and offset), the non-thickened terminal tail cuticle, the presence of three (or two in *T. variopapillatus)* ventromedian cervical papillae with the anterior one in the onchiostyle region and the position of the posterior precloacal supplement at the level of the head of the retracted spicules. Females by the shape of the vulva (a transverse slit) and in both males and females by the pharynx with an offset bulb. *T. variopapillatus* male differs from *T. similis* by having more slender, longer spicules (40-53 μm *vs* 30-44 μm) and the female by the larger rounded triangular to trapezoid vaginal sclerotizations *vs* smaller triangular in *T. similis*, a barrel-shaped vagina *vs* rhomboid-shaped in *T. similis* and two pairs of lateral body pores *vs* one pair.

Remarks

De Waele in Almeida *et al.* (1989) reported that, four out of 44 males in a population from Belgium had two ventromedian cervical papillae whilst the other specimens had three. In an Italian population all males had three ventromedian cervical papillae (Roca & Lamberti, 1984) whilst in the type population half of the specimens had two, and half had three ventromedian cervical papillae.

According to Roca & Lamberti (1984a), the Italian population morphologically and biometrically fitted the original description with the exception of the offset lip region. However, the b-value (9.2-10.5) also differs from the type population (4.7-5.6).

Occasionally, an indication of a few bristles anteriorly on the spicule shaft was observed.

Distribution and host

T. variopapillatus is a European species with a restricted distribution. It has been recorded from England (one record), central and southern Italy, Belgium (rare), The

Netherlands, Germany and Poland. It was found in soils from the rhizosphere of elder (*Sambucus nigra*), *Populus* sp., oak (*Quercus* sp.), olive, grapevine, maize, roses, and in pastures and grasses along the road side (Rau, 1975; Hooper, 1972; Seinhorst & Van Hoof, 1982; De Waele & Coomans, 1991).

Trichodorus velatus Hooper, 1972

Figs 3.1.M, m; 3.3 (1d); 3.5.A; 3.8 (1d); 3.11.S; 4.22.H-J

Measurements

Holotype male: L= 740 µm, a= 25, b=5.1, T=67%, onchiostyle= 46 µm, spicule length= 34 µm, gubernaculum= 18 µm.

Males: L= 616-864 µm (724 µm ± 59), a= 21-33, b= 3.9-7.2, T= 59-72%, onchiostyle= 44-48 µm (46 µm ± 1.3), spicule length= 31-37 µm (34 µm ± 1.7), gubernaculum= 14-18 µm.

Females: L= 623-842 µm (720 µm ± 53), a= 19-26, b= 4.3-6.0, V= 53-59% (56% ± 1.6), Onchiostyle= 42-50 µm (46 µm ± 2.2) (from Hooper, 1972).

Emended diagnosis

T. velatus male is characterized by three ventromedian cervical papillae anterior to the S-E pore (the anterior one in the region of the onchiostyle), the position of the lateral cervical pore (just posterior to the base of the onchiostyle), the shape of the spicules (curved ventrally, shaft stout, with a wide proximal end, a narrower middle part with some coarse bristles, distal part with a velum and tapered posteriorly), three ventromedian precloacal supplements (the posterior one opposite the anterior end of retracted spicules), tail shape dorsally convex-conoid with a broadly rounded end and non-thickened terminal cuticle, and the female by the shape of the vulva (a transverse slit), the shape of the vaginal sclerotizations (medium-sized, roughly triangular, tips pointed to vulva), the shape of the vagina (rhomboid) and the presence of one pair of postadvulvar lateral body pores. The pharyngeal bulb is usually offset in males and females.

Relationships

T. velatus males most closely resemble *T. azorensis* , *T. beirensis* and *T. lusitanicus*, by the shape of the spicules (wide proximal part, clear constriction at about mid-shaft provided with a few bristles, then widened again, tapered towards distal end), the non-thickened terminal tail cuticle, the offset pharyngeal bulb and the presence of three ventromedian cervical papillae with the anterior one in the region of the onchiostyle (*T. lusitanicus* partim). It differs from these species by the presence of a spicular velum, a character it shares only with *T. carlingi*. The latter species, however, has a different spicule shape with the spicule shaft distally strongly curved, ending on a bifid tip. *T. velatus* females differ from those of *T. azorensis*, *T. beirensis* and *T. lusitanicus* by the smaller roughly triangular vaginal sclerotizations in lateral view *vs* respectively large square-shaped, large rounded triangular to almost oval and large, regular to roughly triangular with obtuse angles. Females can also be distinguished from those of *T. lusitanicus* by the shape of the vulva (a transverse slit *vs* a pore).

Remarks

Some specimens have a slight ventral overlap of the pharyngeal glands.

Distribution and host

T. velatus has been described from Belgium, France, Great Britain, Bulgaria, Germany, The Netherlands, Poland and Florida. It was found in the rhizosphere of strawberry, pea, wheat and in nurseries of Sitka spruce (*Picea sitchensis*), *Picea abies*, *Pinus sylvestris*, *Pinus contorta*, *Larix* sp., Phlox, rose and in deciduous woodland (alder, oak, poplar and *Crataegus* sp.)(Hooper, 1972; Alphey & Boag, 1976; Sturhan, 1978; Katalan-Gateva, 1985; Szczygiel & Brzeski, 1985; Alphey & Taylor, 1986; Mead, 1989a; McGowan, 1989a; De Waele & Coomans, 1991). In Great Britain its distribution is associated with light sandy to sandy loam soils (Hooper, 1972; Alphey & Boag, 1976).

Trichodorus viruliferus Hooper, 1963

Figs 3.2.AA, aa; 3.3 (1e); 3.5.C; 3.8 (1e); 3.11.T; 4.22.K-M, 6.7.A-J

See chapter on economically important species

Trichodorus yokooi Eroshenko & Teplyakov, 1975
Figs 3.1.J; 3.4 (7b); 3.5.O; 3.9 (7b)

Measurements

Holotype male: L= 640 µm, a= 15, b= 3.6, T= 44%, onchiostyle= 72 µm, spicule length= 44 µm, gubernaculum= 22 µm.

Males: L= 570-850 µm (725 µm*; 730 µm*), a= 13-21, b= 3.6-5.3, T= 44-69%, onchiostyle= 57-82 µm (72 µm*; 59.5 µm*), spicule length= 38-46 µm (40.5 µm*; 42 µm*).

Females: L= 560-750 µm (655 µm*), a= 14-22, b= 3.5-5, V= 53-65% (59%*), onchiostyle= 62-77 µm (71.5 µm *) (from Eroshenko & Teplyakov, 1975).

Emended diagnosis

T. yokooi male is characterized by three ventromedian cervical papillae anterior to the S-E pore and behind the onchiostyle region, three ventromedian precloacal supplements, none of them within the region of retracted spicules and by the shape of the spicules (ventrally bent, manubrium long and wide, shaft deeply grooved, tapered distally). The female by the shape of the vaginal sclerotizations in lateral view (medium-sized triangular with tips toward the vulva), the pear-shaped vagina, one postadvulvar lateral body pore on each side. The pharyngeal bulb is offset in males and females.

Relationships

T. yokooi males resemble those of *T. dilatatus*, *T. minzi* and *T. taylori* in general spicule shape i.e. with long, wide manubrium, but differs (except *T. dilatatus*) by the manubrium not being offset by a constriction as in *T. minzi* and *T. taylori*. *T. yokooi* males also differ from these species by having three ventromedian cervical papillae *vs* two in *T. minzi* and *T. taylori* and one papilla in *T. dilatatus*, shorter spicules (38-46 µm *vs* 48-55 µm in T. *dilatatus*, 63-71 µm in *T. minzi* and 57-65 µm in *T. taylori*) and an offset pharyngeal bulb as in *T. dilatatus vs* a dorsal intestinal overlap in *T. minzi* and *T. taylori*. *T. yokooi* females can be distinguished by the shape of the vaginal sclerotizations in lateral view i.e. smaller, triangular *vs* large roughly triangular with tips away from vulva in *T. dilatatus* and large rounded in *T. minzi* and *T. taylori* .

Remarks

Lateral cervical pores in males were not described.

Distribution and host

T. yokooi has been reported from the far east of the former USSR: from Central Sihote-Alinya (near Taiga) and from Sakhalina (near Korsakov). It was found in the rhizosphere of *Abies nephrolepis* and *Picea ajanensis* from the mountain region at the upper course of the river Bolshaya Ussurka, and in a nursery of pine trees (Eroshenko & Teplyakov, 1975).

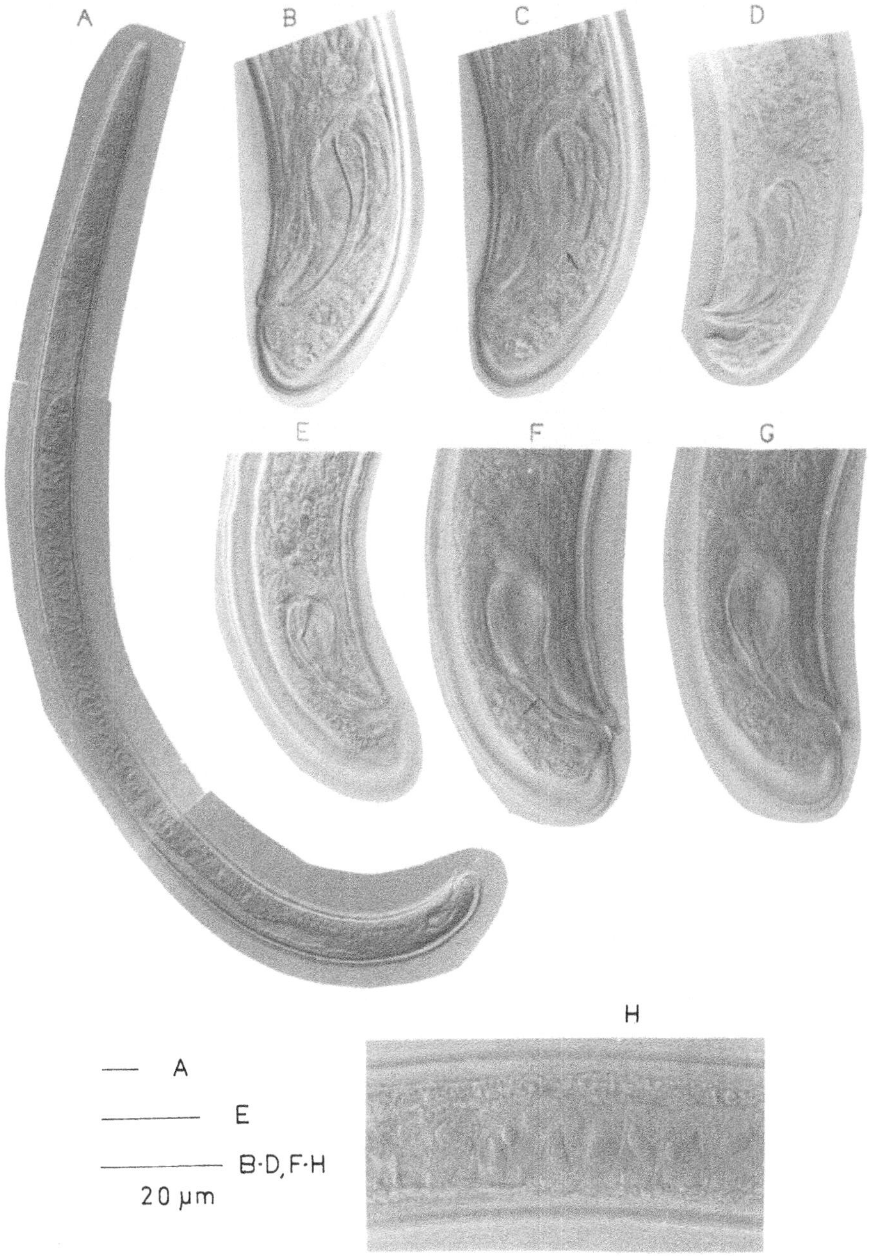

Fig. 4.1. *Trichodorus aquitanensis*. Male: (A) total view (paratype); copulatory apparatus and tail: (B-C) respectively with focus on spicule and on gubernaculum, see arrow (paratype), (D) aberrant spicule (paratype), (E) aberrant spicule (paratype), (F-G) respectively with focus on gubernaculum and on spicule (paratype); (H) sperm (paratype).

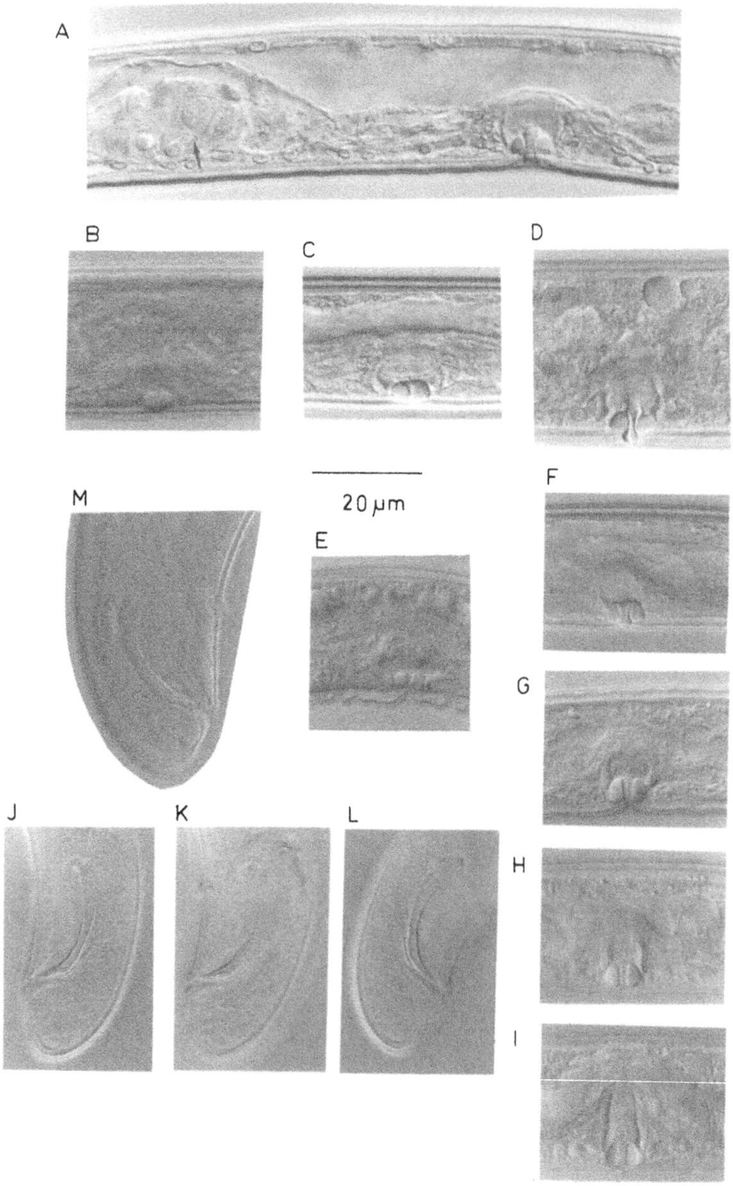

Fig. 4.2. *Trichodorus aquitanensis*. Female: (A) part of anterior genital branch with spermatheca (arrow indicates sperm cell), vagina with secretion plug (paratype); (B-D) vaginal region, D with secretion plug (paratypes). *Trichodorus azorensis*. (E) Female, vaginal region (paratype); Male: (M) copulatory apparatus and tail (paratype). *Trichodorus beirensis*. Female: (F-I) vaginal region, F with secretion plug. Male: (J-K) focus respectively on spicule and on gubernaculum, (L) copulatory apparatus.

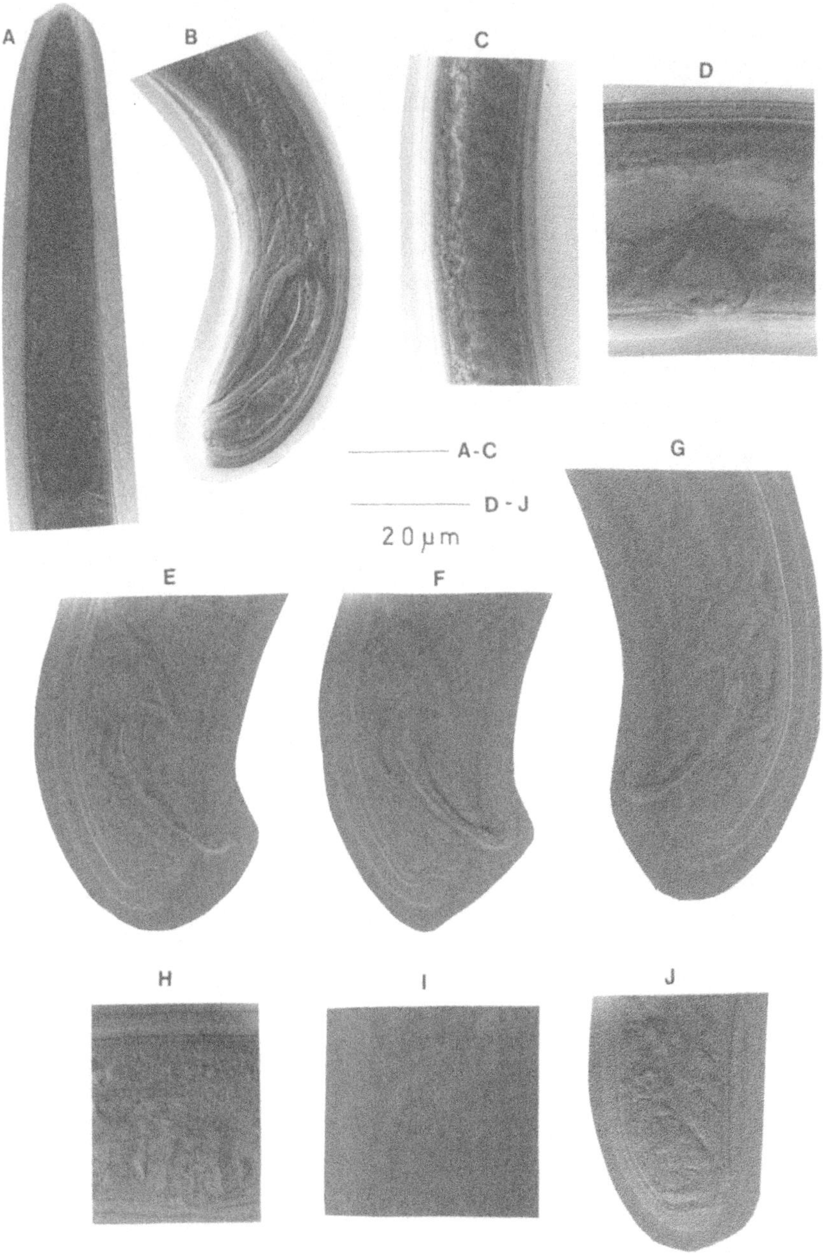

Fig. 4.3. *Trichodorus borneoensis*. Male, paratypes: (A) anterior body region, (B) posterior body region, (C) spermatozoa. Female: (D) vaginal region (paratype). *Trichodorus carlingi*. Male: (E-G) posterior body region and copulatory apparatus (paratypes). Female, paratypes (H) vaginal region, (I) ventral view vulva. Juvenile (J) tail region with spicular primordium (paratype).

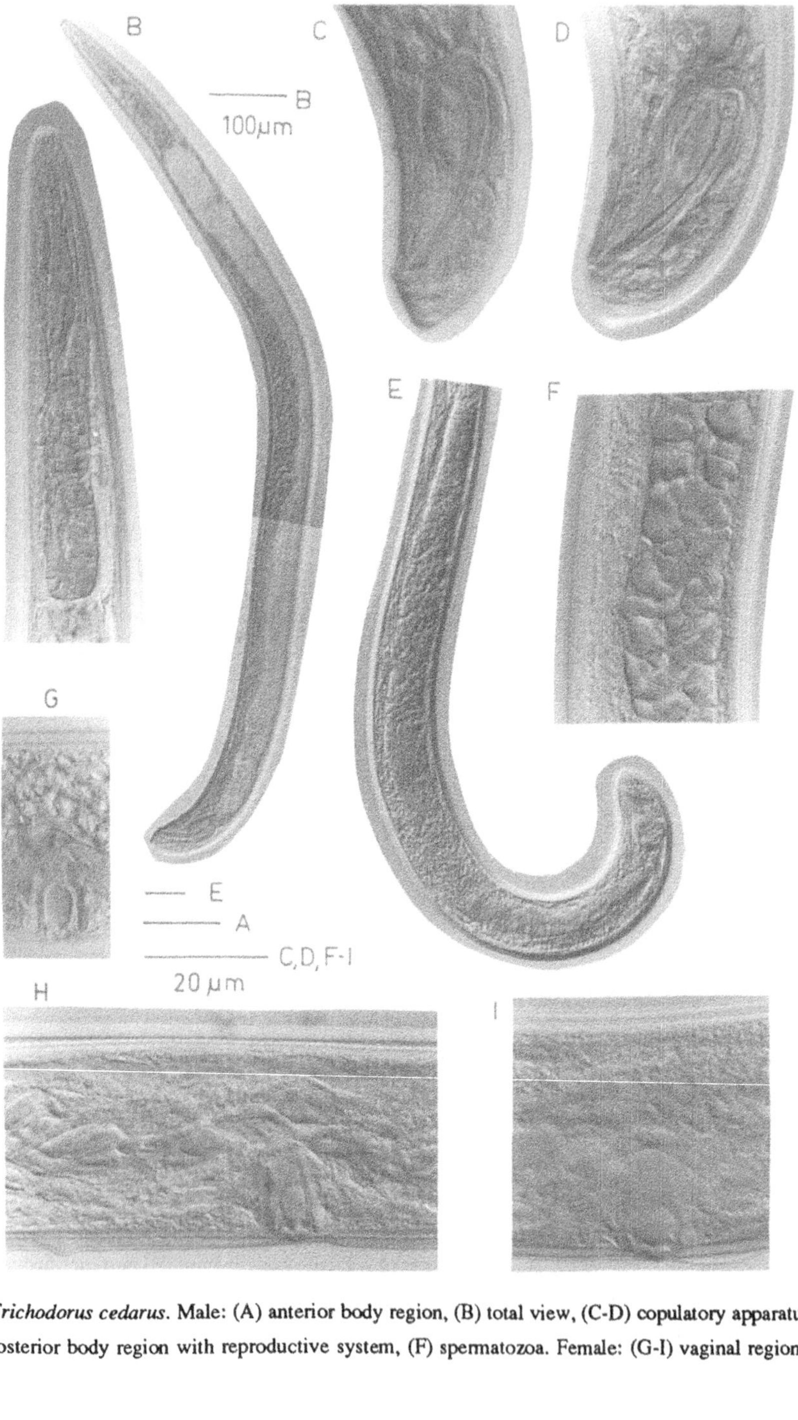

Fig. 4.4. *Trichodorus cedarus*. Male: (A) anterior body region, (B) total view, (C-D) copulatory apparatus and tail, (E) posterior body region with reproductive system, (F) spermatozoa. Female: (G-I) vaginal region.

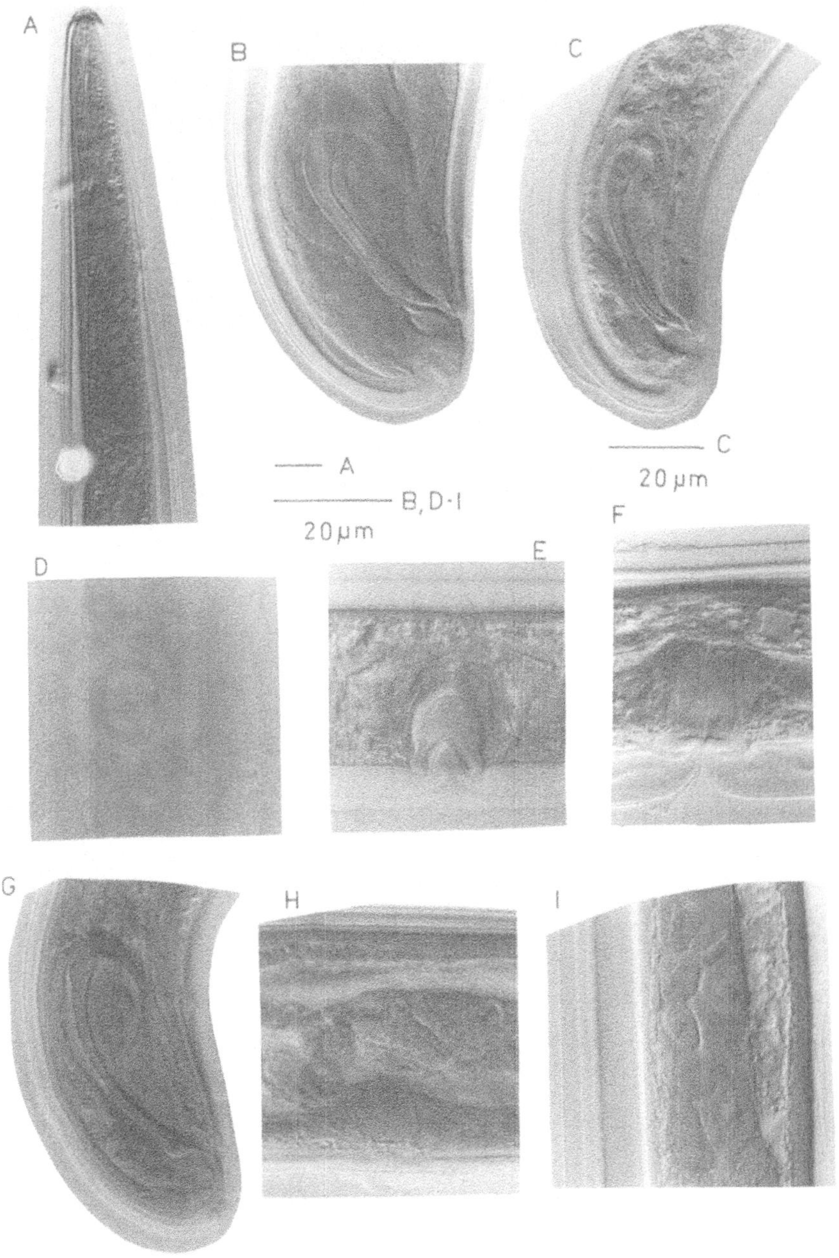

Fig. 4.5. *Trichodorus cottieri*. Male, paratypes: (A) anterior body region, (B-C) copulatory apparatus and tail. Female, paratypes: (D) ventral view vulva, (E-F) vaginal region. *Trichodorus coomansi*. Male, paratype: (G) copulatory apparatus and tail, (I) spermatozoa. Female: (H) vaginal region (paratype).

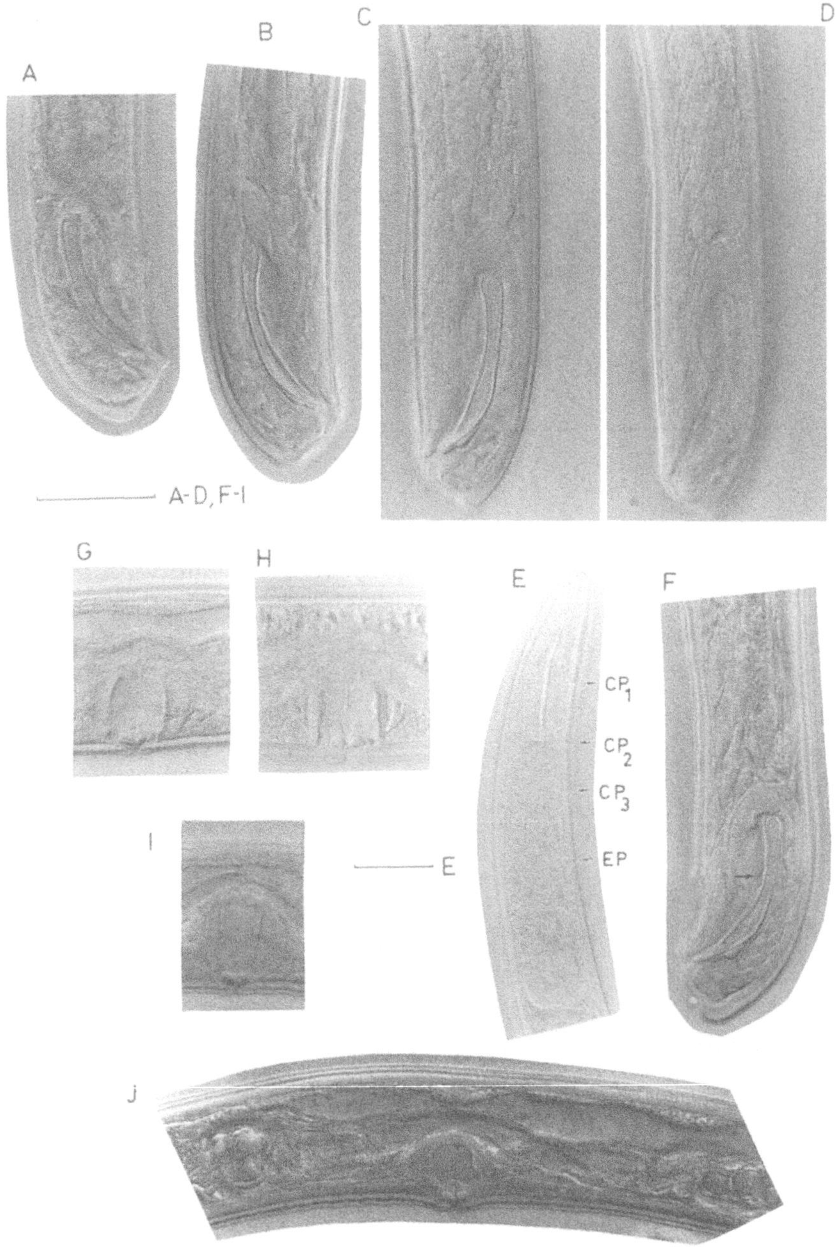

Fig. 4.6. *Trichodorus cylindricus*. Male: (A-B) copulatory apparatus and tail, (C-D) posterior body region, with focus respectively on spicule and on gubernaculum, (E) anterior body region, (F) posterior body region (arrow indicates bristles). Female: (G-H, I) vaginal region, (J) part of genital system with spermathecae, uterus and vaginal region (scale= 20 μm).

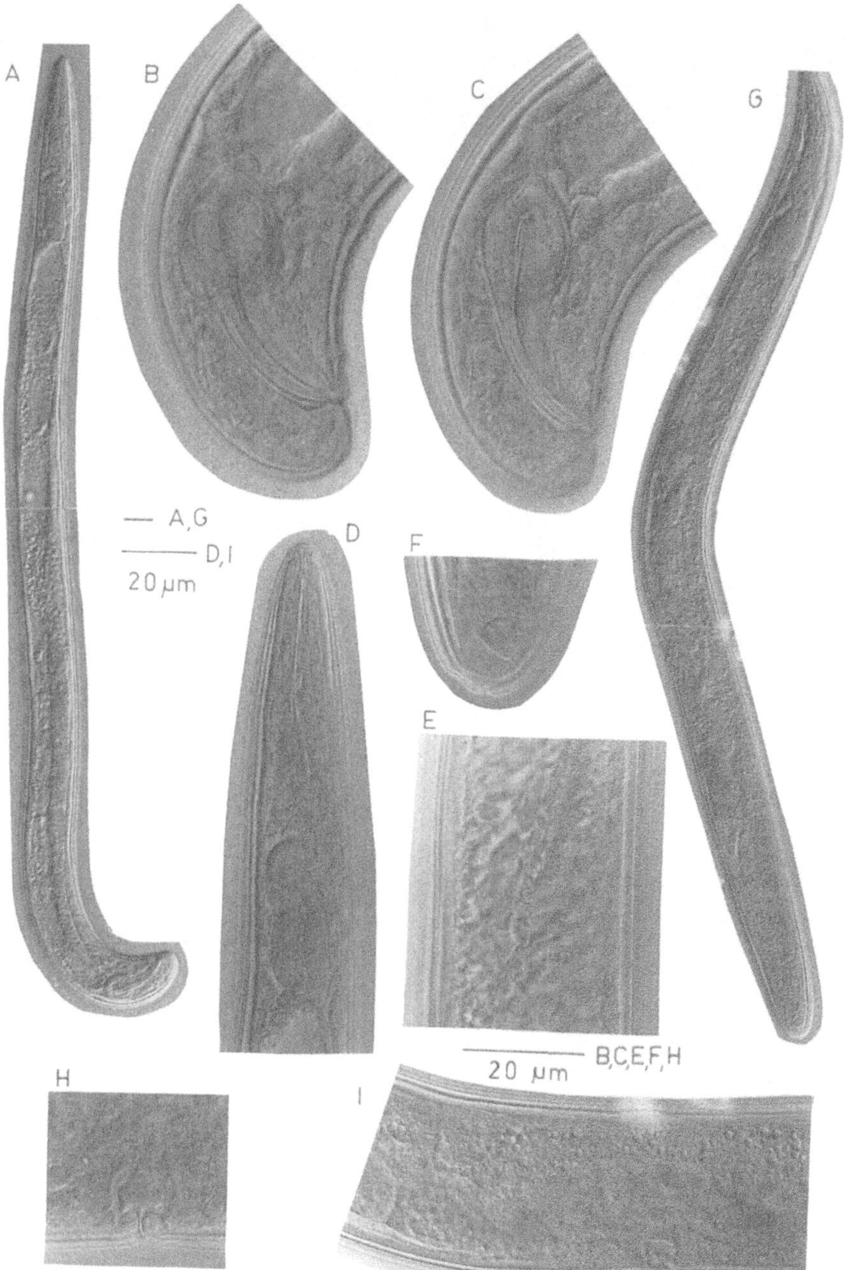

Fig. 4.7. *Trichodorus dilatatus*. Male, paratypes: (A) total view, (B-C) tail region with focus respectively on gubernaculum and on spicule, (D) anterior body region, (E) testis with sperm cells. Female, paratypes: (F) tail region, (G) total view, (H) vaginal region, (I) part of genital system with spermatheca.

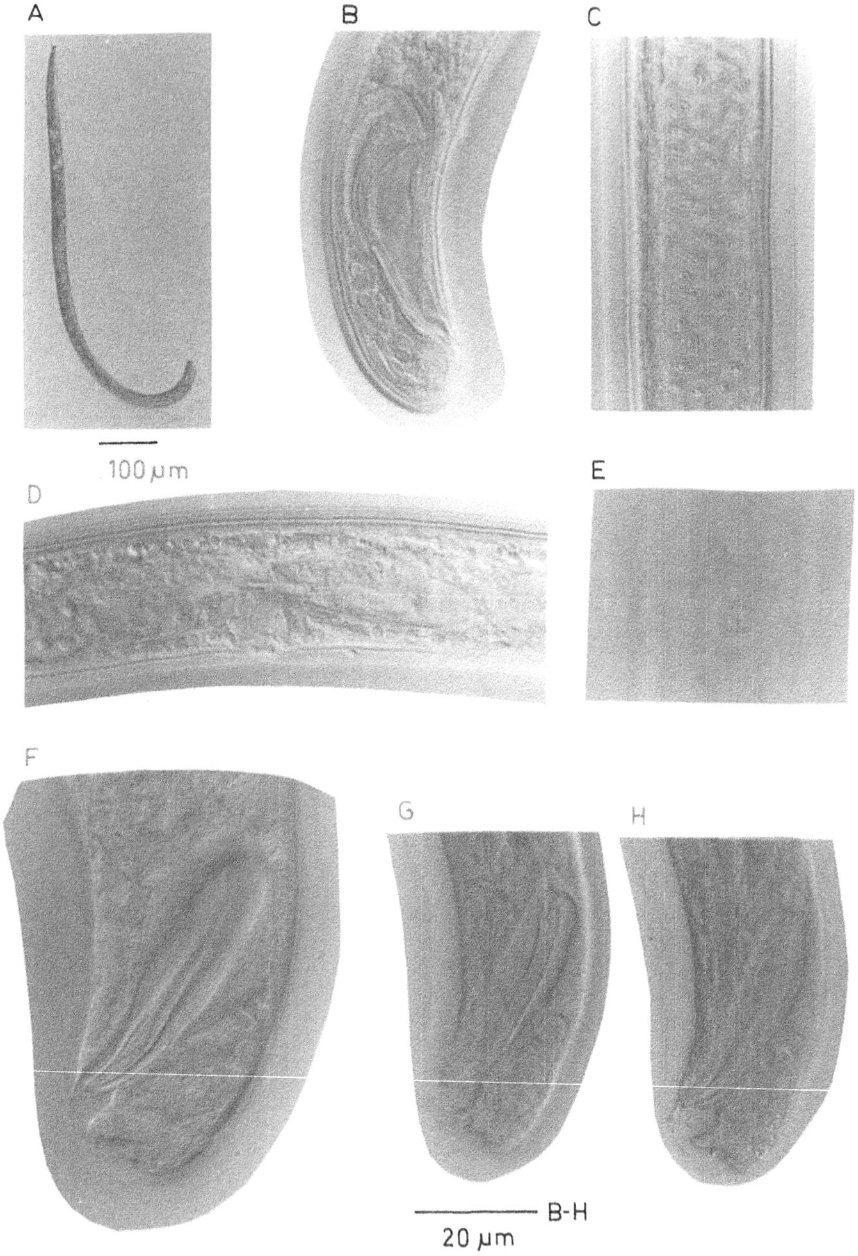

Fig. 4.8. *Trichodorus eburneus*. Male: (A) total view (paratype), (B) copulatory apparatus and tail (paratype), (C) spermatozoa. Female: (D) vaginal region and uterus (paratype), (E) ventral view vulva. *Trichodorus elegans*. Male, paratypes: (F) copulatory apparatus and tail, (G-H) tail region with focus respectively on spicule and on gubernaculum.

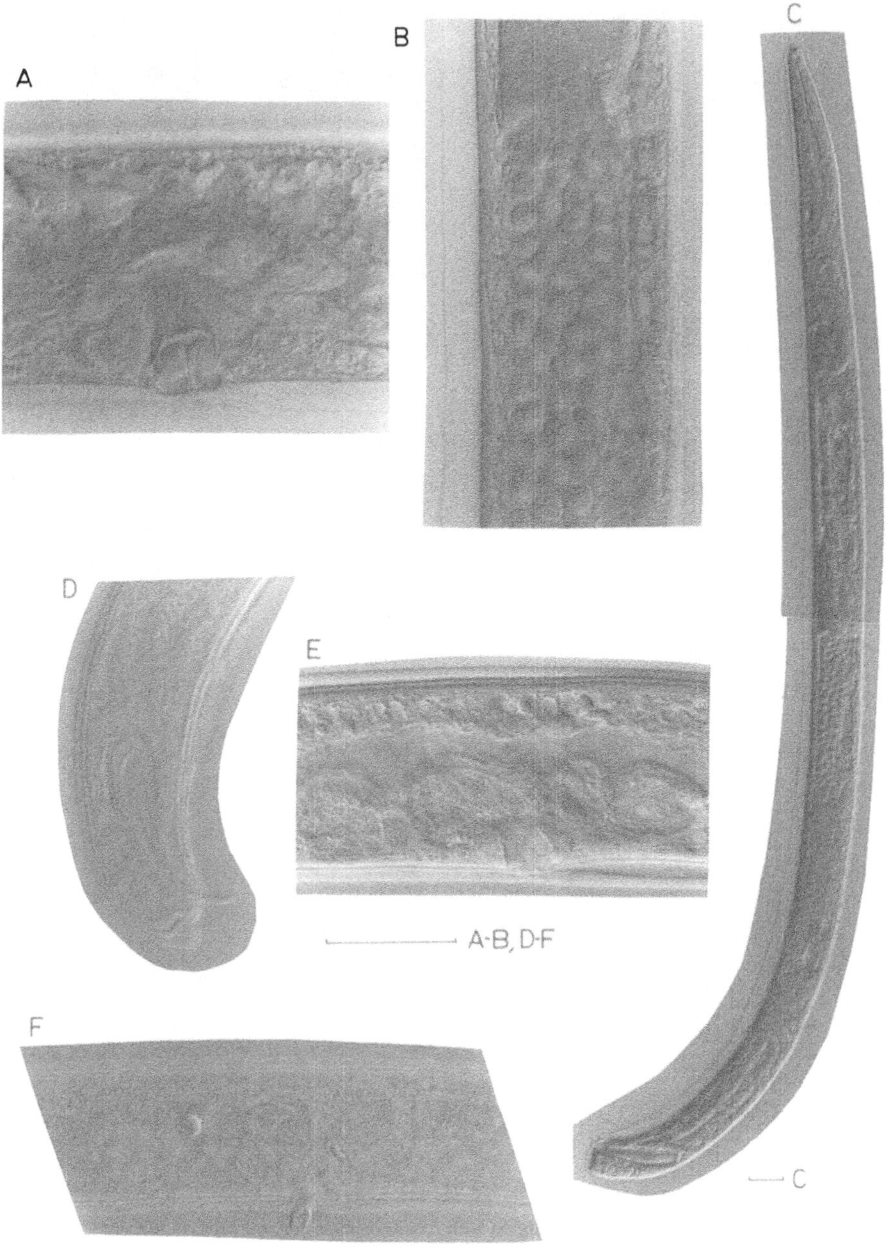

Fig. 4.9. *Trichodorus elegans*. Female: (A) vaginal region (paratype). Male, paratype: (B) spermatozoa, (C) total view. *Trichodorus elefjohnsoni*. Male: (D) copulatory apparatus and tail (paratype). Female, paratypes: (E) vaginal region, (F) vaginal region with secretion plug (scale= 20 μm).

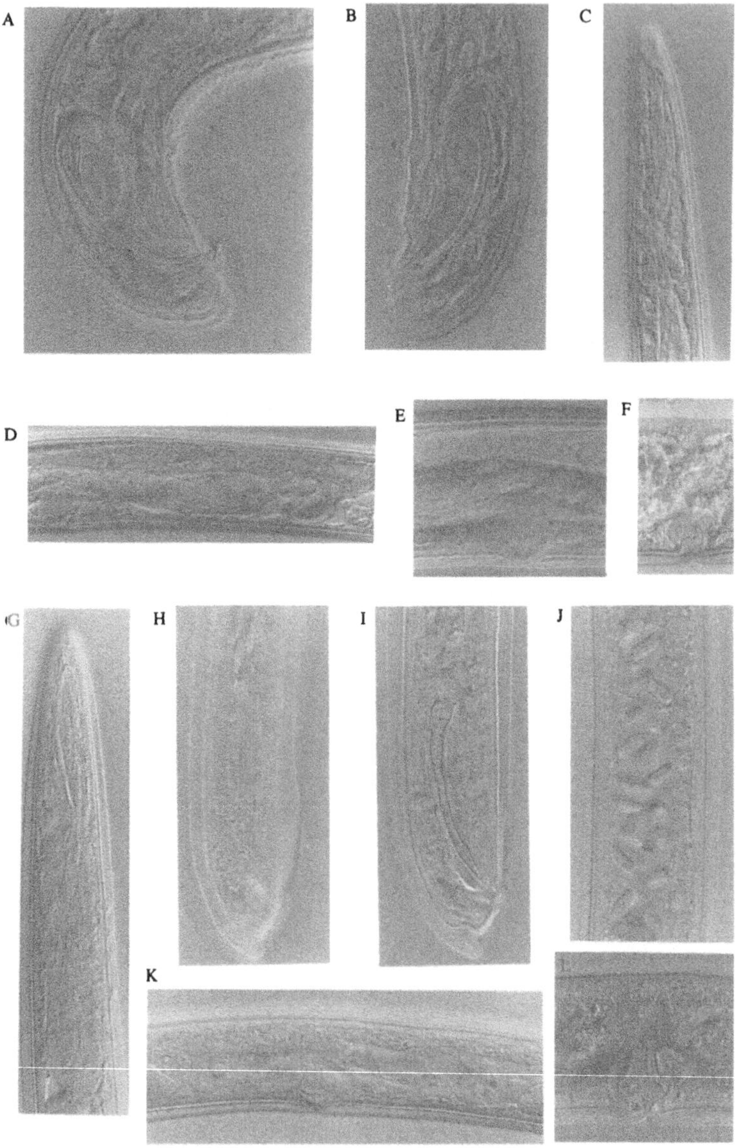

Fig. 4.10. *Trichodorus giennensis*. Male, paratype and holotype: (A-B) copulatory apparatus and tail region (arrow indicates bristle), (C) anterior body region with 2 ventromedian cervical papillae anterior to the S-E pore (holotype). Female, paratypes: (D) reproductive system, anterior branch, (E) vaginal region, lateral view. *T. philipi*. Female, paratype: (F) vaginal region. *T. paracedarus*, paratype specimens. Male: (G) anterior region, (H) posterior region showing the bursa, (I) copulatory apparatus and tail region, (J) sperm cells in testis; Female: (K) reproductive system, (L) vulval region. Scale 20 µm.

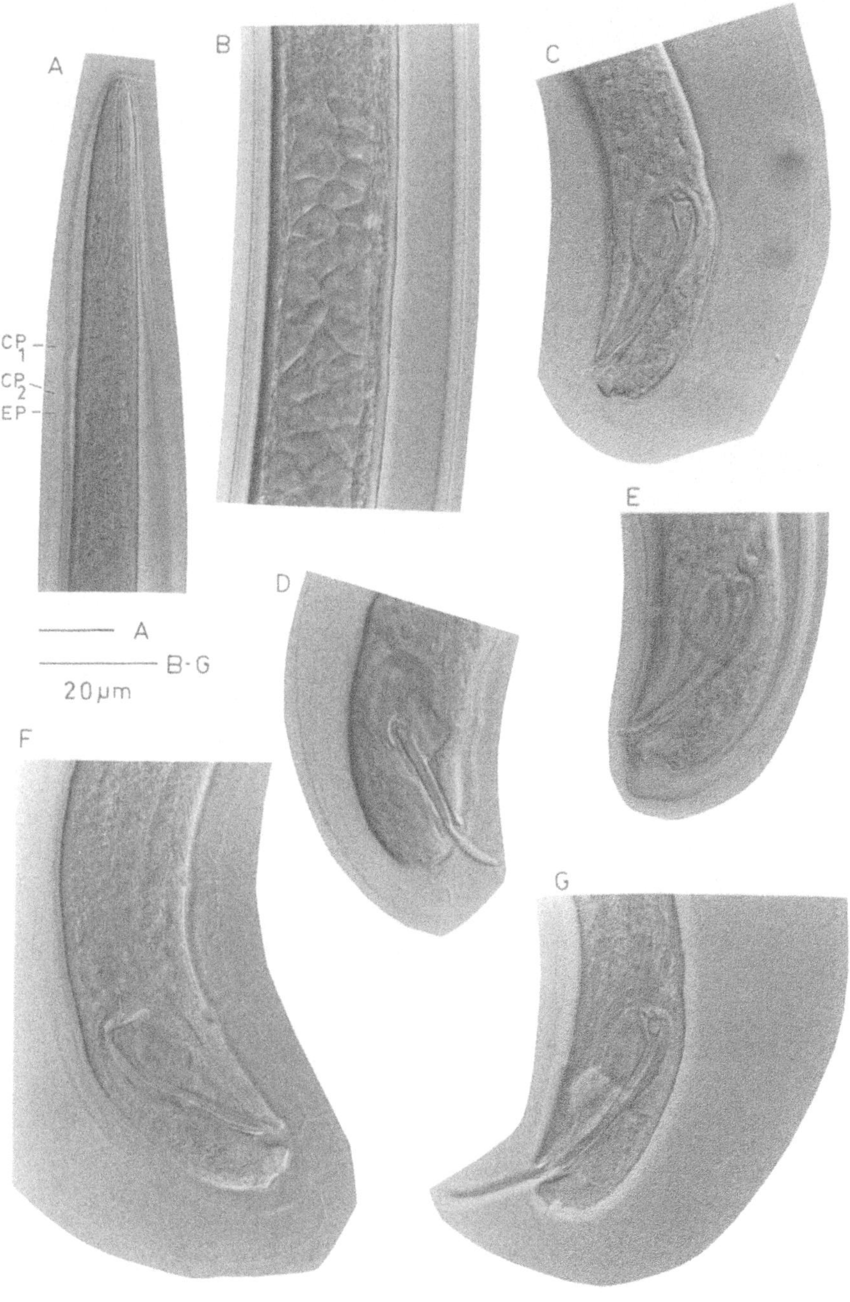

Fig. 4.11. *Trichodorus hooperi*. Male, paratypes: (A) anterior body region, (B) spermatozoa, (C-G) copulatory apparatus and tail, showing a different degree of swelling of the body cuticle.

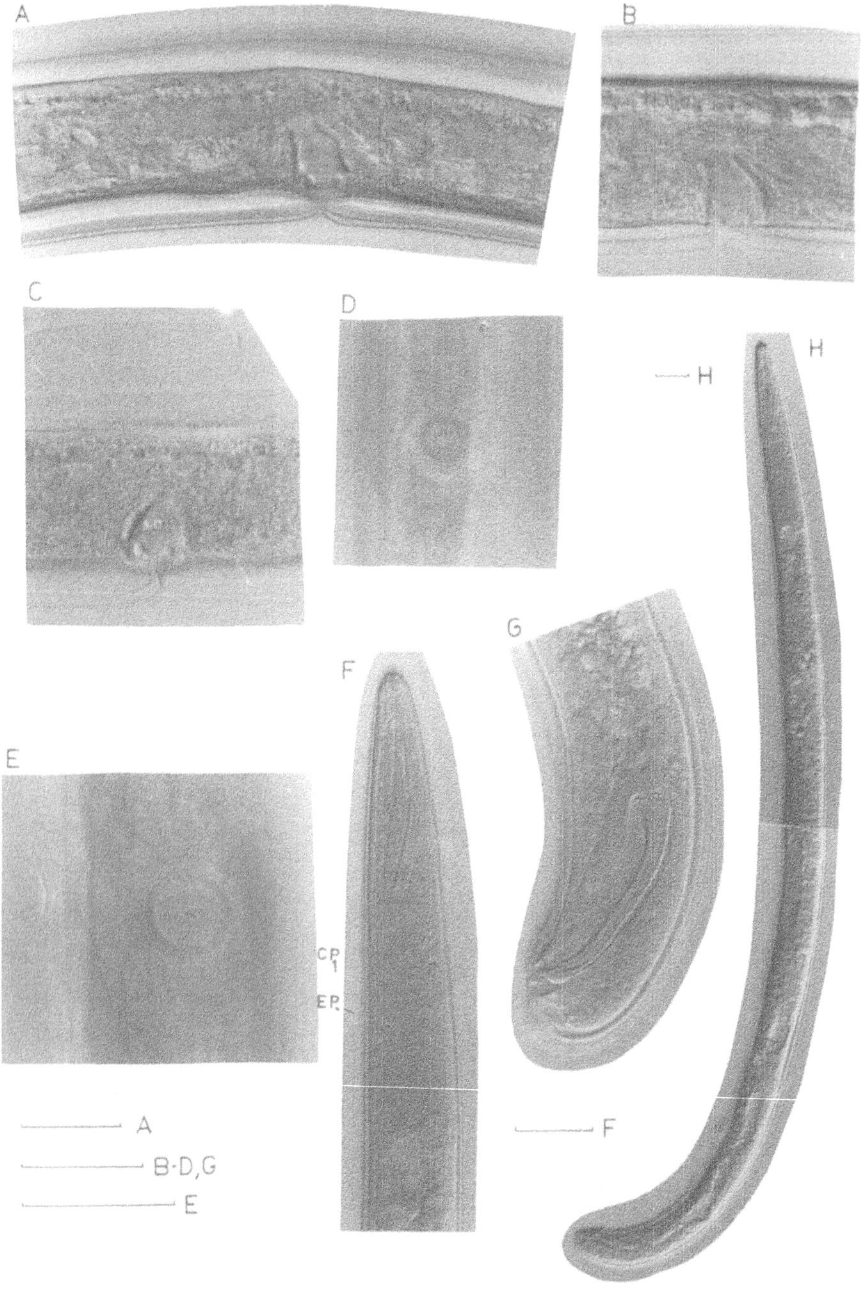

Fig. 4.12. *Trichodorus hooperi*. Female, paratypes: (A) part genital system, vaginal region,(B-C) vaginal region, (D) vulva, ventral view. *Trichodorus intermedius*. Female: (E) vulva, ventral view (paratype). Male, paratypes: (F) anterior body region, (G) tail and copulatory apparatus, (H) total view (scale= 20 μm).

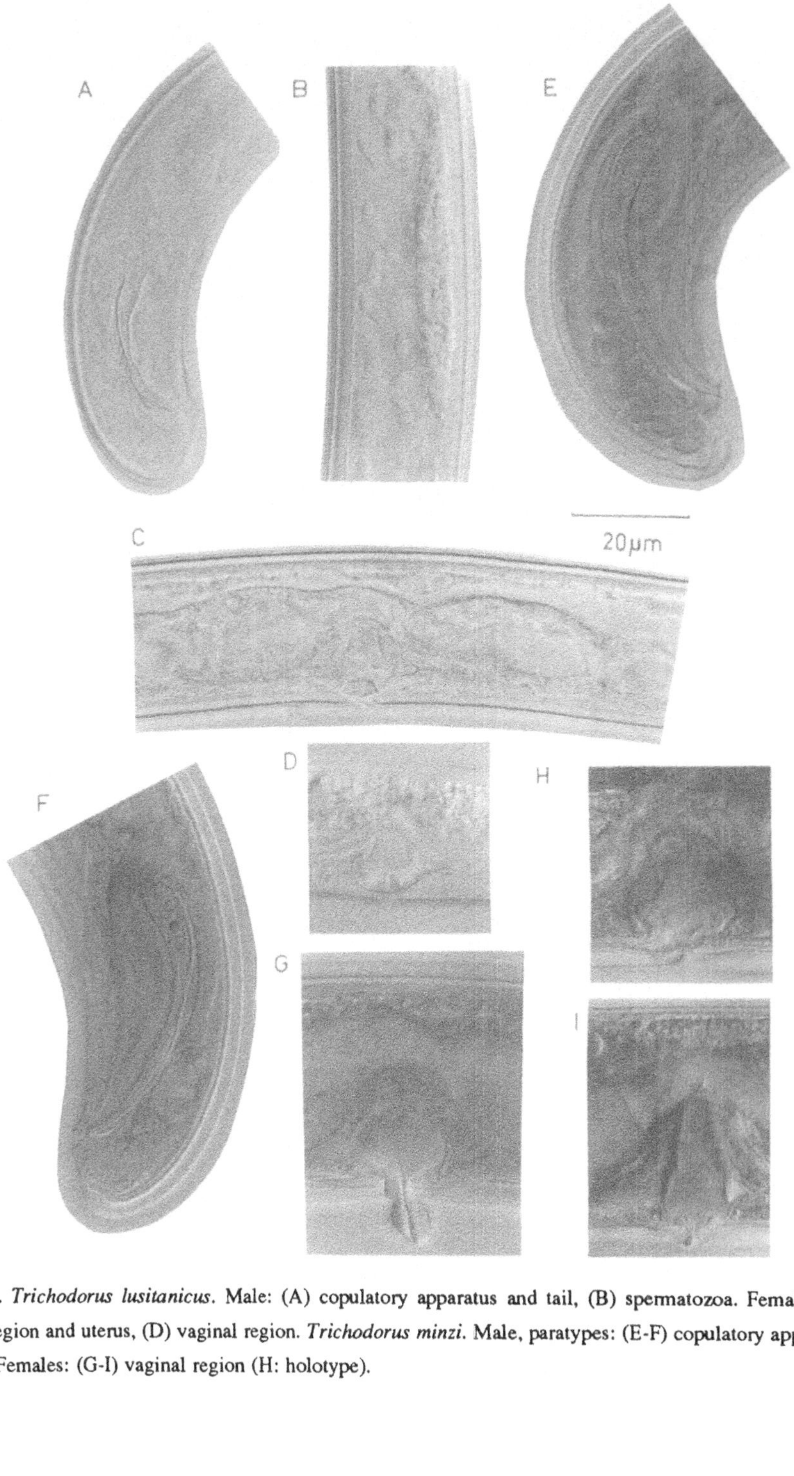

Fig. 4.13. *Trichodorus lusitanicus*. Male: (A) copulatory apparatus and tail, (B) spermatozoa. Female: (C) vaginal region and uterus, (D) vaginal region. *Trichodorus minzi*. Male, paratypes: (E-F) copulatory apparatus and tail. Females: (G-I) vaginal region (H: holotype).

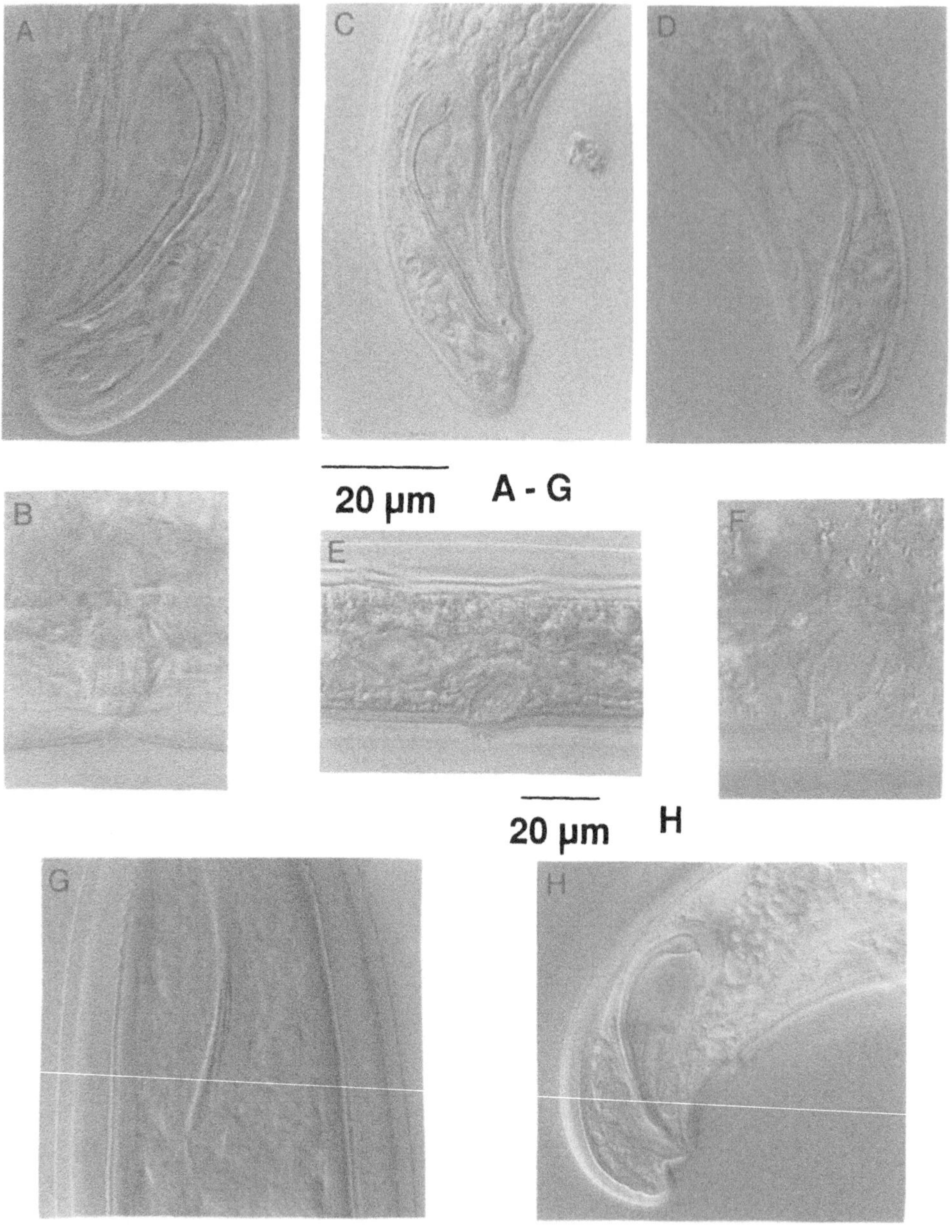

Fig. 4.14. *Trichodorus magnus*. Male: (A) copulatory apparatus and tail (holotype). Female: (B) vaginal region (paratype). *Trichodorus kilianae*. Male, paratypes (C-D) copulatory apparatus and tail. Female: (E) vaginal region, lateral (paratype). *Trichodorus petrusalberti*. Female: (F) vaginal region, lateral (paratype). Male, paratypes: (G) anterior body region with onchiostyle and inner onchium, (H) copulatory apparatus and tail.

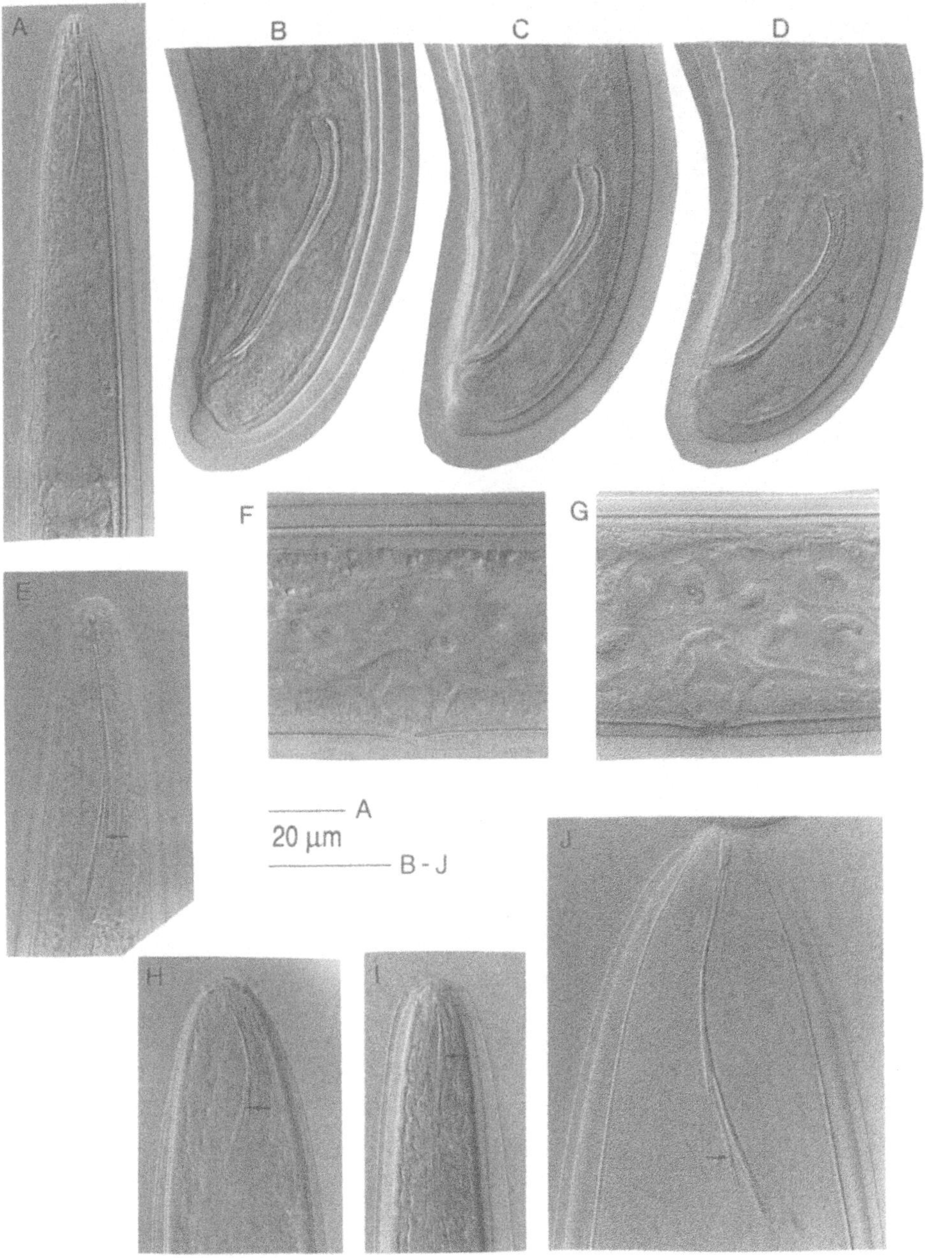

Fig. 4.15. *Trichodorus nanjingensis*. Male, paratypes: (A) anterior body region, (B-D) tail and copulatory apparatus, (E) onchiostyle with inner onchium. Female: (F-G) vaginal region. (H-J) anterior body region of female showing inner onchium in *Paratrichodorus minor*, *P. nanus* and *Trichodorus petrusalberti*, respectively.

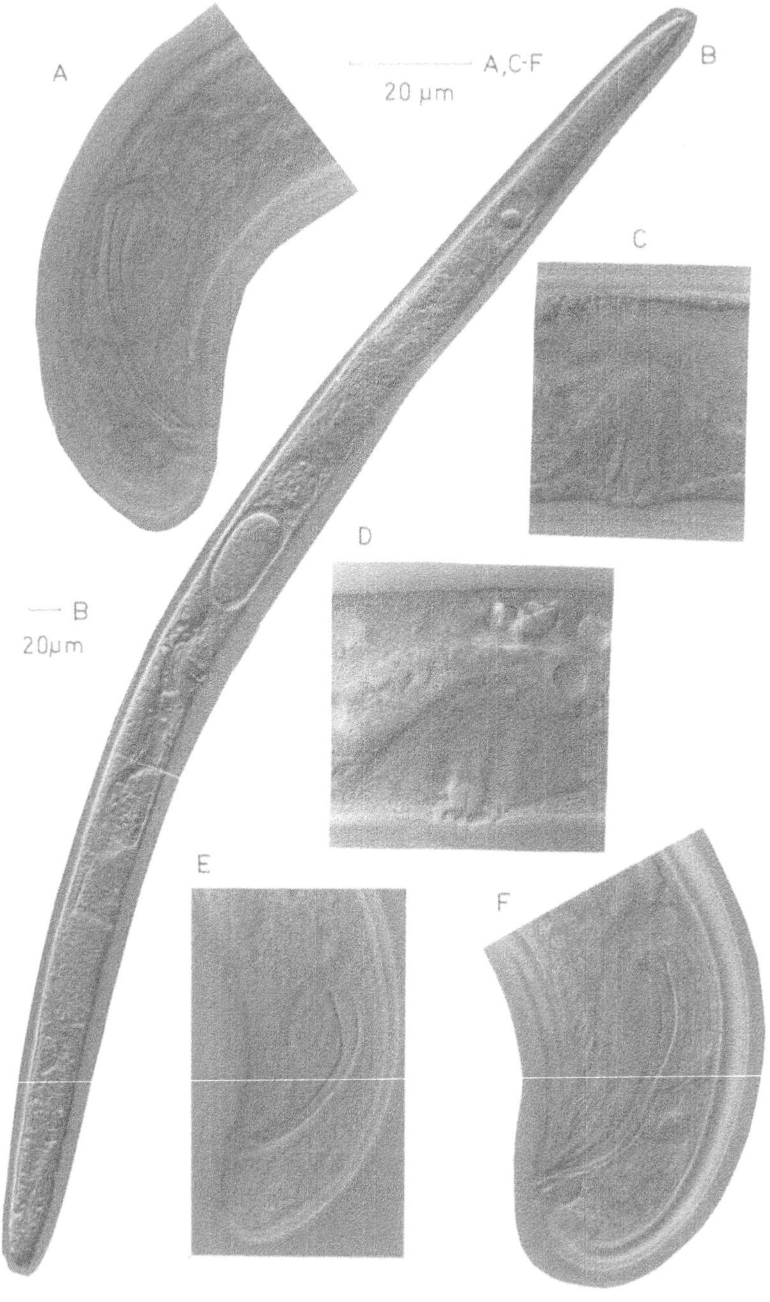

Fig. 4.16. *Trichodorus obscurus*. Male: (A) copulatory apparatus and tail (paratype). Female, paratypes (B) total view, (C-D) vaginal region. *Trichodorus aequalis*. Male from Alaska: (E) copulatory apparatus and tail. *Trichodorus californicus*. Male from Alaska: (F) copulatory apparatus and tail.

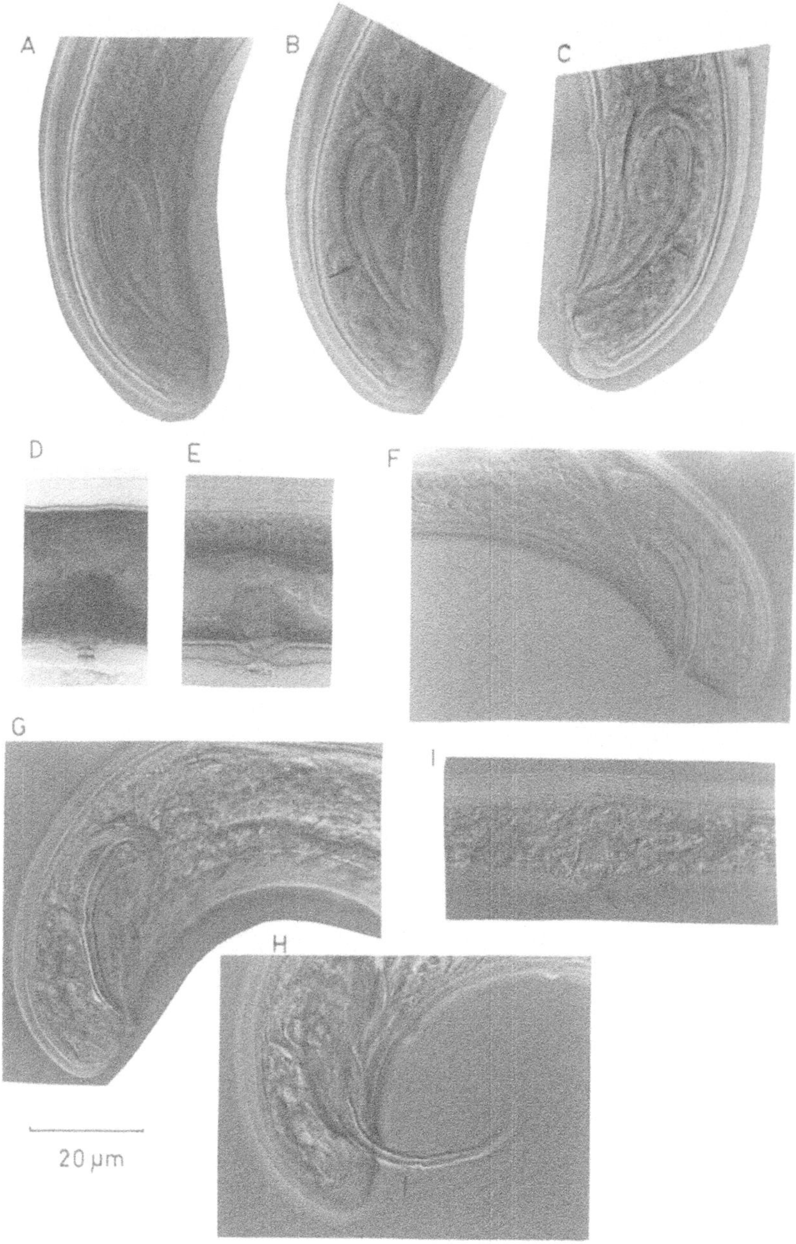

Fig. 4.17. *Trichodorus orientalis*. Male, paratypes: (A-C) copulatory apparatus and tail, B, C respectively right and left side of same specimen, arrow points to constriction. Female, paratypes: (D-E) vaginal region. *Trichodorus paucisetosus*. Male: (F) posterior body region (paratype). *Trichodorus parorientalis*. Male, paratypes: (G-H) copulatory apparatus and tail, arrow indicates bristle. Female: (I) vaginal region (paratype).

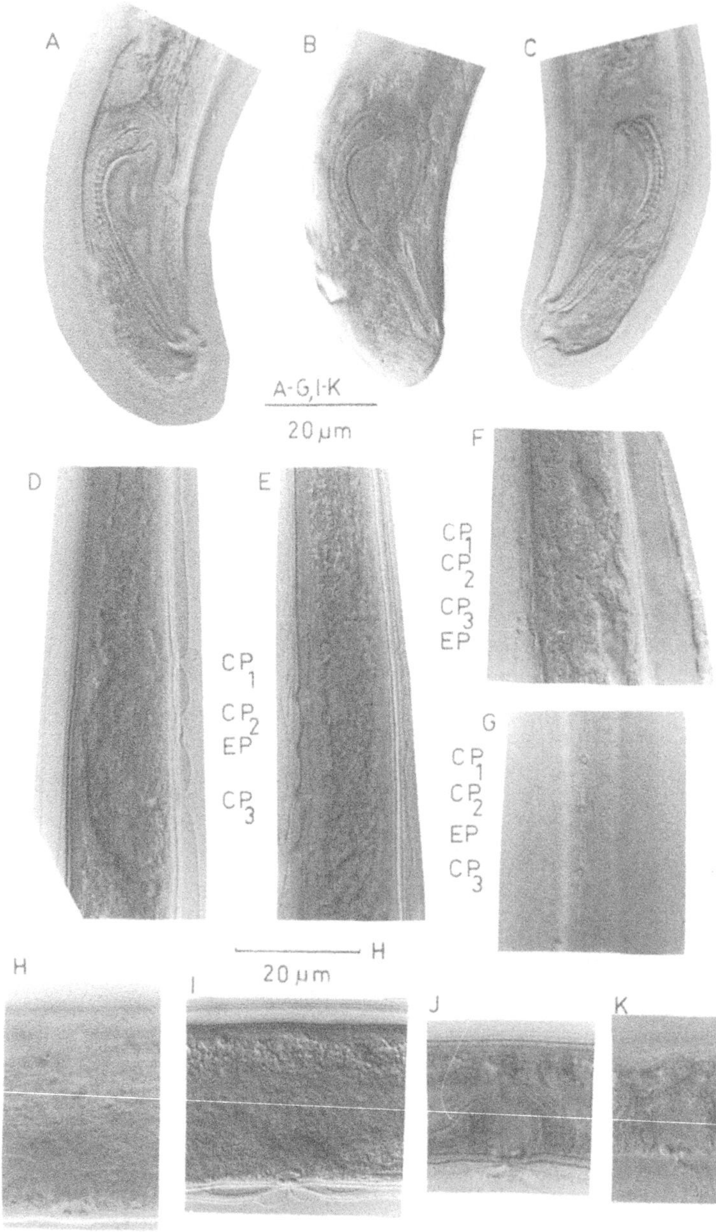

Fig. 4.18. *Trichodorus pakistanensis*. Male: (A-C) copulatory apparatus and tail (B= paratype), (D-G) part of anterior body region, showing the variability in the position of the S-E pore (=EP) in relation to the ventromedian cervical papillae (=CP), (D-F, lateral; G, ventral view). Female: (H-K) vaginal region (H= paratype).

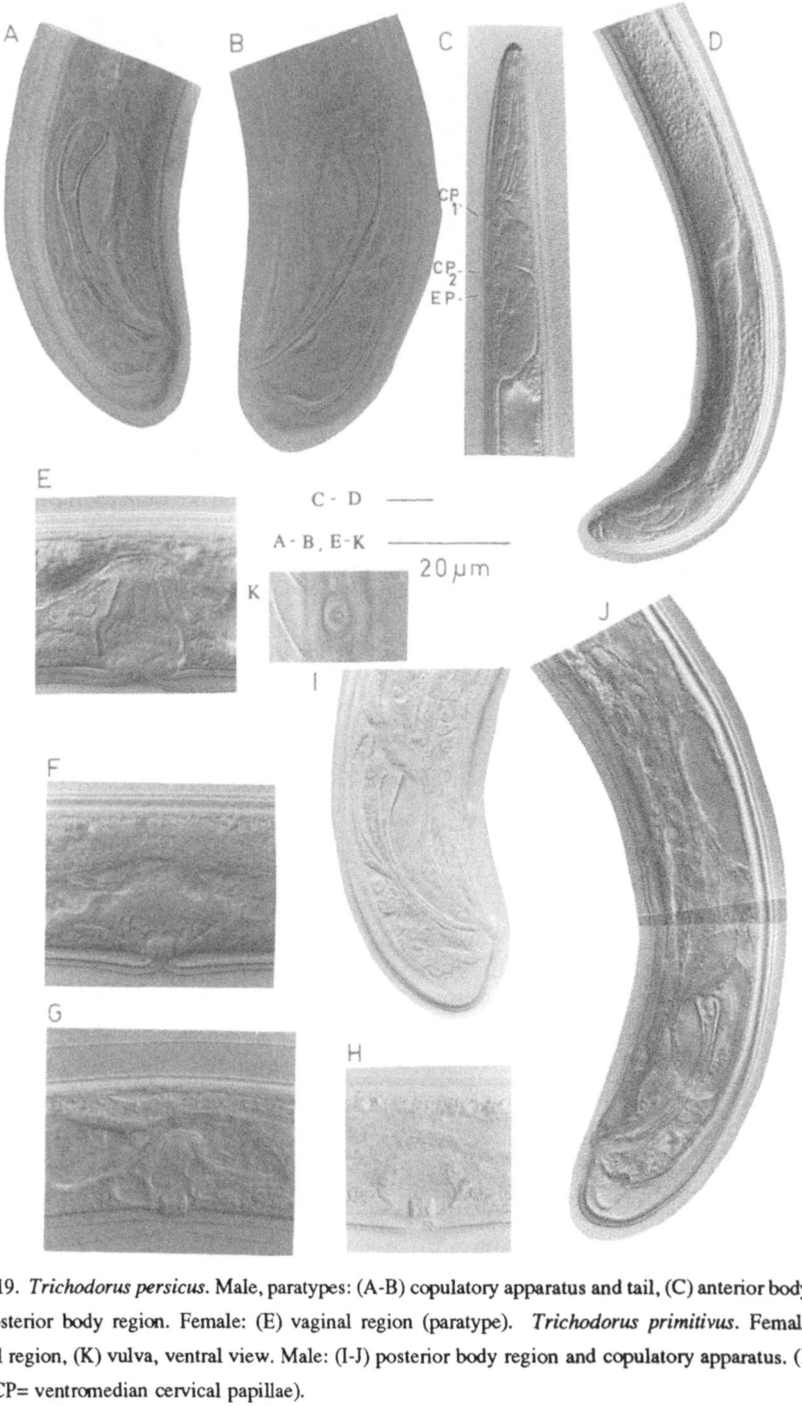

Fig. 4.19. *Trichodorus persicus*. Male, paratypes: (A-B) copulatory apparatus and tail, (C) anterior body region, (D) posterior body region. Female: (E) vaginal region (paratype). *Trichodorus primitivus*. Female: (F-H) vaginal region, (K) vulva, ventral view. Male: (I-J) posterior body region and copulatory apparatus. (EP= S-E pore; CP= ventromedian cervical papillae).

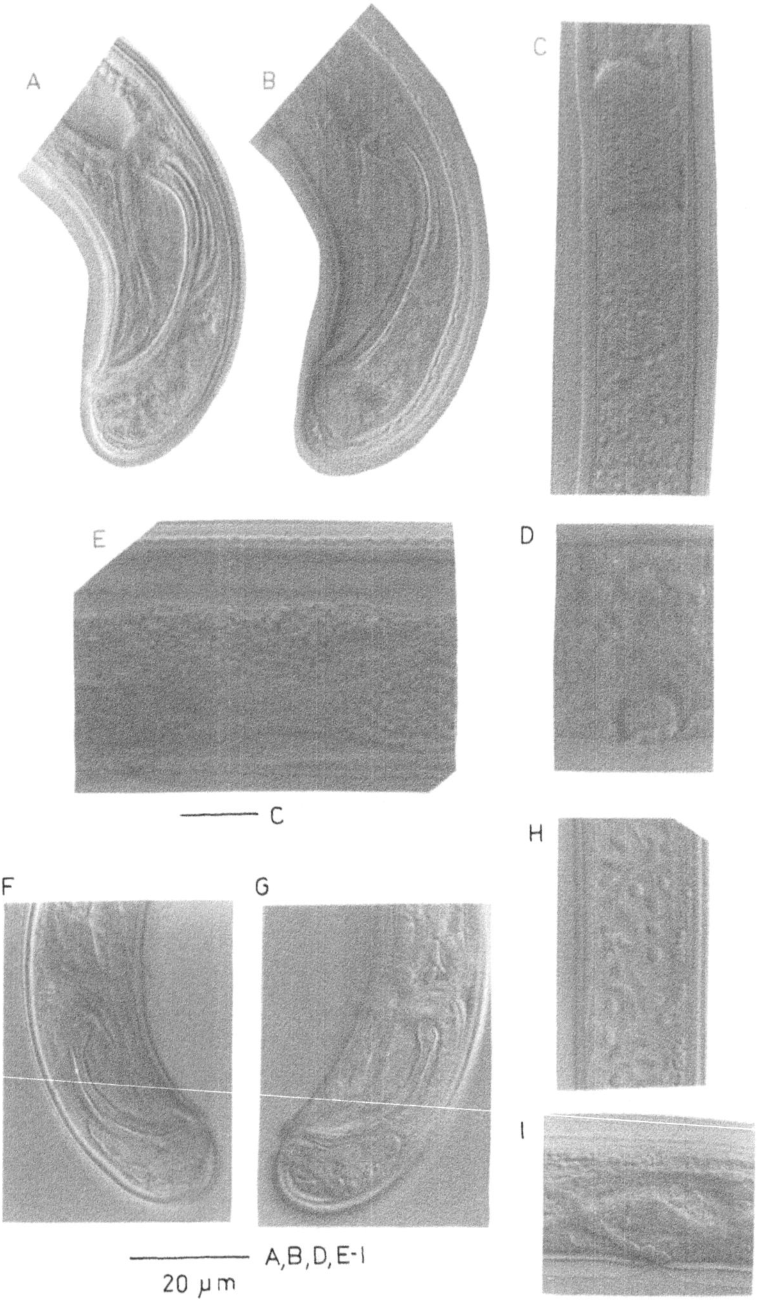

Fig. 4.20. *Trichodorus proximus (=T. obtusus)*. Male: (A-B) copulatory apparatus and tail (B= paratype), (C) spermatozoa (paratype). Female: (D-E) vaginal region (E= paratype). *Trichodorus similis*. Male: (F-G) copulatory apparatus and tail, (H) spermatozoa. Female: (I) vaginal region.

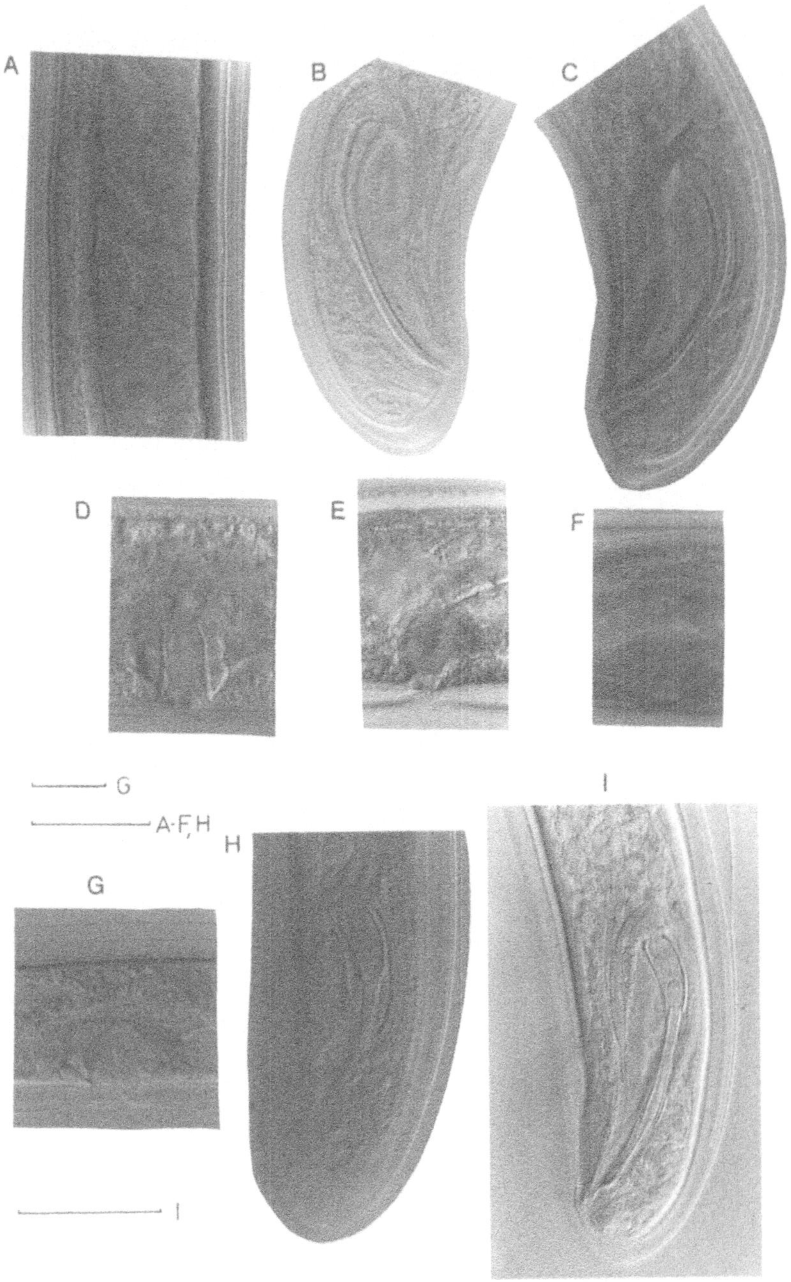

Fig. 4.21. *Trichodorus sparsus*. Male: (A) spermatozoa, (B-C) copulatory apparatus and tail (arrow points to bristles). Female: (D-F) vaginal region. *Trichodorus taylori*. Female: (G) vaginal region (paratype). Male, paratypes: (H-I) copulatory apparatus and tail (scale= 20 µm).

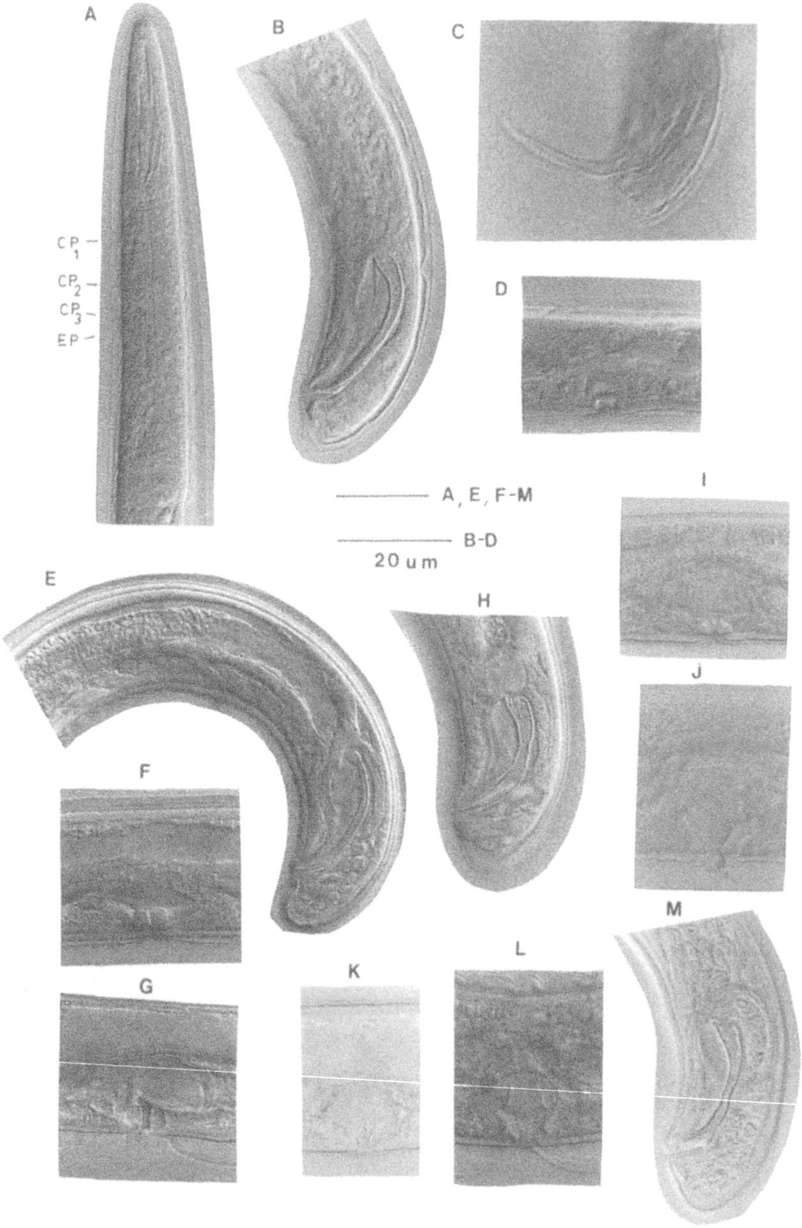

Fig. 4.22. *Trichodorus tricaulatus*. Male, paratypes: (A) anterior body region, (B) posterior body region and copulatory apparatus, (C) copulatory apparatus, bristles on mid-shaft visible. Female: (D) vaginal region (paratype). *Trichodorus variopapillatus*. Male: (E) posterior body region. Female: (F-G) vaginal region (G with plug). *Trichodorus velatus*. Male: (H) copulatory apparatus and tail (paratype). Female: (I-J) vaginal region (paratypes). *Trichodorus viruliferus*. Female: (K-L) vaginal region. Male: (M) copulatory apparatus and tail.

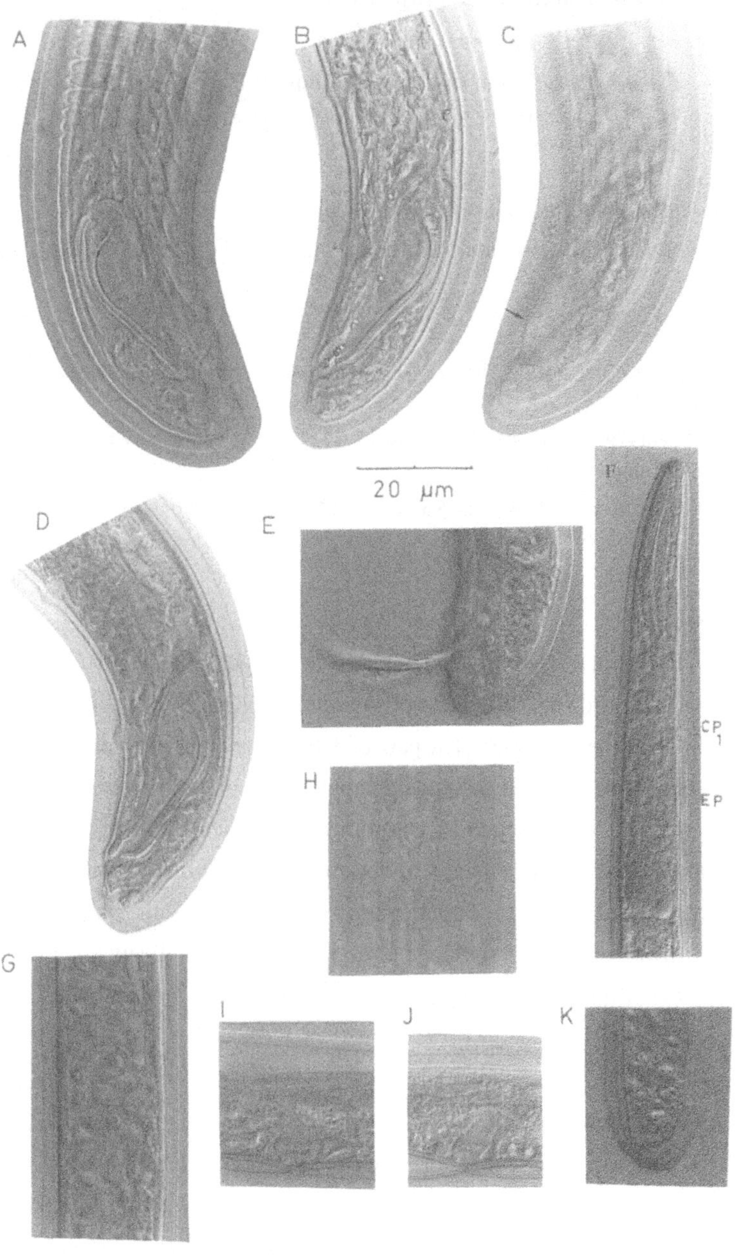

Fig. 4.23. *Trichodorus vandenbergae*. Male: (A-D) copulatory apparatus and tail (B, C respectively with focus on spicule and on "bursa" in same specimen), (E) posterior body region with protruded spicule showing bristles at mid-shaft (paratype), (F) anterior body region (paratype), (G) spermatozoa (paratype). Female: (H) vulva, ventral view, (I-J) vaginal region (paratypes), (K) tail region (paratype).

4.2. The genus *Paratrichodorus* Siddiqi, 1974

Paratrichodorus acaudatus (Siddiqi, 1960) Siddiqi, 1974
Figs 3.13.C; 3.14.V; 3.15.A; 3.17.H'; 4.24.A

syn. *Trichodorus acaudatus* Siddiqi, 1960
syn. *Paratrichodorus (Paratrichodorus) acaudatus* (Siddiqi, 1960) Siddiqi, 1974

Measurements

Holotype male: L= 847 µm, a= 18, b= 6.1, T= 61%, onchiostyle= 78 µm, spicule length= 68 µm, gubernaculum= 12 µm.

Males: L= 750-847 µm (799 µm*), a= 14.5-19.5, b= 5.7-6.1, T= 60-61%, onchiostyle= 75-78 µm (76.5 µm*), spicule length= 67-68 µm.

Females: L= 910-950 µm, a= 15.8-18.2, b= 5.8-6.0, V= 52.4-54.7%, onchiostyle= 82-83 µm (from Siddiqi, 1960).

Emended diagnosis

P. acaudatus is characterized by the onchiostyle length (75-83 µm) and the anterior position of the S-E pore (near the base of the onchiostyle). Males of *T. acaudatus* may be distinguished by the presence of a pair of lateral cervical body pores at the base of the amphids, two ventromedian precloacal supplements, both opposite the distal third of the retracted spicules, the absence of ventromedian cervical papillae, the length (67-68 µm) and shape of the spicules (shaft about straight and striated) and the subterminal position of the cloacal opening. Females are characterized by the shape of the vaginal sclerotizations in lateral view (small, about quadrangular, well separated: 2.5 µm), the shape of the vagina (about quadrangular) and by the almost terminal anus.

Relationships

P. acaudatus most closely resembles *P. atlanticus* by the long onchiostyle (resp. 75-83 µm and 64-82 µm) and long spicules (longer than 2.5 times the anal body width). The males can be distinguished by their somewhat shorter and stouter spicules with less marked manubrium, the number of precloacal ventromedian supplements (2 *vs* 3), the number of postcloacal subventral papillae (1 pair *vs* 2 pairs in *P. atlanticus*) and males and females by the anterior position of the S-E pore (near base of onchiostyle).
Furthermore, *P. acaudatus* males agree with *P. alleni*, *P. allius* and *P. lobatus* by the

combinations of only two precloacal ventromedian supplements, the absence of ventromedian cervical papilla and the presence of ventrally overlapping pharyngeal glands. *P. acaudatus* differs from these species mainly by the longer spicules and onchiostyle, the short tail and the geographic distribution of the species i.e. only recorded from India.

Remarks

Pharyngeal bulb offset in female paratype specimens, with a short ventral overlap in male paratypes. The vulva was described as a minute aperture without distinctive shape (Siddiqi, 1960). No data are available on lateral body pores in females. Sperm cells are small, with a small oval nucleus.

Distribution and hosts

P. acaudatus was described from specimens from Coimbatore, Madras state, South India, associated with soil from the rhizosphere of sugar-cane.

Paratrichodorus acutus (Bird, 1967) Siddiqi, 1974

Figs 3.12.I; 3.17.S; 4.24.B-C; 5.5.C

syn. *Trichodorus acutus* Bird, 1967

syn. *Paratrichodorus (Paratrichodorus) acutus* (Siddiqi, 1960) Siddiqi, 1974

syn. *Nanidorus acutus* (Siddiqi, 1960) Siddiqi, 1974

Measurements

Holotype female: L= 455 µm, a= 19.8, b= 5.6, V= 52.7%, onchiostyle= 23 µm.

Females: L= 338-610 µm (490 µm ± 65; 395 µm + 29.6), a= 14.1-21.3, b= 4.3-6.0, V= 52.3-55.8% (54.0 % ± 1.1), onchiostyle= 22-27 µm (23 µm ± 0.8; 26.0 ± 0.9), distance anterior end to EP= 45-79 µm (60 µm ± 9) (from Bird, 1967; De Waele *et al.*, 1990).

Emended diagnosis

P. acutus females are characterized by their tail shape (anteriorly convex-conoid, then conoid to an acute or slightly rounded end), the absence of lateral body pores, absence of caudal pores (1 exception), the shape of the vaginal sclerotizations (bold dots, close:

1 μm apart), the shape of the vulva (a transverse slit), absence of spermathecae, uteri with thread-like bodies (?sperm), the shape of the vagina (rounded trapezoid) and the pharyngeal bulb offset or with slightly overlapping ventrosublateral pharyngeal glands. S-E pore opposite posterior bulb region or at level of the pharyngo-intestinal junction.

Relationships

P. acutus can be distinguished from all other species of the genus by its tail shape, which is anteriorly convex-conoid, then conoid to an acute or slightly rounded terminus. It most closely resembles *P. nanus* and *P. renifer*.

It has in common with *P. nanus*: a short onchiostyle (resp. 22-27 μm and 21-26.5 μm in *P. nanus*) and vulva shape (a transverse slit), but differs by the absence of males, lack of lateral body pores (2 on each side in *P. nanus*), different sperm structure (thread-like *vs* small with oval nucleus), position of the S-E pore (opposite posterior part or base of pharynx *vs* along anterior region intestine in *P. nanus*), shape of the vaginal sclerotizations in lateral view (dot-like *vs* larger triangular) and a different geographic distribution (USA, South Africa, Madeira *vs* Senegal, western and southern Europe, Tunisia).

P. acutus resembles *P. renifer* by the lack of males, absence of lateral body pores, thread-like "sperm" and vulva shape, but differs by having a somewhat longer onchiostyle (29-34 μm), different shape of the vaginal sclerotizations (reniform in *P. renifer)* and a different geographic distribution (Western Australia, India, Malawi, The Netherlands for *P. renifer*).

Remarks

Males unknown. Juveniles similar to females except for the reproductive system (Bird, 1967).

S-E pore usually at the level of the pharyngo-intestinal junction in South African population (De Waele *et al.*, 1990), about mid-pharyngeal bulb length in a specimen from California and opposite the posterior bulb of the pharynx in type specimens. Pharynx offset in a population from South Africa and a specimen from California. A female from California apparently with a caudal pore, ventrally halfway along the tail.

De Waele *et al.* (1990) described *P. acutus* with spermathecae empty, and referred to a Fig. 2G. From the illustration it appears these authors probably interpreted the well-developed oviduct cells as being the spermathecae.

Distribution and host

P. acutus was described from specimens recovered from Gloxinia (*Sinningia speciosa*) at Ithaca, New York, U.S.A.. It has also been reported from California, Madeira (Portugal) and South Africa (Decraemer & Chaves, 1988; De Waele *et al.*, 1990; nematode collection, California, Riverside, U.S.A.). Another host is leatherleaf fern (*Rumohra adiantiformis)* (Anonymous, 1989a; De Waele *et al.*, 1990).

Paratrichodorus alleni (Andrássy, 1968) Siddiqi, 1974
Figs 3.14.Z; 3.15.B; 3.17.H

syn. *Trichodorus alleni* Andrássy, 1968
syn. *Paratrichodorus (Paratrichodorus) alleni* (Andrássy, 1968) Siddiqi, 1974

Measurements

Holotype male: L= 730 µm, a= 20, b= 5.3, onchiostyle= 44 µm, spicule length= 50 µm, gubernaculum= 9 µm (from Andrassy, 1968).

Emended diagnosis

P. alleni male is characterized by the presence of only two ventromedian precloacal supplements, situated opposite the distal half of the spicules (within the region of the bursa), length (50 µm) and shape of the spicules (cephalated, shaft striated, about straight and equally wide, tapered at distal end with bifid tip), narrow and weakly developed bursa, absence of ventromedian cervical papillae, position of the S-E pore just anterior to pharyngeal bulb, the pharynx with short ventral overlap, small oval sperm cells (2.5-3 µm), and conical tail.

Relationships

P. alleni is described based only on a male and juvenile specimen, therefore, discussion concerning its relationships is restricted to males. *P. alleni* most closely resembles *P. lobatus* by similar metric data for body, onchiostyle and spicule length, by possessing only two precloacal ventromedian supplements, small sperm with oval nucleus, by the absence of ventromedian cervical papilla, a similar tail shape (conical, with small digitate tip) and spicule shape (slender, clearly cephalated). It differs from *P. lobatus* males only by minor differences, some of them lying within the intraspecific range of variability: a shorter bursa (extending anteriorly to halfway the retracted spicules *vs* to the head of spicules in *P. lobatus*), the slightly more anterior position of the S-E pore (opposite the isthmus *vs* along the pharyngeal bulb), longer spicules in relation to the anal body diameter and a different geographic distribution (Congo-Brazzaville *vs* Australia, New Zealand and South Africa for *P. lobatus*).

Remarks

Females unknown.

Distribution and host

P. alleni in known only from the type locality, Brazzaville-Congo where it was found in a sample from the bank of the Congo river.

Paratrichodorus allius (Jensen, 1963) Siddiqi, 1974

Figs 3.13.I; 3.14.D; 3.15.C; 3.17.G, g; 4.24.D-J

syn. *Trichodorus allius* Jensen, 1963

syn. *Paratrichodorus (Paratrichodorus) allius* (Jensen, 1963) Siddiqi, 1974

syn. *Paratrichodorus tansaniensis* Siddiqi, 1974

Measurements

Holotype female: L= 740 µm, a= 16.5, b= 6.4, V= 52%, onchiostyle= 37.5 µm.

Males (n=2): L= 590-680 µm, a= 16.5, b= 6.2, T= 61%, onchiostyle= 42 µm, spicule length= 30.5-30.9 µm, gubernaculum= 10-12 µm (from Jensen, 1963, 1♂; Sturhan, 1989, 1♂).

Females: L= 400-1044 µm (710 µm*; 700 µm*; 654 µm; 567 µm; 434 µm; 550 µm; 607 µm*;882 µm; 946 µm; 580 µm ± 61), a= 11.8-32.5, b= 3.9-10.2, V= 45.8-59.2% (53.5%*; 52.2%*; 53.1%; 52.2%; 58.2%; 53%; 53%*; 56%; 55.5%; 50% ± 2.2), onchiostyle= 35-53 µm (42.5 µm*; 46 µm*; 42 µm; 42.5 µm; 37.5 µm; 38 µm; 39.5 µm*; 45 µm; 48 µm; 47 µm ± 2), anterior end to EP= 80-92 µm (from Jensen, 1963; Siddiqi, 1974; Sturhan, 1989: 3 populations; data for *P. tansaniensis*: Siddiqi, 1974; Vermeulen & Heyns, 1983; Roca & Lamberti, 1984: 2 populations, data of the Italian populations are underlined; De Waele & Cohn, 1992).

Emended diagnosis

P. allius females are characterized by the shape of the vaginal sclerotizations in lateral view (small, about oval, well separated: > 2.5 µm apart, in oblique position), shape of the vagina (small broad, rounded to oval), shape of the vulva (a longitudinal slit of 2 µm), absence of lateral body pores and presence of small sperm in the uteri or at the distal ends of both branches of the genital tract. Males by the shape of the spicules (nearly straight, shaft striated), two ventromedian precloacal supplements within the region of the retracted spicules, small sperm cells with globular nucleus and absence of ventromedian cervical papillae. The S-E pore is located between the nerve ring and anterior end of the pharyngeal bulb in females and males. Pharyngo-intestinal junction usually with ventral overlap of the ventrosublateral pharyngeal glands.

Relationships

P. allius females most closely resemble those of *P. lobatus* and *P. rhodesiensis* by the shape of the vulva (a longitudinal slit), shape of the vaginal sclerotizations in lateral view, small sperm cells with oval nucleus in longitudinal optical section and further with *P. rhodesiensis* by the unusual location of sperm, at the level of the flexure of the genital tracts, and the lack of lateral body pores. It may be distinguished from these species by some differences in the shape of the vagina (rounded, oval *vs* clearly indented in *P. lobatus* and rectangular in *P. rhodesiensis*). The rare males of *P. allius* agree with those of *P. lobatus* by possessing only two ventromedian precloacal supplements, with those of *P. rhodesiensis* by the short caudal alae and with both species by the lack of ventromedian cervical papillae. It differs from the former species by the shorter spicules, and from *P. rhodesiensis* in the number of precloacal supplements.

P. allius has a different geographic distribution from both that of *P. rhodesiensis* (western Africa) and *P. lobatus* (partim) (Australia, New Zealand). Due to its synonymization with *P. tansaniensis*, it may be considered to be present in South Africa (Orange Free State) but in different regions than the few records referring to *P. lobatus* (Transvaal and Cape Province).

Remarks

Decraemer (1988, 1989a, b) described for the first time the unusual sperm location near the beginning of the oviduct, at the level of the flexure of the genital branches, in *P. allius* and *P. rhodesiensis*.

Sturhan (1989b) synonymized *P. tansaniensis* with *P. allius*, because of their similar morphometric data, the comparable shapes of the vagina and vaginal sclerotizations, the males being rare or absent and the location of small sperm cells in the uteri or at the end of the genital branches in both species. He remarked, however, that the Italian populations of *P. tansaniensis* may represent a different species because of some differences in the shape of the vagina (more trapezoid), the more posterior position of the S-E pore (one to three body widths behind the onchiostyle *vs* one half to two body widths), the longer body (779-1044 µm *vs* 400-790 µm) and the apparently slightly larger sperm (3-4 µm *vs* 2 µm long). Sturhan (1989b) described the genital system in the Italian populations as having accumulations of spermatids (?) between "normal" gonocytes indicating hermaphroditic reproduction.

In a population from Israel the females had no lateral body pores and possessed both a short dorsal intestinal overlap and a short ventral pharyngeal overlap. Six out of eight females had sperm cells in the uteri and one female, without sperm in the uteri, had sperm cells accumulated at the tip of both gonadal tracts (De Waele & Cohn, 1992).

A ventral overlap of the pharyngeal glands has been described for the type population and was also observed in populations from Tanzania (Siddiqi, 1974) and Washington State (Sturhan, 1989b).

Exceptionally, faint lateral body pores were observed between the vulva and the anus in a few specimens of the Washington State population; two specimens from the Madeira population also had a faint lateral body pore on the right body side, one to three body widths behind the vulva.

In the population from Washington State, a few specimens had the vagina indented, due to a contraction of the vaginal constrictor muscles (Sturhan, 1989b).

The shape of the vulva in ventral view is always a longitudinal slit, not a transverse slit as described in Roca & Lamberti (1984). The paratype male specimen was originally described without caudal alae; the male specimen from a California population possessed tiny alae (Sturhan, 1989b).

Distribution and host

P. allius (syn. *P. tansaniensis*) has been described from several states in the USA (Oregon, Washington State, California), from Chili, the Madeira Archipelago, Portugal, Italy, Israel, Tanzania and South Africa. It is a polyphagous species found in the rhizosphere of onion (*Allium cepa*), sugar beet, wheat, *Jasminium odoratissimum*, willow tree (*Salix* sp.), *Quercus suber*, avocado, lemon and associated with grapevine, apple and olive in Italy (Jensen, 1963; Vermeulen & Heyns, 1983; De Waele & Cohn, 1992; Roca

& Lamberti, 1984; data from Chili from nematode collection in Riverside, U.S.A.).

Paratrichodorus anemones (Loof, 1965) Siddiqi, 1974
Figs 3.12.N-P; 3.14.A; 3.15.D; 3.17.A, a; 4.25.A-E

syn. *Trichodorus anemones* Loof, 1965
syn. *Paratrichodorus (Atlantadorus) anemones* (Loof, 1965) Siddiqi, 1974
syn. *Atlantadorus anemones* (Loof, 1965) Siddiqi, 1974

Measurements

Holotype male: L= 830 µm, a= 22, b= 5.3, T= 65%, onchiostyle= 48 µm.

Males: L= 710-900 µm (780 µm), a= 18-23, b= 4.8-6.4, T= 63-82%, onchiostyle= 47-52 µm (49 µm), spicule length= 46-53 µm (49.5 µm*), gubernaculum= 10-11 µm.

Females: L= 650-970 µm (790 µm), a= 17-22, b=4.3-5.9, V= 51-59% (55%), onchiostyle= 42-53 µm (49 µm) (from Loof, 1965).

Emended diagnosis

P. anemones males are characterized by the shape of the spicules (long, slender, nearly straight, slightly curved distally, with a characteristic small irregularity in the outline (=kinking part) just anterior to the middle, shaft striated), a narrow bursa, three ventromedian precloacal supplements (posterior two within region of the posterior half of the retracted spicules), one ventromedian cervical papilla just anterior to the S-E pore and one pair of lateral cervical pores slightly more anteriad than the medioventral cervical papilla, large sperm with large sausage shaped nucleus and usually also by a pair of lateral body pores in the anterior half of the body. Females, by the shape of the vagina (often with a constriction at mid-point caused by contraction of the vaginal constrictor muscles), shape of the vaginal sclerotizations in lateral view (small, narrow triangular, oblique and tips 1.5-2 µm apart), pore-like vulva, two spermathecae and usually two pairs of lateral body pores (one prevulvar, one postvulvar). The pharynx usually has a ventral overlap and the S-E pore is usually located just anterior to the pharyngeal bulb in males and females.

Relationships

P. anemones most closely resembles *P. hispanus*, *P. pachydermus* and *P. weischeri* (table

4.5.).

Loof (1965) reported that it was difficult to assign females of mixed populations either to *P. anemones* or to *P. pachydermus*. He distinguished them by differences in the vaginal sclerotizations and the number of lateral body pores, usually two pores on each side in both species, but varying between 0 and 2 in *P. anemones* and between 2 and 5 in *P. pachydermus*. However, the vaginal sclerotizations in both species appear very similar (Decraemer, 1989b) with minor differences lying within the range of intraspecific variability.

Males of both species can be distinguished by differences in tail length, spicule shape and the shape and location of the posterior two precloacal supplements in relation to retracted spicules. Females of both species can be separated on the basis of distribution of sperm in the genital system (usually concentrated near the oviduct in *P. anemones vs* spread throughout the uteri in *P. pachydermus*) and to a lesser extent the shape of the vagina (usually narrower and indented in *P. anemones vs* about rectangular in *P. pachydermus)*.

Remarks

Substantial variability has been observed in the number of lateral body pores in males. Some males (five out of nineteen) had no pores, but the posterior pair at 62-79% of body length is usually present (nine males) and an additional anterior pair at 33-42 % of body length is less frequent (five males with both pairs). More than four pores per male were not observed (Loof, 1965).

The number of lateral body pores in females is usually two pairs (one prevulvar, one postvulvar) but may vary from 0 to 2 on one side of the body, anterior as well as posterior to the vulva. Some females (2 of 32) had no lateral pores, some others had a maximum of 5 or 6 pores (Loof, 1965).

The orientation of the sclerotized pieces occasionally varies under the influence of fixation or due to the physiological condition of the specimens e.g. usually obliquely orientated structures can lie parallel to the body axis (Decraemer, 1989b).

Most females show a concentration of sperm cells near the oviduct, in other specimens sperm may be dispersed throughout the uterus, sometimes with higher concentrations near the oviduct (Decraemer, 1989a, b).

Table 4.5. Comparison between related species of *P. anemones*

	P. anemones	*P. hispanus*	*P. weischeri*	*P. pachydermus*
shape spicules	anteriorly with irregular outline	anteriorly with irregular outline	anteriorly with irregular outline	cephalated,± cylindrical
number of CP	1	1	1	1
location SP1-SP2 in relation to spicule	opposite distal half spicules	opposite distal half but more spread	more spread	more spread
onchiostyle	42-53 µm	43-63.5 µm	43-48 µm	44-60 µm
spicule length	46-53 µm	44-61.7 µm	34-37 µm	40-54 µm
n° lateral body pores on each side in females	2 (0-2)	1	1	2 (2-5)
shape vaginal sclerotizations	small triangular, oblique	small rounded triangular, oblique	small oval, parallel to vagina	small round, oblique
location sperm	in spermathecae or concentration near oviduct	in spermathecae	in spermathecae	spread over uteri
tail length	short	short	short	long

Distribution and host

P. anemones was described from specimens from The Netherlands, and has been reported from France, Great Britain, Poland and Portugal, but occurs usually infrequently (Loof, 1965; Alphey & Boag, 1976, Craveiro & Santos, 1984, Scotto La Massese, 1985; Szczygiel & Brzeski, 1985; Alphey & Taylor, 1986). It has occasionally been recorded from the rhizosphere of barley (*Hordeum vulgare)*, wheat (*Triticum*), *Anemones* sp., hop (*Humulus lupulus*), rye (*Secale cereale)*, *Lolium multiflorum* (Loof, 1965; Craveiro & Santos, 1984; Spaull, 1980)

Paratrichodorus anthurii Baujard & Germani, 1985

Figs 3.12.H, 3.17.I, 4.25.F-H

Measurements

Holotype female: L= 510 µm, a= 16, b=5.5, V= 56%, onchiostyle= 29 µm, anterior end to EP= 106 µm.

Females: L= 380-530 µm (460 µm ± 40), a= 16-21, b= 3.7-5.6, V= 54-58% (57% ± 9), onchiostyle= 28-31 µm

(30 µm ± 8), anterior end to EP= 81-106 µm (from Baujard & Germani, 1985).

Emended diagnosis

P. anthurii female is characterized by the shape of the vaginal sclerotizations in lateral view (small triangular, close: 1 µm), the shape of the vagina (wide, rounded), pore-like vulva, presence of small oval sperm cells (1 by 1.5 µm) accumulated near the oviduct, short onchiostyle (28- 31 µm), absence of lateral body pores, and usually also absence of caudal pores; pharyngeal bulb offset. Males unknown.

Relationships

P. anthurii is comparable to *P. acutus* and *P. renifer* by the absence of males and by possessing a short onchiostyle. It differs from these species by the vulva shape (pore-like *vs* a transverse slit), shape of the vaginal sclerotizations (narrow triangular instead of dot-like in *P. acutus* and reniform in *P. renifer*) and from *P. acutus* by the tail shape.

P. anthurii females have a pore-like vulva, as in eight other species of the genus. Among them only *P. catharinae* possesses similar small sperm, however, they are not concentrated in spermathecae as in *P. anthurii*.

Remarks

The position of the S-E pore in *P. anthurii* is variable: between 71-115% of the length of the pharynx from the anterior end (Baujard & Germani, 1985).

The shape of the vulva was described as a transverse slit, but a study of type specimens (slide MNHN 21745) revealed a pore-like vulva.

Caudal pores were observed in one out of nine female paratype specimens.

According to Sturhan (1985) hermaphroditic reproduction may be assumed.

Distribution and host

P. anthurii is known from the type locality, Morne-Rouge, Martinique where it was found in the rhizosphere of *Anthurium* sp. Esser (1990b) recorded *P. anthurii* in Florida, found in soil from the rhizosphere of *Quercus virginiana* (live oak) (data not checked).

Paratrichodorus atlanticus (Allen, 1957) Siddiqi, 1974

Figs 3.14.W, 3.15.E

syn. *Trichodorus atlanticus* Allen , 1957

syn. *Paratrichodorus (Atlantadorus) atlanticus* (Allen, 1957) Siddiqi, 1974

syn. *Atlantadorus atlanticus* (Allen, 1957) Siddiqi, 1974

Measurements

Holotype male: L= 1200 μm, a= 19, b=7.1, T= 58%, onchiostyle= 82 μm, spicule length= 87 μm, gubernaculum= 18 μm.

Males: L= 930-1200 μm (1065 μm*), a= 12-19, b= 5.3-7.1, T= 54-68%, onchiostyle= 68-82 μm (75 μm*), spicule length= 79-87 μm (83 μm*), gubernaculum= 18-20 μm.

Females: L= 760-1130 μm (945 μm*), a= 10-15, b= 5.5-6.3, V= 50-55% (52.5% *), onchiostyle= 64-70 μm (67 μm*) (from Allen, 1957).

Emended diagnosis

P. atlanticus is characterized by a long onchiostyle (64-82 μm) and the position of the S-E pore at the level of the anterior margin of the pharyngeal bulb to mid-bulb. The male can be distinguished by the long spicules (79-87 μm), a single ventromedian cervical papilla just anterior to the S-E pore, a pair of lateral cervical pores at the level of the S-E pore, three ventromedian precloacal supplements (posterior two supplements opposite the posterior third of the retracted spicules, i.e. within bursa range), two pairs of postcloacal papillae, sperm with sausage shaped nucleus, and a bilobed flap extending over the cloacal opening. It is characterized in the female by the shape of the vulva (pore-like), the presence of spermathecae and one pair of ventro-submedian and one pair of ventrosublateral postadvulvar body pores.

Relationships

P. atlanticus belongs to the group of longest species of the genus, showing similarities with *P. acaudatus* and *P. macrostylus,* especially in the male. It can be distinguished from these species by several characters in males and females as presented in table 4.6.:

Remarks

P. grandis, another species with a long body was not included in the comparison (table 4.6.) because of its different spicules (short: 39-43 μm and with different shape).

Data are not available on vaginal sclerotizations in lateral view in female as the paratype specimen examined is in a dorsoventral position and the original illustration also shows the vulvar region in ventral view.

Females were originally described with the vulva being a longitudinal slit. Study of a paratype specimen revealed it to be pore-like. Moreover, Siddiqi (1974, 1980a) indicated *P. atlanticus* as type species of *Atlantadorus* which is characterized by a pore-like vulva.

Table 4.6.

	P. atlanticus	*P. acaudatus*	*P. macrostylus*
L	♂: 930-1200 µm ♀: 760-1130 µm	♂: 750-847 µm ♀: 910-950 µm	♂: 1160-1630 µm ♀: 1350-1730 µm
onchiostyle	♂: 68-82 µm ♀: 64-70 µm	♂: 75-78 µm ♀: 82-83 µm	♂: 133-180 µm ♀: 160-188 µm
spicule length	79-87 µm	67-68 µm	67.5-87.6 µm
n° postcloacal papillae	2	1	1
n° precloacal supplements	3	2	3
number of CP	1	0	1
shape spicules	slender with marked manubrium	stouter without marked manubrium	stouter without marked manubrium
V: shape	pore	?	a longitudinal slit
geographic distribution	Florida	India	Romania

Distribution and host

P. atlanticus is known from the type locality, Winter Haven in Florida. It was found in soil from the roots of sand pine (Allen, 1957).

Paratrichodorus catharinae Vermeulen & Heyns, 1983

Figs 3.13.K; 3.14.T; 3.15.F; 3.17.J; 4.26.A-K

syn. *Paratrichodorus (Atlantadorus) catharinae* Vermeulen & Heyns, 1983, n.syn.

Measurements

Holotype male: L= 691 µm, a= 13.4, b= 4.7, onchiostyle= 49 µm, spicule length= 52 µm, gubernaculum= 11 µm, anterior end to EP= 76 µm.

Males: L= 523-742 µm (630 µm; 560 µm), a= 9.8-22, b= 3.5-4.7, onchiostyle= 40-51 µm (46 µm; 44 µm), spicule length= 52-55 µm (54 µm; 40 µm), gubernaculum= 9.2-12 µm, anterior end to EP= 51-87 µm, distance CP1-EP= 9-27 µm.

Females: L= 546-766 µm (611 µm; 643 µm), a= 10.9-34 µm (15.2; 26), b= 4.1-6.6, V= 51-56% (52%; 53%), onchiostyle= 34-50 µm (46 µm; 42 µm), anterior end to EP= 45-85 µm (from Vermeulen & Heyns, 1983; Vermeulen & Heyns, 1984).

Emended diagnosis

P. catharinae males are characterized by a single ventromedian cervical papilla anterior to the S-E pore (near base of onchiostyle or more posteriorly), shape of the spicules (straight, striated except at extremities, slightly cephalated), three ventromedian precloacal supplements with the two posterior supplements within region of the retracted spicules, caudal alae narrow, reaching anteriorly to the head of the retracted spicules, tail shape (short, conical, dorsally convex), shape of sperm cells (small, oval, 3 by 3.5 µm, nucleus oval about 2 µm long) and protruding anterior cloacal lip. Females by the pore-like vulva, shape of the vaginal sclerotizations in lateral view (dot-like, 2 µm apart), shape of the vagina (wide, rounded), spermathecae absent and two lateral body pores (one prevulvar, one postvulvar). S-E pore near base of the onchiostyle. Overlaps at pharyngo-intestinal junction variable (either with anteriodorsal intestinal overlap or with ventral pharyngeal overlap) in male and female.

Relationships

P. catharinae most closely resembles *P. sacchari* and both species are described only from South Africa. They have comparable metric data for males and females (e.g. L= 523-766 µm *vs* 589-733 µm; onchiostyle= 34-50 µm *vs* 38-48 µm; spicule length= 52-55µm *vs* 46-52 µm in *P. sacchari*). Furthermore, they possess the same number of ventromedian precloacal supplements (3) and ventromedian cervical papilla (1) in males and a wide, rounded vagina and pore-like vulva in females. They differ in the structure of the sperm: small sperm in *P. catharinae vs.* large sperm with sausage shaped nucleus in *P. sacchari*. Minor differences were observed in males in spicule shape and the location of the posterior two precloacal supplements (slenderer spicules in *P. sacchari*; posterior two supplements more dispersed within the region of the retracted spicules in

P. catharinae) and in females in the shape of the vaginal sclerotizations (dot-like and well separated in *P. catharinae vs* minute triangular with tips close to one another in *P. sacchari*). These observations are based on type specimens of *P. catharinae* and *P. sacchari*. A few other specimens of *P. catharinae* were examined, the type specimens being badly fixed (Vermeulen & Heyns, 1983; De Waele & Kilian, 1992; own observations). Both species are very similar, but were differentiated by their authors mainly on the structure of vagina (with refractive folds in *P. sacchari*) and vaginal sclerotizations (Vermeulen & Heyns, 1983). *P. catharinae* is also close to *P. meyeri* but differs by the more anterior position of the S-E pore, the slenderer spicules, the shape of the vulva and vaginal sclerotizations (see *P. meyeri*).

Remarks

The paratype specimens are flattened, with extremely swollen cuticles, therefore morphometric data such as L, a-value have to be interpreted with caution. The specimens examined (slide 1543) showed a ventral overlap of the pharyngeal glands.

De Waele & Kilian (1992) have reexamined some paratype specimens, giving additional morphometric and morphological information, without comments. Their redescription differs from that of the original in the following characters: 1) males and females with anterior dorsal intestinal overlap *vs* a ventral overlap of the pharynx in original description and illustration; 2) position of the S-E pore from anterior end in female paratypes is 83-85 (84) µm *vs* 45-75 (64) µm originally; two lateral body pores on each side (one prevulvar, one postvulvar) *vs* one postvulvar in original description of females.

The second population of *P. catharinae* found by Vermeulen & Heyns (1984) showed some morphometric differences from the type population: larger a-value, shorter onchiostyle (34-49 µm *vs* 41-51 µm (type population), shorter spicules (40 µm *vs* 52-55 µm) and slightly shorter gubernaculum (9.2 *vs* 11-12 µm).

Lateral cervical pores, not described previously, have been observed.

The specimens described as *Paratrichodorus* sp.1 in Vermeulen & Heyns (1983) also belong to *P. catharinae*. They were described as close to *P. catharinae* and *P. lobatus* but different from the first species in a ratio, L, length of spicules (42 µm *vs* 52-55 µm in type population), length of gubernaculum and possibly absence of ventromedian cervical papilla. These differences are probably due to fixation and variability, as suggested by the authors when describing another population (Vermeulen & Heyns,

1984).

Distribution and host

P. catharinae is recorded only from South Africa (Crystal waters, Rustenburg, Transvaal; Oranje Free State; Cape Province) except for one record from Florida (Esser, 1991)(record not checked). It was found in soil samples from the roots of garden flowers, grasses, pumpkin, maize, pear tree and cherry laurel (Vermeulen & Heyns, 1983; 1984; Esser, 1991a).

Paratrichodorus delhiensis (Khan, Saha & Lal, 1993) n. comb.

Figs 3.14.L; 3.15.G

syn. *Atlantadorus delhiensis* Khan, Saha & Lal, 1993, n.syn.

Measurements

Holotype male: L= 410 μm, a= 22.7, b= 4.9, T= 55.8%, onchiostyle= 31 μm, spicule length= 34 μm, gubernaculum= 6.5 μm.

Males: L= 410-520 μm (465 μm), a= 20.5-26.5, b= 4.9-6.3, T= 52.2-55.8%, onchiostyle= 31-35 μm (33 μm), spicule length= 30-34 μm (32 μm).

Females: L= 440-595 μm (520 μm), a= 19.8-25.8, b= 5.5-7.0, V= 50.1-56.6% (53.4%), onchiostyle= 31-36 μm (33.5 μm) (from Khan *et al.*, 1993).

Emended diagnosis

P. delhiensis males are characterized by the absence of ventromedian cervical papillae, the shape of the spicules (largely straight but slightly curved at distal end, manubrium marked and small knob-like), two ventromedian precloacal supplements with the posterior one close to the cloacal opening, the anterior one outside the region of retracted spicules. Females by the pore-like vulva, a short vagina (less than 1/3rd of vulval body width) and apparently without spermathecae. The S-E pore is situated opposite the middle of the basal pharyngeal bulb and the pharyngeal bulb is offset in males and females.

Relationships

P. delhiensis males lack ventromedian cervical papillae as eight other species of the genus. Among them *P. acaudatus*, *P. alleni*, *P. allius* and *P. lobatus* also have two ventromedian precloacal supplements as *P. delhiensis*, with only the posterior supplement within the region of retracted spicules (except for *P. allius* with both supplements in the region of the retracted spicules). *P. delhiensis* further agrees with *P. allius* by having short spicules (30-34 µm), shorter than in the other three species. *P. delhiensis* differs from the four previous species by the short body length and the small onchiostyle (31-35 µm). *P. delhiensis* females have a pore-like vulva as eleven other species of the genus, among them *P. anthurii, P. faisalabadensis* and *P. psidii* have a small onchiostyle (smaller than 34 µm). Since no data are available on the shape of the vaginal sclerotizations and the structure and location of sperm cells, the relationship of *P. delhiensis* females can not be determined.

Remarks

Originally, the relationship of *P. delhiensis* was only given for males. Khan *et al.* (1993) considered *P. delhiensis* close to *P. porosus* and *P. anemones* without explanation.

No information is available on the shape of the sperm in males nor on the shape of the vaginal sclerotizations, the shape of the vagina and the presence of lateral body pores in females.

Apart from morphometric data and the shape of the vulva, the description of females do not include enough data to allow identification to species level.

Distribution and host

P. delhiensis is known from the type locality in New Delhi, India, where it was found in the rhizosphere of mulberry (*Morus alba*) (Khan *et al.*, 1993).

Paratrichodorus faisalabadensis Nasira & Maqbool, 1994

Figs 3.13.P; 3.14.F; 3.15.H

Measurements

Holotype female: L= 650 µm, a= 21, b= 4.9, V= 54%, onchiostyle= 34 µm, distance anterior end to EP= 72 µm.

Females: L= 540-650 µm (580 µm ± 40), a= 18.6-23.3, b=4.2-5.4, V= 51-54% (52.7% ± 1.42), onchiostyle= 30-34.4 µm (33.3 µm ± 1.05), distance anterior end to EP= 56-76 µm.

Males: L= 520-660 µm (570 µm ± 20), a= 16.8-21.2, b= 4.7-5.6, T= 54-61.3%, onchiostyle= 32.8-35.2 µm (34.3 µm ± 1.33), spicule length= 30-33.6 µm (31.9 µm ± 1.81), gubernaculum= 7.0-8.8 µm, distance anterior end to EP= 65.6-68.0 µm, anterior end to CP1= 34-39.2 µm, distance anterior end to CP2= 49.2-57.6 µm, distance CP2-EP= 11.2-17.4 µm (from Nasira & Maqbool, 1994).

Emended diagnosis

P. faisalabadensis females are characterized by a pore-like vulva, the shape of the vaginal sclerotizations in lateral view (short rods, about 2 µm apart), the shape of the vagina (small trapezoid) and by the shape, size and position of sperm (medium-sized rounded, located throughout the uteri with possible concentration near the oviducts). Males by two ventromedian cervical papillae anterior to the S-E pore and with the anterior papilla near the level of the onchiostyle base, three ventromedian precloacal supplements (the posterior one near the cloacal opening, the second one at the level of the spicule head when spicules are retracted, the anterior supplement lying out of reach of the retracted spicules), the length (30-33.6 µm) and shape of the spicules (largely straight except for a slightly curved distal end, manubrium slightly marked and shaft striated except at both extremities) and by the medium-sized sperm cells with rounded nucleus. Males and females with a short onchiostyle (30-35 µm) and an anteriodorsal intestinal overlap of the pharynx. S-E pore posterior to the nerve ring and opposite the isthmus or anterior region of the pharyngeal bulb.

Relationships

P. faisalabadensis males are similar to those of *P. mirzai* and *P. psidii* by having medium-sized sperm cells with rounded nucleus, two ventromedian cervical papillae, three ventromedian precloacal supplements, short spicules (33-33.6 µm in *P. faisalabadensis*, 30-40 µm in *P. mirzai* and 30-31 µm in *P. psidii*) and a short onchiostyle (32.8-33.5 µm in *P. faisalabadensis*, 29-36 µm in *P. mirzai* and 30-32 µm in *P. psidii*). They differ from *P. mirzai* in males, by the arrangement of the precloacal supplements with the SP2 at the level of the spicule manubrium *vs* clearly anterior to the region of the retracted spicules in *P. mirzai*. *P. faisalabadensis* females are similar to those of *P. psidii* by the structure and location of sperm, the pore-like vulva and short onchiostyle, but they differ by the position and shape of the vaginal sclerotizations in lateral view (short rods, well separated in *P. faisalabadensis vs* dot-like, less separated in *P. psidii*) and by the shape of the vagina (trapezoid vs rounded to oval in *P. psidii*). Females of *P. faisalabadensis* further differ from *P. mirzai* by the shape of the vulva (pore-like vs a longitudinal slit in *P. mirzai*).

Remarks

Nasira & Maqbool (1994) further differentiated *P. faisalabadensis* from *P. psidii* by body length (520-660 μm *vs* 400-520 μm, spicule length (30-33.6 μm vs 30-31 μm) and spicule shape, distance of S-E pore from anterior end and onchiostyle length. Former differences fall in the range of intraspecific variability. Moreover, the type specimens of *P. psidii* are not well fixed, with strongly swollen body cuticle and retracted head region. Both species essentially differ in the shape of the vagina and vaginal sclerotizations, but these differences seem rather small. *P. faisalabadensis* is very close to *P. psidii* described from the same locality. It could be that both species in fact represent a single species. No information is available on the presence of lateral body pores in *P. faisalabadensis*. Type specimens were not available.

Distribution and host

P. faisalabadensis was described from the type locality Faisalabad, Pakistan where it has been collected from the rhizosphere of guave (*Psidium guajava*) (Nasira & Maqbool, 1994).

Paratrichodorus grandis Rodriguez-M. & Bell, 1978

Figs 3.12.M; 3.14.G-H; 3.15.I-J; 3.17.D, d; 4.26.I-J

syn. *Paratrichodorus (Atlantadorus) grandis* Rodriguez-M & Bell, 1978
syn. *Atlantadorus grandis* (Rodriguez-M & Bell, 1978) Siddiqi, 1974

Measurements

Holotype male: L= 920 μm, a= 24.0, b=6.3, T= 57.8%, onchiostyle= 70 μm, spicule length= 42 μm, gubernaculum= 13 μm, anterior end to EP= 113 μm, distance CP1-EP= 45 μm.

Males: L= 790-1040 μm (900 μm ± 80), a= 22.6-27, b= 5.0-7.4, T= 51.9-60.3%, onchiostyle= 64-72 μm (70 μm ± 2.2), spicule length= 39-43 μm (40.0 μm ± 1.4), gubernaculum= 12-16 μm.

Females: L= 900-1040 μm (960 μm ± 40), a= 23-30, b= 5.1-7.0, V= 56.4-64% (59.0% ± 2.1), onchiostyle= 65-72 μm (68.0 μm ± 2.2), anterior end to EP= 102-127 μm (from Rodriguez-M & Bell, 1978).

Emended diagnosis

P. grandis males are characterized by a single ventromedian cervical papilla at the level of the onchiostyle base and well anterior to the S-E pore which is located in the posterior isthmus region, a pair of lateral cervical pores at the level of the ventromedian cervical papilla, the shape of the spicules (largely straight but slightly curved at distal end, manubrium marked, shaft striated), three ventromedian precloacal supplements with the posterior one clearly within the region of the retracted spicules and the middle one at the head or just anterior to the spicules, one pair of sublateral postcloacal papillae, the cloacal opening covered by a bilobed flap and caudal alae narrow, extending from slightly anterior to the second supplement. Females by the pore-like vulva, shape of the vaginal sclerotizations in lateral view (small rounded triangular, close), shape of the vagina (rounded trapezoid, with or without slight indentation), lateral body pores postvulvar and large sperm with sausage-shaped nucleus, dispersed throughout the uteri (no spermathecae). The pharyngeal base usually with dorsal intestinal overlap in males and females.

Relationships

P. grandis is most similar to *P. pachydermus*, having a pore-like vulva, large sperm cells with sausage-shaped nucleus dispersed over the uteri, a comparable shape of the vagina and vaginal sclerotizations (small triangular) in females and three ventromedian precloacal supplements and one ventromedian cervical papilla in males. They differ by their longer onchiostyle (64-72 µm *vs* 44-60 µm in *P. pachydermus*), more anterior position of the ventromedian cervical papilla , shorter tail and differently shaped spicules.

Remarks

According to the authors of the species, lateral body pores of *P. grandis* vary in number and position but always occur posterior to the vulva, usually two on the right side and one on the left side; but one paratype has three lateral body pores close together on the right side; they may be advulvar.

The second precloacal supplement varies in position in relation to the retracted spicules, from opposite to just anterior to the spicule head (Decraemer, 1989b).

Distribution and host

P. grandis is known from the type locality in California where it was found in soil around manzanita (*Arctostaphylos* sp.) (Rodriguez-M & Bell, 1978).

Paratrichodorus hispanus Roca & Arias, 1986
Figs 3.12.Q; 3.14.B; 3.15.K; 3.17.E, e; 4.27.A-D

Measurements

Holotype male: L=976 µm, a= 20.5, b= 6.5, T= 65%, onchiostyle= 58.2 µm, spicule length= 55.8 µm, gubernaculum= 14.7 µm, anterior end to EP= 109.4 µm.

Males: L= 691-1103 µm (956 µm ± 85; 856 µm ± 87.9), a= 13-26, b= 4.0-8.5, T= 57-73%, onchiostyle= 49.0-63.5 µm (59.4 µm ± 1.96; 54.0 µm ± 5.41), spicule length= 44.0-61.7 µm (56 µm ± 2.76; 48.5 µm ± 3.14), gubernaculum= 11.5-16.4 µm, distance anterior end to EP= 79.5-122.3 µm, distance anterior end to CP1= 74.5-109.5 µm.

Females: L= 660.5-1100 µm (985 µm ± 57; 808.0 µm ± 104.16), a= 17.1-26.5, b= 4.5-7.3, V= 51-61% (54.7% ± 1.77; 55.0% ± 3.02), onchiostyle= 43.0-63.5 µm (59.6 µm ± 2.01; 51.0 µm ± 4.97), distance anterior end to EP= 81.0-123.5 µm (from Roca & Arias, 1986; Decraemer *et al.*, 1993).

Emended diagnosis

P. hispanus males are characterized by their single ventromedian cervical papilla just anterior to the S-E pore, position of the lateral cervical pore (on each side at the level of the ventromedian cervical papilla, i. e. at the level of the isthmus), shape of the spicules (almost straight, distal third ventrally curved, head slightly larger, shaft anteriorly with irregular outline, finely striated except at tapered distal end), three ventromedian precloacal supplements with the posterior two supplements opposite the distal half of the retracted spicules, large subcylindrical sperm cells with granular sausage-shaped nucleus and caudal alae extending from halfway spicule level to the postcloacal papillae. Females by the size, shape and orientation of the vaginal sclerotizations (small rounded triangular, oblique, close), pore-like vulva, shape of the vagina (about trapezoid, with or without indentation), the presence of spermathecae and a single pair of prevulvar lateral body pores. The pharynx has an anterior dorsal intestinal overlap in males and females; the S-E pore is located opposite the anterior end of the pharyngeal bulb.

Relationships

P. hispanus resembles *P. anemones* and *P. weischeri* (see table 4.5. under relationships of *P. anemones*). *P. hispanus* females also resemble *P. grandis* by their pore-like vulva, comparable shape of vagina and vaginal sclerotizations, similar sperm shape, but differ mainly by having a shorter onchiostyle (43-63.5 µm *vs* 65-72 µm in *P. grandis*), presence of spermathecae (absent in *P. grandis*) and males by their different shape of spicules.

Remarks

The specimens from Sierra Morena showed some variability with the type population in morphometric data: slightly shorter mean body length, somewhat shorter onchiostyle length in female (43-59 µm *vs* 56.4-63.5 µm in type population), a more anterior position of the S-E pore and somewhat shorter spicules (44-54.5 µm *vs* 53-61.7 µm in the type population) (Decraemer *et al.*, 1993).

Distribution and host

P. hispanus is known only from Spain (Toledo, Sierra Morena) where it was found in the rhizosphere of wheat and *Fraxinus angustifolia* (Roca & Arias, 1986; Decraemer *et al.*, 1993).

Paratrichodorus lobatus (Colbran, 1965) Siddiqi, 1974
Figs 3.13.E-F; 3.14.S; 3.15.L; 3.17.K, k; 4.28.A-H

syn. *Trichodorus lobatus* Colbran, 1965
syn. *Paratrichodorus (Paratrichodorus) lobatus* (Colbran, 1965) Siddiqi, 1974
syn. *Trichodorus clarki* Yeates, 1967

Measurements

Holotype female: L= 630 µm, a= 14.2, b= 5.9, V= 54.4%, onchiostyle= 47.2 µm.

Females: L= 535-1051 µm (943 µm*; 858 µm; 866 µm; 785 µm; 605 µm*; 727 µm; 677.5 µm; 793 µm ± 115.4), a= 10.3-32.0, b= 3.6-8.5, V= 47.7-60.1% (53.9%*; 53.3%; 55%; 54%; 51.8%*; 54.3%; 56.0%; 52.0% ± 3.1); onchiostyle= 44.5-58.0 µm (50.0 µm; 57 µm; 49 µm; 51 µm; 47.5 µm; 48 µm; 48.7 µm*; 50.5 µm ± 5.1), anterior end to EP= 62.0-107.0 µm

Males: L= 550-1095 µm (813 µm*; 899 µm; 869 µm; 778 µm; 593 µm; 655 µm*; 612.5 µm*; 719.0 µm ± 102.9), a= 10.2-35.0, b= 3.6-9.9, T= 45.3-77.7%, onchiostyle= 42-56 µm (46 µm*; 52 µm; 47 µm; 49 µm; 46 µm; 53 µm*; 46.3 µm*; 57.5 µm ± 3.0), spicule length= 46-70 µm (56.5 µm*; 67 µm; 56 µm; 57.5 µm; 55 µm; 58.5 µm*; 54 µm*; 60.5 µm ± 5.5), gubernaculum= 8.5-16 µm, anterior end to EP= 60-134 µm (from Colbran, 1965; Yeates, 1967; Vermeulen & Heyns, 1983, 1984; De Waele *et al.*, 1990; Decraemer & Reay, 1991: 3 populations).

Emended diagnosis

P. lobatus is characterized by a pharynx with a short, moderate or most usually more than one body width long ventral overlap in males and females. Females are distinguished by the vulva shape (a longitudinal slit), shape of the vaginal sclerotizations (small oval, well separated: > 2.5 μm), shape of the vagina (wide, rounded trapezoid when relaxed), usually one pair of lateral postvulvar body pores and sperm cells usually dispersed throughout the uteri (at most with some concentration near the oviduct). Males by the shape of their spicules (cephalated, shaft striated), caudal alae well developed, small sperm cells with oval-shaped nucleus (may appear fibrillar), number of precloacal supplements and their position in relation to the retracted spicules (two precloacal supplements opposite the distal half of the retracted spicules), absence of ventromedian cervical papillae and lateral cervical pores, and the protruding anterior cloacal lip. S-E pore opposite anterior half pharyngeal bulb in male and female.

Relationships

P. lobatus is closely related to *P. teres*. They have a comparable habitus, a similar overlapping of the pharyngeal glands and small sperm cells in males and females. *P. lobatus* males have a similar spicule shape and the females have a similar shape of the vagina and vaginal sclerotizations as in *P. teres*. Both species are distinguished by a different sex ratio (males abundant in *P. lobatus vs* males absent or rare in most populations of *P. teres*) and a related difference in the amount of sperm, a slight difference in sperm structure (nuclei not always clear and cells appearing more fibrillar in *P. teres*), small differences in spicule length (longer: mean length between 55 and 67 μm in *P. lobatus vs* mean length between 45 and 48.5 μm in *P. teres*; the measurements may overlap when several populations of these two species are compared), usually three precloacal supplements in *P. teres vs* mainly two supplements in *P. lobatus*, and by different geographical distributions (*P. lobatus* common in Australia, South Africa, *P. teres* common in Europe) (Decraemer & Reay, 1991).

Remarks

Spermathecae have only been described in the type population; in the other populations the sperm cells are dispersed throughout the uteri.

Females of a population from rice in South Africa have mainly two pairs of lateral body pores (one prevulvar, one postvulvar pair) (De Waele *et al., 1990)*.

In eleven out of fourteen males of a population from rice in South Africa, the pharyngeal overlap was less than one body width long (De Waele *et al., 1990)*. The tail may be provided with a small digitate tip.

Among specimens of *P. lobatus* from Australian populations, exceptionally two specimens were observed with three distinct precloacal supplements (Decraemer & Reay, 1991). One male specimen from the population described as *T. clarki* Yeates (1967) possesses two lateral cervical pores anterior to the S-E pore.

All Australian specimens, previously identified as *P. teres*, belong to *P. lobatus* (Decraemer & Reay, 1991).

Distribution and host

P. lobatus is common in Australia, but has also been observed in New Zealand, Chili and South Africa. It was found associated with eucalyptus forest, *Ammophila arenaria* (European beach grass), *Prunus mahaleb* (mahaleb), *Citrus jambhiri*, *Citrus* sp., unthrifty cotton, pumpkin, grasses, maize, pear, rice and almond (Colbran, 1965; Yeates, 1967; Vermeulen & Heyns, 1983, 1984; De Waele *et al.*,1990; Decraemer & Reay, 1991; nematode collection Riverside, U.S.A.).

Paratrichodorus macrostylus Popovici, 1989
Figs 3.13.Q; 3.14.X; 3.15.M; 4.29.A-G

Measurements

Holotype: L= 1560 µm, a= 23.9, b= 5.7, T= 54.5%, onchiostyle= 163 µm, spicule length= 80 µm, gubernaculum= 15.5 µm, anterior end to EP= 168 µm, distance CP1-EP= 7.5 µm

Males: L= 1160-1630 µm (1510 µm; 1505 µm*; 1610 µm*; 1320 µm), a= 16.5-26.7, b= 4.2-7.3, T= 53-63.3%, onchiostyle= 133-180 µm (156.8 µm; 134.5 µm*; 170 µm*; 154 µm), spicule length= 67.5 µm-87.6 µm (75.7 µm; 81.5 µm*; 85 µm*; 73 µm), gubernaculum= 11-19.5 µm, anterior end to EP= 148-188 µm, distance CP1-EP= 7.5-15 µm

Females: L= 1350-1730 µm (1560 µm; 1720 µm; 1450 µm), a= 15.8-24.3, b= 4.6-6.6, V= 50.6-65.5% (57.9%; 57%; 54.3%), onchiostyle= 160-188 µm (166 µm; 160 µm; 171 µm), anterior end to EP= 153-182.5 µm (from Popovici, 1989).

Emended diagnosis

P. macrostylus is characterized by its very long onchiostyle (133-188 µm), long body (1160-1730 µm), long (15-20 µm) fusiform sperm cells with elongated striated nucleus and slightly overlapping pharyngeal glands. Males can be distinguished by the anterior position of the ventromedian cervical papilla (at the level of the onchiostyle base, i.e.

immediately anterior to the S-E pore), the anterior position of the S-E pore (shortly posterior the onchiostyle base), the position of a pair of lateral cervical pores (about halfway the isthmus, shortly posterior to the nerve ring), one pair of postcloacal papillae, the length (67.5-87.5 µm) and shape of the spicules (slightly ventrally curved, shaft striated except at extremities), the position of the three ventromedian precloacal supplements (posterior two opposite the distal third of retracted spicules, anterior supplement about one spicule length anterior to retracted spicules) and short caudal alae (extending anteriorly to distal third of retracted spicules). Females by the shape of the vulva (a longitudinal slit), shape of the vaginal sclerotizations in lateral view (rod-like, well separated, oblique), vagina wide, rounded trapezoid with eight refractive folds, one pair of prevulvar lateral body pores and by the presence of two spermathecae, each filled with sperm.

Relationships

P. macrostylus male is most similar to *P. atlanticus* by the long body (respectively 1160-1630 µm and 930-1200 µm), number and position of the ventromedian precloacal supplements (three, with posterior two close to the cloacal opening) and the length of the spicules (respectively 67.5-87.6 µm and 79-87 µm). It differs mainly by the length of the onchiostyle (133-180 µm *vs* 68-82 µm), the shape of the spicules (not cephalated *vs* cephalated in *P. atlanticus*), the position of the ventromedian cervical papilla (one just anterior to the S-E pore and at the base of the onchiostyle *vs* at level mid-bulb in *P. atlanticus*) and in females by the vulva shape (a longitudinal slit *vs* pore-like).

Remarks

The long onchiostyle which is ventrally curved usually has a knick at the onchium/onchiophore junction.

Nerve ring with anterior position (from just at, to just posterior to the onchiostyle base); in its anteriormost position at the same level as the S-E pore.

Lateral cervical pores associated with inner canal provided with granulae.

In four male paratypes, the dorsal pharyngeal gland nucleus lies in the posterior half of the bulb.

Sperm cells elongated, their nucleus long, oval with the central tube surrounded by longitudinal fibrils giving a striated appearance.

Cloacal aperture covered by a digitate flap (Fig. 4.29 E); in the original description this was apparently interpreted as a minute postcloacal papilla.

One pair of prevulvar lateral body pores observed in three female paratype specimens (Popovici, 1989).

Distribution and host

P. macrostylus is known only from the type population from the Gilau Mountains, Romania where it was found in acid black soil in a beech-spruce forest (Popovici, 1989).

Paratrichodorus meyeri De Waele & Kilian, 1992

Figs 3.13.N; 3.14.U; 3.15.O; 4.29.H-J

Measurements

Holotype female: L= 838 µm, a= 13.9, b= 7.2, V= 49.4%, onchiostyle= 48µm, anterior end to EP= 111 µm

Females: L= 510-896 µm (689 µm ± 138), a= 9.4-19.3, b= 3.8-7.0, V= 47.9-53.2 % (50.6% ± 2.3), onchiostyle= 45-53 µm (48 µm ± 3), anterior end to EP= 72-113 µm

Males: L= 510-865 µm (700 µm ± 101), a= 9.3-19.4, b= 6.3-7.7, T= 65-76.1%, onchiostyle= 44-54 µm (47 ± 3), spicule length= 48-58 µm (53 µm ± 3), gubernaculum= 9-14 µm, anterior end to EP= 74-121 µm, distance CP1-EP= 7-14 µm (from De Waele & Kilian, 1992).

Emended diagnosis

P. meyeri males are characterized by having a single ventromedian cervical papilla anterior to the S-E pore and opposite the anterior or middle part of the pharyngeal bulb, a pair of lateral cervical pores at level nerve ring, i.e. anterior to the ventromedian cervical papilla, the number and position of ventromedian precloacal supplements (three, the posterior two within the region of the retracted spicules), the shape of the spicules (cephalated, slightly curved distally, striated, except at both extremities), anterior cloacal lip with two finger-like projections and one pair of postcloacal subventral papillae. Females by the vulva shape (a longitudinal slit), shape of the vaginal sclerotizations in lateral view (small triangular pieces, well separated: 2 µm apart), shape of vagina (wide, rounded), number of lateral body pores (two to five on each side, but not always opposite each other, one or two prevulvar) and the absence of spermathecae. Furthermore, in males and females by the length of the onchiostyle (44-54 µm), the presence of a ventral pharyngeal gland overlap and an anterior dorsal overlap by the intestine, the position of the S-E pore opposite mid-bulb and small sperm cells with oval nucleus.

Relationships

P. meyeri most closely resembles *P. pachydermus* and *P. catharinae*. All three species have comparable spicules (in length and shape), about equally long onchiostyle, three ventromedian precloacal supplements with the same position of the posterior two supplements in relation to retracted spicules, a single ventromedian cervical papilla, and more than one lateral body pore in females.

P. meyeri differs from both species by vulva shape (a longitudinal slit *vs* a pore). Furthermore, from *P. pachydermus* by the size and shape of sperm (small cells with oval nucleus *vs* large cells with sausage-shaped nucleus), a shorter tail in males and from *P. catharinae* by somewhat more slender spicules, different shape of vaginal sclerotization (small triangular *vs* dot-like) and a more posterior position of the S-E pore.

Remarks

The following exceptions have been observed: one female with only an anterior intestinal overlap, one female with only a ventral pharyngeal overlap and two males without ventromedian cervical papilla.

Lateral cervical pores were described as being absent but were observed in paratype males.

Distribution and host

P. meyeri is known from the type locality in Waterberg district, Transvaal, South Africa where it was found in sandy soil around the roots of tobacco (*Nicotiana tabacum*) (De Waele & Kilian, 1992).

Paratrichodorus minor (Colbran, 1956) Siddiqi, 1974

Figs 3.12.C, D, G; 3.14.Y; 3.15.N; 3.17.L, R, l, r; 4.27.E-F; 6.1.

syn. *Trichodorus minor* Colbran, 1956
syn. *Paratrichodorus (Nanidorus) minor* (Colbran, 1956) Siddiqi, 1974
syn. *Nanidorus minor* (Colbran, 1956) Siddiqi, 1974
syn. *Trichodorus christiei* Allen, 1957
syn. *Paratrichodorus christiei* (Allen, 1957) Siddiqi, 1974
syn. *Paratrichodorus (Nanidorus) christiei* (Allen, 1957) Siddiqi, 1974

syn. *Nanidorus christiei* (Allen, 1957) Siddiqi, 1974
syn. *Paratrichodorus obesus* (Razjivin & Penton, 1975) Rodriguez-M & Bell, 1978
syn. *Trichodorus obesus* Razjivin & Penton, 1975
syn. *Paratrichodorus (Nanidorus) obesus* (Razjivin & Penton, 1975) Rodriguez-M & Bell, 1978

See chapter on economically important species.

Paratrichodorus mirzai (Siddiqi, 1960) Siddiqi, 1974

Figs 3.13.O; 3.14.I; 3.16.A; 3.17.F'; 4.30.A-C

syn. *Trichodorus mirzai* Siddiqi, 1960
syn. *Paratrichodorus (Paratrichodorus) mirzai* (Siddiqi, 1960) Siddiqi, 1974
syn. *Trichodorus musambi* Edward & Misra, 1970

Measurements

Holotype male: L= 518 µm, a= 18, b= 4.5, T= 63%, onchiostyle= 29 µm, spicule length= 32.5 µm, gubernaculum= 11 µm

Males: L= 450-630 µm (519 µm*; 530 µm; 500 µm; 525 µm*), a= 14.5-20, b= 4.2-5.8, T= 53-70%, onchiostyle= 29-36 µm (29 µm; 34 µm; 30 µm*), spicule length= 30-34 µm (32.5 µm; 33 µm; 30 µm; 33 µm), gubernaculum= 7-13 µm, anterior end to EP= 72 µm, distance CP1-CP2= 22 µm, distance CP2-CP3**= 7-10 µm

Females: L= 420-640 µm (545 µm*; 540 µm; 550 µm; 495 µm*), a= 13-20.7, b= 3.0-6.3, V= 50.0-58.0% (54.8%*; 55.8%; 54.1%; 52%*), onchiostyle= 29-37 µm (32.5 µm*; 35 µm; 31.3 µm; 30.5 µm*), anterior end to EP= 65 µm (from Siddiqi, 1960; 1962; Yokoo, 1964; Edward & Misra, 1970).

** According to Edward & Misra (1970) *P. mirzai* has three ventromedian cervical papillae in male, the posterior one at the level of the S-E pore.

Emended diagnosis

P. mirzai males are characterized, by the number and position of the ventromedian cervical papillae (two: the anterior papilla immediately posterior to the onchiostyle base, the second one being immediately behind the nerve ring), S-E pore just posterior to the second papillae, the length (30-34 µm) and shape of the spicules (relatively straight, cylindrical, tapered distally, cephalated, shaft striated), sperm structure (rounded sperm cells each with a rounded nucleus), number and position of ventromedian precloacal supplements (three, posterior one opposite posterior half of retracted spicules, other two supplements outside the region of the spicule), two pairs of postcloacal papillae and weak caudal alae. Females by the vulva shape (a longitudinal slit), shape of the vaginal sclerotizations in lateral view (small, dot-like, well separated), shape of vagina (about quadrangular), sperm either absent or located in the uteri and usually two lateral body pores (one prevulvar, one postvulvar) on each side. The pharyngeal bulb is offset or with a slight ventral overlap.

Relationships

P. mirzai males differ from those of the other species of the genus (except *P. paramirzai*, *P. faisalabadensis* and *P. psidii*) by the larger number of ventromedian cervical papillae (2) and by sperm structure (medium-sized sperm cells with rounded nucleus). They differ from *P. faisalabadensis* and *P. psidii* males by the position of the second precloacal supplement (shortly anterior to retracted spicules *vs* at the level of the spicule head)

P. mirzai females resemble *P. paramirzai*, *P. lobatus* and *P. teres* by the combination of vulva shape (a longitudinal slit), lack of spermathecae and similar shape of vaginal sclerotizations (dot-like, well separated pieces). *P. mirzai* differs from the three former species by the characters enumerated in table 4.7. *P. mirzai* females further differ from those of *P. faisalabadensis* and *P. psidii* by the shape of the vulva (a longitudinal slit *vs* pore-like).

Remarks

P. mirzai was described as having three ventromedian cervical papillae in males, but the S-E pore could not be located. Siddiqi (1962) again mentioned three ventromedian cervical papillae and S-E pore not distinguishable; he remarked that the most posterior papilla may be the S-E pore but that it was not possible to ascertain its true nature. Decraemer (1989b) observed only two ventromedian cervical papillae in *P. mirzai*, the "third one" appeared to be the S-E pore. Edward & Misra (1970) described three ventromedian cervical papillae and the excretory pore at the level of the posteriormost papilla. No data are available on lateral cervical pores.

Lateral body pores in females showed some intraspecific variability: usually two pores on each side (one prevulvar, one postvulvar) but also 1-3 pores on each side have been observed (rarely located advulvar; if so, than prevulvar) (Decraemer, 1989b).

Spermathecae were described only by Edward & Misra (1970) but were not distinct in the illustration given. Other descriptions of the species do not mention the presence of spermathecae nor sperm cells in the genital branches. From my own observations, sperm cells are rounded, medium-sized, with a round nucleus. They were concentrated in each branch near the vagina region and there were no marked spermathecae.

Edward & Misra described males of *P. musambi* (= *P. mirzai*) without a bursa, with only one pair of caudal postcloacal papillae and only two ventromedian precloacal supplements. *P. musambi* was synonymized with *P. mirzai* by Siddiqi (1974) without comment.

Males are nearly as abundant as females (Siddiqi, 1960).

Table 4.7.

	P. mirzai	*P. paramirzai*	*P. lobatus*	*P. teres*
L	♂: 450-630 µm ♀: 420-640 µm	♂: 590-710 µm ♀: 580-750 µm	♂: 550-1095 µm ♀: 535-1051 µm	♂: 537-1029 µm ♀: 616-1325 µm
onchiostyle	♂: 29-36 µm ♀: 29-37 µm	♂: 40-44 µm ♀: 41-44.5 µm	♂: 42-56 µm ♀: 44.5-58 µm	♂: 41-64 µm ♀: 41-61 µm
spicules	30-34 µm	33-38 µm	46-70 µm	36-57.5 µm
number of SP	3	3	2	3
location SP in spicule range	SP1	SP1	SP1 and SP2	SP1 and SP2
number of CP	2	2	0	0
n° postcloacal papillae	2 pairs	1 pair	1 pair	1 pair
sperm shape	cell medium sized, rounded, nucleus rounded	cell large, nucleus sausage-shaped	cell small, nucleus oval, may be fibrillar	cell small, nucleus oval, fibrillar

Distribution and host

P. mirzai was first reported from Aligarh (U.P.) , India (Siddiqi, 1960). Other records are from Madras State, South India, Allahabad (U.P.) India, Western Japan (Edward & Misra, 1970; Yokoo, 1964) and Florida (Mead, 1971, 1974, 1987a). It occurred in soil around the roots of cabbage (*Brassica oleracea*), sugar cane (*Saccharum afficinarum*) and citrus (*Citrus sinensis*) in India, green grass in Japan and (slash pine) *Pinus elliotti* in Florida.

Paratrichodorus nanus (Allen, 1957) Siddiqi, 1974
Figs 3.12.A, B; 3.14.CC; 3.16.B; 3.17.M; 4.30.D-E

syn. *Trichodorus nanus* Allen, 1957
syn. *Paratrichodorus (Nanidorus) nanus* (Allen, 1957) Siddiqi, 1974
syn. *Nanidorus nanus* (Allen, 1957) Siddiqi, 1974

Measurements

Holotype female: L= 520 µm, a= 16, b= 5.4, V= 53

Females: L= 430-600 µm (520 µm*; 540 µm; 455 µm*; 480 µm*), a= 12.8-23, b= 5.1-7.8, V= 49-56% (51.5%*; 54%; 54%; 54%), onchiostyle= 21-26.5 µm (22.5 µm*; 22 µm; 25.3 µm; 26 µm), anterior end to EP= 79-148 µm

Males: L= 430-550 µm (-; 510 µm; 430 µm;-), a= 17-25, b= 5.9-7.3, onchiostyle= 21-26 µm (22 µm; 26 µm), spicule length= 42-50 µm (43 µm; 50 µm) (from Allen, 1957; Siddiqi, 1963; Baujard, 1983: 2 populations of ♀, 1 of ♂).

Emended diagnosis

P. nanus is characterized by the posterior position of the S-E pore (clearly posterior to the pharyngo-intestinal junction, along the anterior part of the intestine), small onchiostyle (21-26.5 µm) with inner onchium, pharyngeal bulb offset and small sperm cells, oval to rounded. Furthermore, females are distinguished by the shape of the vulva (a transverse slit), the shape of the vaginal sclerotizations in lateral view (rounded triangular pieces, close to one another), sperm cells dispersed throughout the uteri or at the end of the branches of the genital tract, the shape of the vagina (about rectangular), two lateral body pores on each side. It is distinguished in males, by the length (more than 2.5 times the anal body diameter) and shape of the spicules (almost straight, shaft slender, striated), a single ventromedian precloacal supplement close to the cloacal opening, the absence of ventromedian cervical papillae and well developed caudal alae.

Relationships

P. nanus most closely resembles *P. minor* in the following characters: males: rare, with a single ventromedian precloacal supplement, slender spicules (more than 2.5 times the anal body width), absence of ventromedian cervical papillae, a relatively long tail; females: vulva a transverse slit, vagina about rectangular and small sperm cells with oval nucleus (in *P. minor* sperm may also appear fibrillar).

Further, *P. nanus* has a very short onchiostyle (21-26.5 µm) similar to the minimum values in the range of *P. minor* (26-47 µm). *P. nanus* together with *P. minor* and *P. porosus* are the only species of the genus where an inner onchium has been observed in adults.

P. nanus differs from *P. minor* by the far posterior position of the S-E pore (along the anterior part of the intestine *vs* at the level of the pharyngeal bulb or pharyngo-intestinal junction in *P. minor*), the shape of the vaginal sclerotizations (rounded triangular *vs* thin rods parallel to the body axis), the presence of two lateral body pores on each side in females (absent in *P. minor*), the offset pharyngeal bulb (more variable: overlapping pharyngeal glands, offset bulb or pharyngeal and intestinal overlap in *P. minor*) and the shorter spicules in males (42-50µm *vs* 48-73µm).

Remarks

Specimens from Senegal differ from those from The Netherlands (Allen, 1957) and Tunisia (Siddiqi, 1963) in having a slightly longer onchiostyle (25-26.5 µm *vs* 21-23 µm) and slightly more anterior position of the S-E pore (79-100 µm *vs* 120-148 µm) (Baujard, 1983). Siddiqi (1963) described females with thin-walled oval spermathecae containing much sperm. Sturhan (1989a) observed sperm in the uteri although males are rare, but rarely found sperm in the uteri and at the distal ends of the branches of the genital tract at the same time. He considered *P. nanus* to reproduce by syngonic hermaphroditism.

Distribution and host

P. nanus was originally described from The Netherlands and subsequently reported from Tunisia, Senegal, and from Europe (Belgium, France, Germany, Great Britain, Italy and Portugal). It was found in soil around roots of rye, strawberry, vine, peanut and soja (Allen, 1957; Siddiqi, 1963; Baujard, 1983). It has also been recorded in association with *Rumorha adiantiformis* in Florida (Anonymous, 1989a).

Paratrichodorus orrae Decraemer & Reay, 1991

Figs 3.13.J; 3.14.O; 3.16.C; 4.30.F-H

Measurements

Holotype male: L= 655 µm, a= 18.2, b= 4.7, T= 55.6, onchiostyle= 44 µm, spicule length= 44 µm, gubernaculum= 12 μm, anterior end to EP= 95 μm, distance CP1-EP= 32 μm

Males: L= 470-680 μm (596 μm ± 74.3), a= 13.2-18.4, b= 4.6-5.5, T= 50.7-62.4, onchiostyle= 35-47 μm (43 μm ± 4.5), spicule length= 35-47 μm (42 μm ± 4.7), gubernaculum= 10-13 μm, anterior end to EP= 68-103 μm, distance CP1-EP= 16-39 μm (31.6 μm ± 7.9)

Females: L= 455-790 μm (645 μm ± 94.8), a= 11.9-18.6, b= 4.4-6.1, V= 53.5-59 %(56.0 ± 1.9), onchiostyle= 38-47 μm (43.7 μm ± 3.0), anterior end to EP= 78-109 μm (from Decraemer & Reay, 1991).

Diagnosis

P. orrae is characterized by the onchiostyle length (35-47 μm) and by the large sperm cells with long sausage-shaped nucleus. In the male by the presence of a single ventromedian cervical papilla (posterior to the onchiostyle base and well anterior to the S-E pore, located opposite anterior part of pharyngeal bulb), by the shape of the spicules (distal third slightly curved ventrally, manubrium subcylindrical, narrowing at junction with calamus which gradually widens before tapering distally to a narrow rounded tip; corpus striated except at both extremities; inner cytoplasmic core with marked inclusions), three ventromedian precloacal supplements (posterior two supplements dispersed within the region of the retracted spicules) and by weakly developed caudal alae. In the female by the shape of the vagina (short, about trapezoid, with or without slight indentation), the inconspicuous, well separated oval vaginal sclerotized pieces in lateral view, the dispersion of sperm throughout the uteri, a small longitudinal slit-like vulva and a pair of postadvulvar lateral body pores. Pharyngeal bulb offset or intestine with anteriodorsal overlap in male and female.

Relationships

P. orrae closely resembles *P. grandis* and *P. sacchari* in location of the single medioventral cervical papilla with regard to the onchiostyle base and the S-E pore, in number and position of the three precloacal supplements, in the shape of the copulatory apparatus, the weak development of the caudal alae and the presence of a protruding anterior cloacal lip. It also resemble *P. grandis* in sperm structure and spicule length (37-47 μm *vs* 39-48 μm in *P. grandis*) and agrees with *P. sacchari* in morphometric data such as body length and onchiostyle. *P. orrae* differs from both species in the female longitudinal slit-like vulva instead of a pore-like vulva. It can be distinguished from *P. grandis* males and females by the shorter onchiostyle (64-72 μm in *P. grandis*) and the smaller body length (790-1040 μm in *P. grandis* males). *P. orrae* males have shorter spicules than *P. sacchari* (46-52 μm). *P. orrae* also resmbles *P. queenslandensis* (see further).

P. orrae shows some similarity to *Trichodorus* species because of the presence of weakly developed caudal alae and a ventrally curved posterior end in males (fixed specimens), and the presence of a pair of postadvulvar lateral body pores. However, it represents a true *Paratrichodorus* species due to a much swollen cuticle in fixed specimens, weakly developed copulatory muscles and capsule of suspensor muscles, spicule shape and a single ventromedian cervical papilla in males; and by a short vagina, inconspicuous vaginal sclerotizations and sperm dispersed throughout the uteri in females.

Remarks

Among the ten male and nine female specimens found, two males and one female were obviously smaller than the other specimens, possessing also a shorter onchiostyle: 35-36 µm (♂), 38 µm (♀) against 42-47 µm (♂) and 40-47 µm (♀); shorter spicules: 35-36 µm against 40-47 µm; the medioventral cervical pore nearer to the S-E pore: at 16-23 µm distance against 32-36 µm and a slightly more anterior position of the S-E pore in male.

Distribution and host

P. orrae is known only from its type locality, the southern bank of the King Edward River, about 1 km upstream from the Mitchell Plateau Road crossing, in the north-west of Western Australia. It was found on sloping ground, slightly wooded with *Eucalyptus latifolia* and *Grevillea pyramidalis* as dominant plant species; other vegetation included *Bossiaea bossiaeoides*, *Melaleuca* sp. (honey myrtle) and grasses.

Paratrichodorus pachydermus (Seinhorst, 1954) Siddiqi, 1974

Figs 3.12.R-T; 3.14.N; 3.16.D; 3.17.C, c; 4.31.A-D, 6.2.

syn. *Trichodorus pachydermus* Seinhorst, 1954

syn. *Paratrichodorus (Atlantadorus) pachydermus* (Seinhorst, 1954) Siddiqi, 1974

syn. *Atlantadorus pachydermus* (Seinhorst, 1954) Siddiqi, 1974

See chapter on economically important species

Paratrichodorus paramirzai Siddiqi, 1991

Figs 3.13.D; 3.14.M; 3.16.E

Measurements

Holotype male: L= 630 µm, a= 28, b= 5.7, T= 60%, onchiostyle= 41 µm, spicule length= 34 µm, gubernaculum= 11 µm

Males: L= 590-710 µm (660 µm ± 34), a= 21-29, b= 4.9-6.6, T= 48-65%, onchiostyle= 40-44 µm (41.8 µm ± 1.16), spicule length= 33-38 µm (35.6 µm ± 1.15), gubernaculum= 10-12.5 µm, anterior end to CP1= 39-49 µm, anterior end to CP2= 57-80 µm, distance CP2-EP= 3-10 µm

Females: L= 580-750 µm (680 µm ± 38), a= 22-30, b= 4.9-6.9, V= 51-57% (53.7% ± 2.45), onchiostyle= 41.0-44.5 µm (42.5 µm ± 0.8), anterior end to EP= 60-70 µm (from Siddiqi, 1991).

Emended diagnosis

P. paramirzai males are characterized by the number and position of the ventromedian cervical papillae (two papillae: the anterior one near base retracted onchiostyle; both papillae anterior to the S-E pore), a pair of lateral cervical pores at the level of the anterior ventromedian cervical papilla, length (33-38 µm) and shape of spicules (almost straight, cephalated, shaft striated), anterior lip of cloacal aperture with two elongate lobes, sperm with large sausage-shaped nucleus, three ventromedian precloacal supplements with the posterior one near the cloacal opening, the second one opposite the head of retracted spicules and caudal alae narrow. Females, by the vulva shape (a longitudinal slit), the shape of the vaginal sclerotizations in lateral view (inconspicuous, dot-like, well separated), sperm scattered throughout the uteri, shape of vagina (about quadrangular), usually one postvulvar lateral body pores on each side. S-E pore located opposite the nerve ring or just posterior to it and the pharyngeal bulb usually offset or slightly overlapping the intestine in male and female.

Relationships

P. paramirzai most closely resembles *P. mirzai* which was also found at the type locality. *P. paramirzai* is longer (580-750 µm *vs* 420-640 µm) and more slender (a= 21-30 *vs* 13-20.7) and has a longer onchiostyle (40-44.5 µm *vs* 29-37 µm). According to Siddiqi (1991) *P. paraporosus* is a somewhat similar species from around rice roots in Bihar State, India, but differs from *P. paramirzai* in having males with 28-36 (32) µm long spicules and four supplements of which 3 are within the region of the retracted spicules and females have 4-5 ventromedian body pores located posterior to the vulva.

Remarks

Lateral body pores variable: most females have a pair of lateral pores (not opposite each other) at 57-80 µm posterior to the vulva; four out of seventeen females had two pairs of postvulvar lateral pores at 55-85 µm and 151-158 µm from the vulva; one female had one of the posterior pairs located about two body widths from the tail tip; three females had one pair of lateral pores 56-78 µm anterior to the vulva, beside the postvulval pair (Siddiqi, 1991).

Distribution and host

P. paramirzai was described from the type locality Andhra Pradesh, India where it was found in soil around the roots of groundnut and around rhizomes of turmeric (Siddiqi, 1991).

Paratrichodorus paraporosus Khan, Jairajpuri & Ahmad, 1989

Figs 3.13.G; 3.14.K; 3.16.F

Measurements

Holotype females: L= 630 µm, a= 22, b= 5.2, V= 60%, onchiostyle= 39 µm

Females: L= 620-750 µm (700 µm), a= 22-27, b= 4.6-5.8, V= 57-59% (57%), onchiostyle= 39-42 µm (40 µm), anterior end to EP= 62-81 µm

Males: L= 570-690 µm (630 µm), a= 22-28, b= 4.6-5.8, T= 44-69%, onchiostyle= 39-41 µm (40 µm), spicule length= 28-36 µm (32 µm), gubernaculum= 6-10 µm, anterior end to EP= 71-85 µm, anterior end to CP1= 39-42 µm (from Khan, Jairajpuri & Ahmad, 1989).

Emended diagnosis

P. paraporosus females are characterized, by the shape of the vulva (a longitudinal slit), shape of the vaginal sclerotizations in lateral view (small, inconspicuous, well separated pieces), two spermathecae filled with sperm, shape of vagina (wide, rounded) and four to five ventromedian body pores posterior to the vulva. Males by the single ventromedian cervical papilla at the level of the distal end of the onchiostyle, anterior to the S-E pore which is located between the nerve ring and the anterior end of the pharyngeal bulb, four ventromedian precloacal supplements (posterior two clearly within the region of retracted spicules, third one opposite the head of retracted spicules, anterior one at three spicular lengths from the cloacal aperture), and by the length (28-36 µm) and shape of the spicules (about straight, cephalated, shaft striated). The pharyngeal glands slightly overlap the intestine in males and females.

Relationships

P. paraporosus most closely resembles *P. porosus*, mainly by possessing ventromedian body pores in the female instead of the common lateral body pores as in the other species of the genus. It has a similar body length but differs by having a shorter onchiostyle (39-42µm *vs* 39-58µm), apparently small sperm cells *vs* large sperm cells with large sausage-shaped nucleus. *P. paraporosus* males differ by the number of ventromedian precloacal supplements (four *vs* two), a somewhat longer tail with digitate tip (tail very short and rounded in *P. porosus*), shorter gubernaculum (6-10µm *vs* 12-13µm) and males common *vs* rare in *P. porosus*. Females can be distinguished from those of *P. porosus* by the vulva shape (a longitudinal slit *vs* pore-like), presence of spermathecae *vs* absent in *P. porosus*, position of ventromedian body pores (only postvulvar *vs* pre- and postvulvar in *P. porosus*).

Remarks

Sperm type not mentioned in original description.

Vermeulen & Heyns (1983) described four female specimens, closely resembling *P. porosus*, but with a spermatheca present in one of the specimens and with a quite large variability in number and position of ventromedian vulval body pores (1-4 prevulvar, and postvulvar; e.g., one female with one prevulvar and four postvulvar pores).

Distribution and host

P. paraporosus was described from the type locality in the district Madhubani, Bihar state, India where it was found in soil around roots of paddy (*Oryza sativa*) (Khan *et al.*, 1989).

Paratrichodorus porosus (Allen, 1957) Siddiqi, 1974

Figs 3.12.U; 3.14.J; 3.16.G-H; 3.17.F; 4.31.E-H, 6.3.

syn. *Trichodorus porosus* Allen, 1957

syn. *Paratrichodorus (Atlantadorus) porosus* (Allen, 1957) Siddiqi, 1974

syn. *Atlantadorus porosus* (Allen, 1957) Siddiqi, 1974
syn. *Trichodorus bucrius* Lordello & Zamith, 1958

See chapter on economically important species

Paratrichodorus psidii Nasira & Maqbool, 1994
Figs 3.12.F; 3.14.E; 3.16.I

Measurements

Holotype female: L= 440 µm, a= 13.8, b= 4.4, V= 55%, onchiostyle= 32 µm, distance anterior end to EP= 54.2 µm.

Females: L= 400-520 µm (460 µm ± 40), a= 13-18, b= 3.5-5.3, V= 51-56% (53.9 ± 1.49), onchiostyle= 30.4-33.6 µm (31.9 µm ± 1.34), distance anterior end to EP= 49.6-60 µm.

Males: L= 430-510 µm (480 µm ± 10), a= 13.8-18.0, b= 3.4-5.1, T= 52-58%, onchiostyle= 30-32 µm (30.6 ± 0.89), spicule length= 30-31 µm (30.6 ± 0.54), gubernaculum= 7-8 µm, distance anterior end to CP1= 30.4-35.6 µm, distance anterior end to EP= 52-59 µm (from Nasira & Maqbool, 1994).

Emended diagnosis

P. psidii females are characterized by the vulva shape (pore-like), shape of vaginal sclerotizations (dot-like, close), shape of vagina (rounded), number of lateral body pores (two posterior to the vulva) and medium-sized sperm with rounded nucleus, concentrated near the oviduct. Males by having two ventromedian cervical papillae anterior to the S-E pore and with the anterior one near the level of the onchiostyle base, lateral cervical pores absent, the number and position of the ventromedian precloacal supplements (three supplements with the posterior supplement near the cloacal opening, the middle one at the level of the spicule manubrium and the anterior supplement clearly anterior to the retracted spicules), the shape of the spicules (almost straight with slight ventral curvature in distal third, no marked manubrium, mid-shaft with fine transverse striation) and tail with one pair of postcloacal subventral papillae. Males and females with anterior dorsal overlap of the pharynx by the intestine and S-E pore opposite posterior part of isthmus.

Relationships

P. psidii males are similar to those of *P. faisalabadensis* and *P. mirzai* by having medium-sized cells with rounded nucleus, two ventromedian cervical papillae, three ventromedian precloacal supplements, short spicules and short onchiostyle. They differ from *P. mirzai* in males by the position of the second precloacal supplement (at the level of the head of the retracted spicules *vs* anterior to the spicules in *P. mirzai*). *P. psidii* females are similar to those of *P. faisalabadensis* by the pore-like vulva, but differ by the shape of the vagina (rounded to oval *vs* trapezoid) and by the shape of the vaginal sclerotizations in lateral view (dot-like, close *vs* short rods, well separated in *P. faisalabadensis*). Females of *P. psidii* also differ from *P. mirzai* by the shape of the vulva (pore-like *vs* a longitudinal slit).

Remarks

Females have been described with "spermatheca present, sperm scattered throughout the length of the uteri, but occasionally they appear accumulated in an elongated oval spermatheca distally" (Nasira & Maqbool, 1994). The original drawings show the reproductive system with an accumulation of sperm near the oviduct in both branches.

Distribution and host

P. psidii was described from the type locality Faisalabad, Pakistan where it was found in soil around the roots of guave (*Psidium guajava*) and around the roots of citrus (*Citrus* sp.) (Nasira & Maqbool, 1994).

Paratrichodorus queenslandensis Decraemer & Reay, 1991

Figs 3.13.H; 3.14.P; 3.16.L; 4.32.A-C

Measurements

Holotype male: L= 595 µm, a= 18.6, b= 4.3, T= 63.0, onchiostyle= 52 µm, spicule length= 38 µm, gubernaculum= 9.5 µm, anterior end to EP= 75 µm, distance CP1-EP= 4 µm

Males: L= 440-705 µm (547 µm ± 55.3; 500 µm ± 29.8; 610 µm ± 67.9; 515 µm), a= 13.4-23.0, b= 3.3- 4.8, T= 46.2- 73.7, onchiostyle= 49- 60 µm (52.5 µm ± 1.8; 53 µm ± 1.9; 58.5 µm ± 1.2; 51 µm), spicule length= 36- 43 µm (39 µm ± 1.7; 39 µm ± 1.8; 39.5 µm ± 2.3; 38 µm), gubernaculum= 9.5-12 µm, anterior end to EP= 62-89 µm, distance CP1-EP= 2-23 µm (7 µm ± 7.1; 3.5 µm ± 1.8; 3.4 µm ± 0.5; 2 µm)

Females: L= 460-655 µm (510 µm ± 60.8; 505 µm ± 19.4; 627 µm*), a= 12.4-18.3, b= 3.3-4.7, V= 51.5- 60% (58.1 % ± 2.2; 57.5% ± 3.7; 60.0%*), onchiostyle= 46-55 µm (52.6 µm ± 3.0; 51.3 µm ± 1.0; 52µm*), anterior end to EP= 64-94 µm (from Decraemer & Reay, 1991).

Diagnosis

P. queenslandensis male is characterized by one ventromedian cervical papilla just anterior to the S-E pore, by large sperm with sausage-shaped nucleus, the spicule length (36-43 µm) and spicule shape (straight to slightly bent, proximally with widened knob-like manubrium, followed by a slender, striated calamus), three precloacal supplements (posteriormost supplement within the region of the posterior third of retracted spicules, the second supplement opposite the anterior end of the spicules or just anterior to them) and weakly developed caudal alae. The female by the longitudinal slit-like vulva, the shape of the vagina (short, wide trapezoid with discernable vulvar constrictor muscles), the minute oval vaginal sclerotizations (2 µm apart), the absence of spermathecae and two postvulvar lateral body pores (one in advulvar position). S-E pore well anterior to pharyngeal bulb which is usually offset or with a minute anteriodorsal overlap of the intestine.

Relationships

P. queenslandensis shows a close similarity to *P. orrae* but differs from it: in males - in slightly shorter spicules with a different shape, in a straight posterior body end in fixed specimens (curved in *P. orrae*), in the position of the single ventromedian cervical papilla in relation to the S-E pore (mean distance 7 µm against 31.6 µm in *P. orrae*); in both sexes - in the longer onchiostyle (46-52 µm against 35-47 µm in *P. orrae*), and in females - in the presence of two postvulvar lateral body pores on each side of the body instead of a single postadvulvar lateral pore as in *P. orrae* (Decraemer & Reay, 1991).

Remarks

In two specimens from Elliott River Forest only one pair of lateral body pores was observed.

Distribution and host

P. queenslandensis has a distribution limited to Queensland, Australia, with Eungella National Park as type locality. It was found in soils collected from tropical rainforest with the plant species: *Acmena smithii*, *Lophostemon suaveolens*, *Archontophoenix cunninghamiana*, *Livistona drudei*, *Argrodendron actinophylla*, *Alsophila australis*, *Toona australis* and from Wallum association (woodland) with dominant plant species including *Melaleuca quinquenerva*, *Banksia oblongifolia*, *B. aemula* and occasionally *Eucalyptus umbra.*

Paratrichodorus renifer Siddiqi, 1974
Figs 3.12.E; 3.17.T 4.32.D-F

syn. *Paratrichodorus (Nanidorus) renifer* Siddiqi, 1974
syn. *Nanidorus renifer* Siddiqi, 1974

Measurements

Holotype female: L= 480 μm, a= 22, b= 5.1, V= 56%, onchiostyle= 31.5 μm
Females: L= 420-560 μm (500 μm; 535 μm*), a= 19-25, b= 4.6-6.9, V= 54-59 % (56.5%; 55.5 %*), onchiostyle= 29-34 μm (32.5 μm; 31 μm*) anterior end to EP= 83-90 μm (from Siddiqi, 1974; Jana & Baqri, 1984).

Emended diagnosis

P. renifer females are characterized, by the vulva shape (a transverse slit), shape of the vaginal sclerotizations in lateral view (kidney-shaped pieces, close), a few thread-like bodies (?sperm) in the uteri, the shape of the vagina (about quadrangular to rectangular with slight midvagina indentation), lateral body pores and caudal pores absent, S-E pore opposite base pharynx and pharyngeal bulb offset, rarely with ventrally overlapping glands. Male unknown.

Relationships

P. renifer is most similar to *P. acutus*, *P. nanus*, *P. minor* and *P. anthurii* from which it may be distinguished as described in table 4.8. Furthermore, *P. renifer* also differs from *P. acutus* in tail shape and from all four species by the shape of the kidney-shaped vaginal sclerotization.

Remarks

Jana & Baqri (1984) described the pharyngeal base as "somewhat flattened", their illustration showing a minute anterior overlap by the intestine.Nasira & Maqbool (1994) described a clear ventral overlap of the pharyngeal glands. Sturhan (1989a) confirmed the absence of distinct sperm in the genital system.

Table 4.8.

	P. renifer	*P. acutus*	*P. anthurii*	*P. minor*	*P. nanus*
L	420-560	338-610µm	380-530µm	440-1530µm	430-600µm
onchiostyle	29-34µm	22-27µm	28-31µm	26-47µm	21-26.5µm
males	absent	absent	absent	rare	rare
shape vulva	transverse slit	transverse slit	pore	transverse slit	transverse slit
shape sperm	thread-like structures	thread-like structures	small oval	thread-like, or small oval	small oval
location sperm	uteri	uteri	spermatheca	uteri	uteri
pharynx end	usually offset	usually offset	offset	variable	usually offset
n° lateral body pores in ♀	0	0	0	0	2 on each side

Distribution and host

P. renifer was described from Limbe, Malawi and has also been recorded from Upper Swan and Middle Swan, Western Australia, and Lebon Tea Estate, Lebong, district Darjeeling, India. It was found in soil around the roots of potato (*Solanum tuberosum*), grapevine and tea (*Thea sinensis*) (Siddiqi, 1974; Jana & Baqri, 1984). It has also been recorded from Florida, in association with *Ilex glabra* (Anonymous, 1986) and from Pakistan from soil around the roots of pine trees (*Pinus* sp.) and the roots of grass (*Cynodon dactylon*) (Nasira & Maqbool, 1994). In 1990, it was recorded for the first time in the UK during a pre-export inspection of a consignment of pot-grown deciduous azaleas (*Rhododendron japonicum*) in Hampshire (Cotten & Hooper, 1991). Recently, *P. renifer* was recorded from China associated with imported azaleas from Europe (Ye, 1995).

Paratrichodorus rhodesiensis (Siddiqi & Brown, 1965) Siddiqi,1974
Figs 3.12.J-L; 3.14.BB; 3.16.K; 3.17.N, n; 4.32.G-H

syn. *Trichodorus rhodesiensis* Siddiqi & Brown, 1965

syn. *Paratrichodorus (Paratrichodorus) rhodesiensis* (Siddiqi & Brown, 1965) Siddiqi,

1974

Measurements

Holotype male: L= 690 µm, a= 25, b= 5.7, T= 68%, onchiostyle= 41 µm, spicule length= 43 µm, gubernaculum= 12 µm

Males: 360-690 µm (645 µm*; 360 µm), a= 16-26, b= 4.2-6.2, T= 56-69%, onchiostyle= 39-43 µm (40 µm*; 43 µm), spicule length= 42-44 µm (41 µm*), gubernaculum= 12-14 µm, anterior end to EP= 49.5 µm+

Females: L= 380-800 µm (760 µm*; 430 µm ± 55; 410 µm*; 410 µm ± 57), a= 14-26, b= 4.1-7.1, V= 50.7-59% (52.5%*; 55 % ± 1.1; 58.5%; 55% ± 1.4), onchiostyle= 37.5-44 µm (41 µm*; 40 µm ± 1.1; 42 µm*; 41 µm ± 1.7), anterior end to EP= 64-90 µm (from Siddiqi & Brown, 1965; Baujard, 1983: 3 populations of ♀, 1 of ♂).

Emended diagnosis

P. rhodesiensis males are characterized by the absence of ventromedian cervical papillae (one exception) and lateral cervical pores, the length (42-44 µm) and shape of the spicules (slightly cephalated, ventrally curved, shaft striated), the three ventromedian precloacal supplements (the posterior two within the region of the retracted spicules, the second supplement near the spicule head), one pair of postcloacal subventral papillae and short caudal alae (extending to a level between the posterior two precloacal supplements). Females by the rounded shape of the vagina, shape of the vaginal sclerotizations in lateral view (small, oval pieces, 1.5-2 µm apart), vulva shape (a longitudinal slit) and small sperm cells with oval nucleus dispersed throughout uteri sometimes with concentration near oviduct; sperm location at flexure of the genital branches common and lateral body pores absent. Furthermore, *P. rhodesiensis* is characterized by the slightly overlapping ventro-sublateral pharyngeal glands, the position of the S-E pore (at about the level of the beginning of the pharyngeal bulb).

Relationships

P. rhodesiensis most closely resembles *P. allius*, especially in females by the unusual sperm location at the flexure of the genital branches, body length (380-800µm *vs* 400-1044µm) and length of onchiostyle (37.5-44µm *vs* 35-53µm), vulva shape (a longitudinal slit), rounded to oval shape of vagina, sperm structure, absence of lateral body pores. Males by the absence of ventromedian cervical papilla. It differs from *P. allius* mainly in males having longer (42-44µm *vs* 30.5-30.9µm) spicules which are different in shape (slender, elongated *vs* stouter), number of ventromedian precloacal supplements (3 *vs* 2) and males are common in *P. rhodesiensis vs* rare in *P. allius.*

Remarks

Baujard (1983) reported specimens of *P. rhodesiensis* from East Africa as being much smaller (360-570 µm *vs* 600-800 µm) than type specimens from Southern Rhodesia, the only male (from Ivory Coast) had a ventromedian cervical papilla (= erroneously described as lateral, see Fig.1J) just posterior to the base of the onchiostyle, well anterior to the S-E pore, and all females had the uteri filled with sperm.

Lateral body pores were not observed in females (Siddiqi & Brown, 1965).

Distribution and host

P. rhodesiensis was described from Low Veldt, Southern Rhodesia, and subsequently from Casamance, Senegal, Ivory Coast and Cap Vert Isles. It has been found in association with sugar-cane (*Saccharum officinarum*) and in forests with *Tectona grandis* and *Ceiba pentandre* (from Siddiqi & Brown, 1965; Baujard, 1983).

Paratrichodorus sacchari Vermeulen & Heyns, 1983
Figs 3.12.V; 3.14.Q, 3.16.J; 4.27.G-I

syn. *Paratrichodorus (Atlantadorus) sacchari* Vermeulen & Heyns, 1983, n.syn.

Measurements

Holotype male: 694 µm, a= 17.5, b= 5.8, onchiostyle= 44 µm, spicule length= 52 µm, gubernaculum= 11.8 µm, anterior end to EP= 80 µm, distance CP1-EP= 8 µm

Paratype male: L= 666 µm, a= 18, b= 7.6, onchiostyle= 38 µm, spicule length= 46 µm, anterior end to EP= 73 µm, distance CP1-EP= 9 µm

Paratype females: L= 589-733 µm, a= 13.3-15.8, b= 4.9-6.1, V= 53%, onchiostyle= 46-48 µm, anterior end to EP= 78-80 µm (from Vermeulen & Heyns, 1983).

Emended diagnosis

P. sacchari males are characterized having a single ventromedian cervical papilla (just anterior to the S-E pore and opposite the isthmus), the length (46-52 µm) and shape of the spicules (nearly straight, cephalated, striated except at head and distal end), three ventromedian precloacal supplements (the posterior two opposite the posterior half of the retracted spicules), caudal alae shallow (reaching to the anterior half of retracted spicules), anterior cloacal lip with two finger-like projections and large sperm with sausage shaped nucleus. Females by the vulva shape (pore-like), shape of vaginal sclerotizations in lateral view (minute triangular pieces, their tips close to one another), spermathecae not always differentiated, the shape of the vagina (wide, rounded, with refractive folds) and by one pair of postvulvar lateral body pores.

Relationships

P. sacchari is similar to *P. catharinae* (see relationships *P. catharinae*).

Remarks

Overlapping at base pharynx variable: one female paratype with an anterior dorsal overlap by the intestine, the other female paratype with both an anterio-dorsal overlap by the intestine and a ventral overlap by the pharyngeal glands; both male type specimens with a ventral overlap of the intestine and a dorsal overlap of the pharynx (De Waele & Kilian, 1992).

Lateral cervical pores have been observed in a male paratype.

One female paratype with two spermathecae, the other paratype specimen with sperm in the uterus near the vagina and spermathecae not differentiated. Both female paratypes possess three medioventral body pores just posterior to the vulva; their inner canal is less marked than in lateral body pores.

Distribution and host

P. sacchari was described from four specimens from the type locality near Doringkop, Natal, South Africa where it was found in a soil sample from a sugar cane field (Vermeulen & Heyns, 1983).

Paratrichodorus teres (Hooper, 1962) Siddiqi, 1974

Figs 3.13.L, M; 3.14.R; 3.16.M; 3.17.Q, q; 4.33.A-F, 6.4.

syn. *Trichodorus teres* Hooper, 1962

syn. *Paratrichodorus (Paratrichodorus) teres* (Hooper, 1962) Siddiqi, 1974

syn. *Trichodorus flevensis* Kuiper & Loof, 1962

See chapter on economically important species

Paratrichodorus tunisiensis (Siddiqi, 1963) Siddiqi, 1974
Figs 3.13.B; 3.14.AA; 3.16.N; 3.17.P; 4.33.G-I

syn. *Trichodorus tunisiensis* Siddiqi, 1963
syn. *Paratrichodorus (Paratrichodorus) tunisiensis* (Siddiqi, 1963) Siddiqi, 1974

Measurements

Holotype female: L= 1060 µm, a= 26, b= 6.2, V= 50%, onchiostyle= 50 µm, anterior end to EP= 105 µm
Females: L= 707-1100 µm (970 µm; 777 µm), a= 18.7-29, b= 6.0-10.3, V= 50-58.5% (53%; 56%), onchiostyle= 48-51 µm (49 µm; 49 µm), anterior end to EP= 105-123 µm
Males: L= 714-900 µm (830 µm; 763 µm), a= 20.2-26, b= 5.8-10.8, T= 48-55%, onchiostyle= 47-52 µm (48.5 µm*; 49 µm), spicule length= 45-50 µm (49 µm; 48 µm), gubernaculum= 11-14 µm, distance CP1-EP= 10.5-16 µm (from Siddiqi, 1963; Roca & Lamberti, 1984).

Emended diagnosis

P. tunisiensis females are characterized by the shape of the vulva (a longitudinal slit), shape of the vaginal sclerotizations in lateral view (small, rod-like, oval, well separated, oblique or parallel to the vagina), sperm small oval, dispersed throughout uteri with higher concentrations near oviduct, the shape of the vagina (wide rectangular or high rectangular with slight indentation) and two postvulvar lateral body pores on each side. Males by the single ventromedian cervical papilla anterior to the S-E pore and at the level of, or posterior to the nerve ring, three ventromedian precloacal supplements (the posterior near the cloacal opening, the second about mid-point of retracted spicules or slightly more anterior), the spicule shape (slightly cephalated, shaft cylindrical tapering at distal end, spicules striated except at both ends) and one pair of postcloacal papillae. Furthermore, the species is distinguished by ventrally overlapping pharyngeal glands and the position of the S-E pore opposite the beginning of the pharyngeal bulb.

Relationships

P. tunisiensis females have the following characters in common with *P. allius*: vulva a longitudinal slit, small sperm cells, similar vaginal sclerotizations and vagina shape, and more or less comparable length of onchiostyle (resp. 48-51µm and 35-53µm). Females differ from those of *P. allius* by having two lateral body pores *vs* absent or one pore and males differ by the number of precloacal supplements (three *vs* two in *P. allius*), different shape of spicules (slender, only distally curved *vs* stout and about straight in *P. allius*) and males common *vs* rare in *P. allius*.

Remarks

Italian populations largely agree with the type population except for having a shorter body (707-876 µm *vs* 800-1100 µm), small bristles on the spicules (Roca & Lamberti, 1984) and two pairs of postcloacal papillae *vs* one in type population. This is the only record of spicules ornamented with bristles within the genus. The bristles (minute and close to one another) in the illustration are extended over a short part of the shaft in the posterior half of the spicules and have a different appearance compared with those observed in *Trichodorus* species (observations not checked).

Vaginal sclerotizations described as thin short rods parallel to the vagina in lateral view were observed as oblique in a paratype female. Siddiqi (1963) described sperm in the spermathecae. Small sperm with oval nucleus were observed dispersed throughout uteri with higher concentrations near the oviduct but without any marked spermathecae. Roca & Lamberti (1984) described two oval spermathecae without reference to the presence of sperm.

Distribution and host

P. tunisiensis was described from Teboulba, Tunisia and it is widespread in the central and southern regions of Italy. It was found in sandy soils around the roots of *Sesbania* sp., and in the rhizosphere of grapevine, olive, artichoke, oak, hazelnut, poplar, apple, peach, walnut, alder, willow, elm, clover and pasture (Siddiqi, 1963; Roca & Lamberti, 1984).

Paratrichodorus weischeri Sturhan, 1985

Figs 3.13.A; 3.14.C; 3.16.O-P; 3.17.B, b; 4.33.J-K

syn. *Paratrichodorus (Paratrichodorus) weischeri* Sturhan, 1985, n.syn.

Measurements

Holotype male: L= 730 µm, a= 19, b= 5.3, T= 63%, onchiostyle= 46 µm, spicule length= 35 µm, gubernaculum= 11 µm, anterior end to EP= 85 µm

Males: L= 560-800 µm (710 µm), a= 18-24, b= 4.7-6.1, T= 57-72%, onchiostyle= 43-48 µm (45 µm), spicule length= 34-37 µm (35 µm), gubernaculum= 9.5-11.5 µm, anterior end to EP= 71-86 µm, distance CP1-EP= 1-14 µm

Females: L= 590-790 µm (710 µm), a= 16-22, b= 4.7-6.2, V= 56-62 % (58.5 %), onchiostyle= 43-48 µm (45 µm), anterior end to EP= 70-85 µm (from Sturhan, 1985, data from fixed specimens).

Emended diagnosis

P. weischeri males are characterized by having a single ventromedian cervical papilla (anterior to the S-E pore and posterior to the onchiostyle, rarely at the level of the onchiostyle base), the position of the lateral cervical pore (on both sides located at the level of the onchiostyle base to about one body width posterior to it), the presence of an additional lateral body pore (about seven body widths anterior to the posterior body end), the number and location of ventromedian precloacal supplements (three supplements, the posterior one near the cloacal opening, the second at the level of the head of retracted spicules), the shape of the spicules (shaft proximally slightly narrower than extreme end, slightly less sclerotized and usually kinking), the caudal alae weak and anteriorly reaching to between the posterior precloacal supplements, anterior cloacal lip with two flaps and a pair of large subventral postcloacal papillae. Females by the shape of the vulva (a transverse slit), shape of the vaginal sclerotizations in lateral view (small oval pieces, slightly separated: 1.5-2 µm, parallel to the vaginal lumen), shape of the vagina (usually with clear mid-vagina indentation), sperm concentrated near oviduct but spermathecae not clearly marked, usually one lateral postvulvar body pore on each side and two pairs of caudal pores. Further more, *P. weischeri* is characterized by the intestine overlapping the pharynx dorsally, S-E pore opposite isthmus and large sperm cells with sausage-shaped nucleus.

Relationships

P. weischeri males are similar to those of *P. anemones* and *P. hispanus* in having a similar spicule shape , narrower kinking anterior part and large sperm cells with sausage shaped nucleus. They differ by having shorter spicules (34-37 µm *vs* 46-53 µm in *P. anemones* and 44-61 µm in *P. hispanus).* Furthermore, they could be considered similar to *P. atlanticus, P. mirzai, P. tunisiensis* (Italian population) and *P. pachydermus* (according to some authors) by possessing "two" pairs of postcloacal papillae. However, the anterior pair is considered as pores in *P. pachydermus* and *P. weischeri* (see Sturhan (1985).

P. weischeri females have vaginal sclerotized pieces parallel to the vagina, as in *P. paramirzai* and some specimens of *P. tunisiensis*, but differ from them e.g. by the shape of the vulva (a transverse slit *vs* a longitudinal slit).

Remarks

Sturhan (1985) compared the morphometric data of fixed and living male and female specimens in order to determine the influence of fixation. In non-fixed specimens, the body length was longer (840-1100 µm *vs* 560-800 µm), as well as the length of the pharynx, the reproductive tract(s) and the tail length in males. Furthermore, through shortening of the body, the S-E pore had a more anterior position and the distance between the ventromedian supplements became smaller. The body width, b-, V- and T-values and the length of onchiostyle and spicules were not or were only slightly influenced by fixation.

Male tails may show an additional pair of ventrosublateral pores near the cloacal aperture as in *P. pachydermus* (Sturhan, 1985).

Number of lateral body pores variable in females: 0-2 postvulvar lateral body pores, rarely 1-2 lateral body pores anterior to the vulva (Sturhan, 1985).

Some abnormalities were observed: males with four precloacal supplements, others with a single spicule or with one spicule less developed, one male without ventromedian cervical papillae and a female with many cuticle pores at the posterior end (Sturhan, 1985).

Distribution and host

P. weischeri was described from the type locality in Westfalen, Germany, where it was found in loamy sand soil sample, 100 m from the river Weser (Sturhan, 1985). Its distribution seems restricted to the borders of the river, where it was found in the rhizosphere of fruittrees, grass and riverbank vegetation (personal communication from Sturhan, 1994).

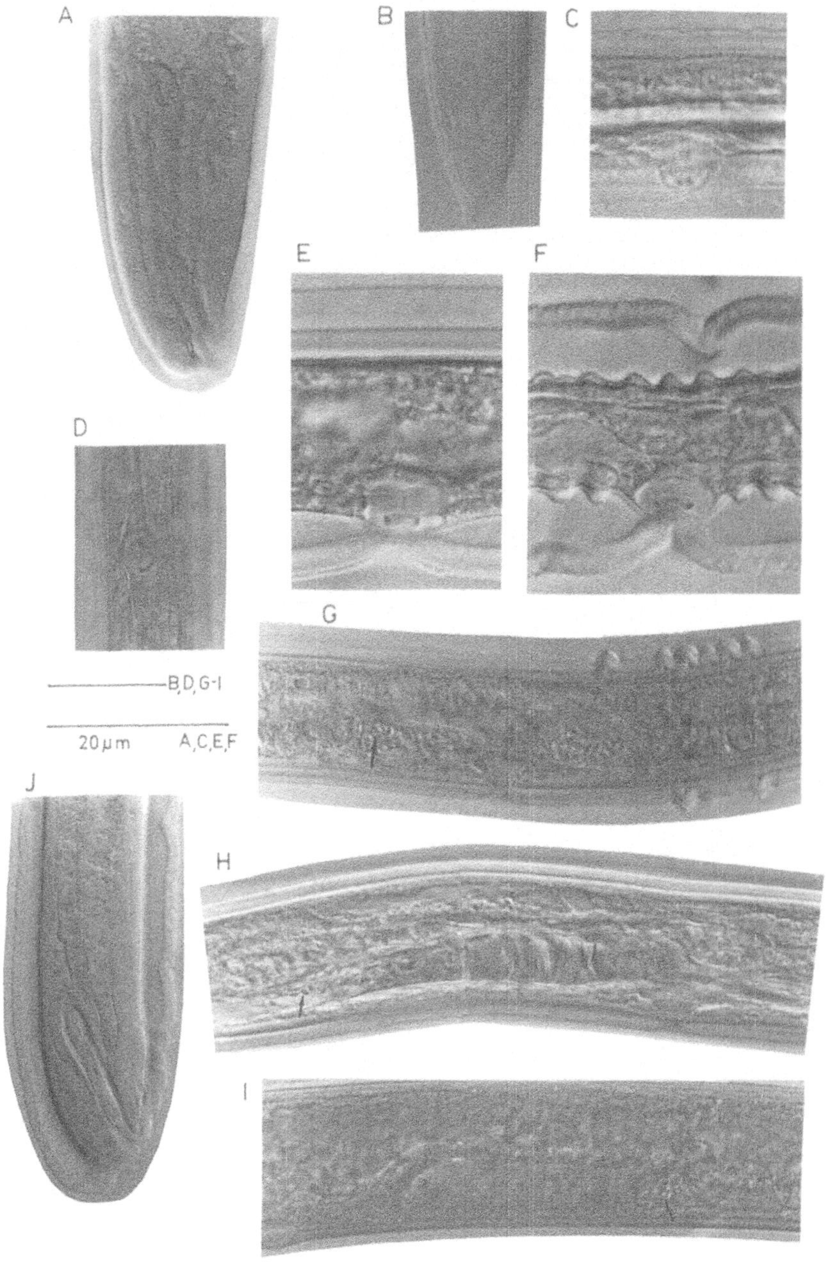

Fig. 4.24. *Paratrichodorus acaudatus*. Male: (A) copulatory apparatus and tail (paratype). *Paratrichodorus acutus*. Female: (B) tail region (paratype), (C) vaginal region. *Paratrichodorus allius*. Females: (D) ventral view vulva, (E-F) vaginal region, (G) vaginal region and uteri with sperm (arrow), (H-I) one branch of genital system with unusual sperm (arrow) location. Male: (J) copulatory apparatus and tail (paratype).

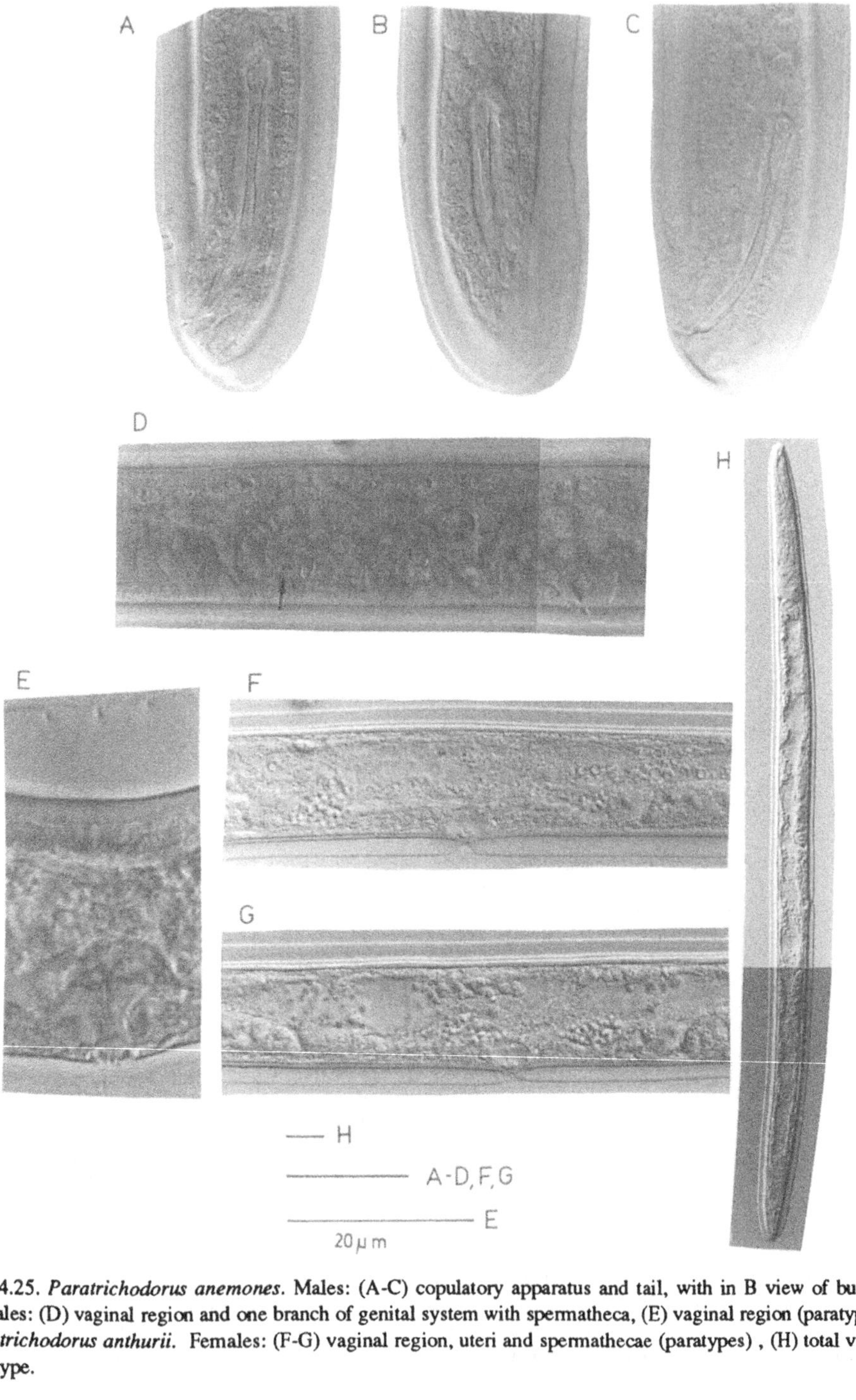

Fig. 4.25. ***Paratrichodorus anemones***. Males: (A-C) copulatory apparatus and tail, with in B view of bursa. Females: (D) vaginal region and one branch of genital system with spermatheca, (E) vaginal region (paratype). ***Paratrichodorus anthurii***. Females: (F-G) vaginal region, uteri and spermathecae (paratypes) , (H) total view paratype.

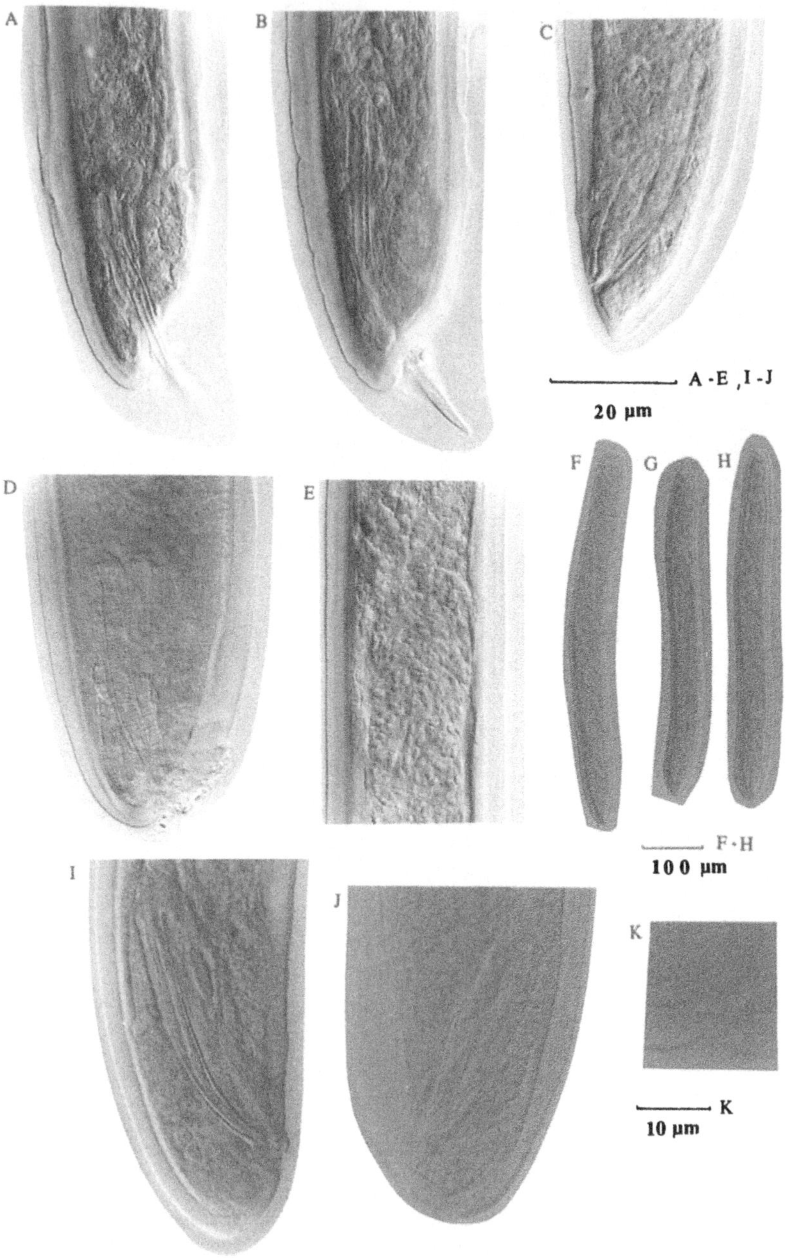

Fig. 4.26. *Paratrichodorus catharinae*. Males: (A-D) copulatory apparatus and tail with A, B resp. right and left spicule of a paratype, C (holotype), D (paratype), (E) sperm, (F-H) total view, paratypes. Females: (K) vaginal region, paratype. *P. grandis*. Males: (I-J) copulatory apparatus and tail, resp. with accent on bursa and on cloacal lip (paratypes).

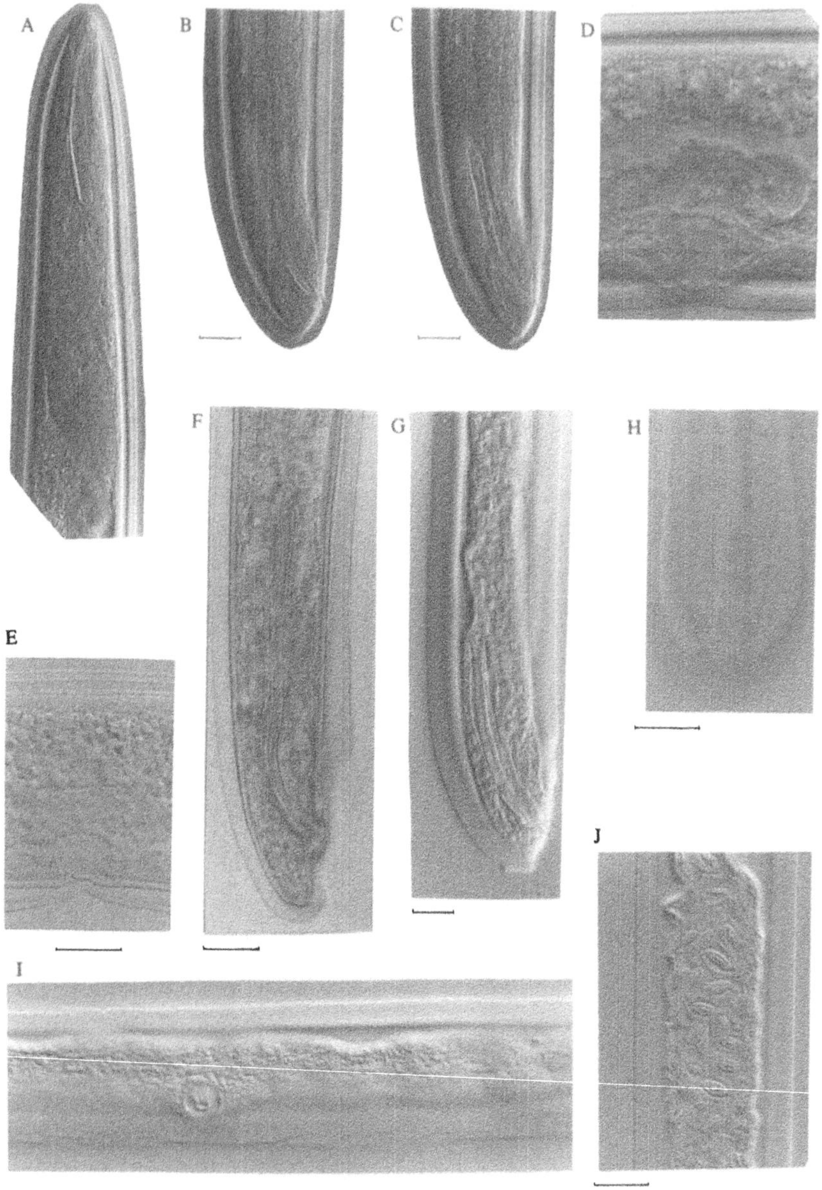

Fig. 4.27. *Paratrichodorus hispanus*. Males: (A) anterior body region (paratype), (B-C) copulatory apparatus and tail, resp. with accent on gubernaculum and spicule (paratype). Female: (D) vaginal region. *P. minor*. Female: (E) vaginal region. Male: (F) copulatory apparatus and tail. *P. sacchari*. Males: (G) copulatory apparatus and tail (holotype), (H) ventral view of tail with bilobate anal lip (arrow)(paratype), (J) sperm (paratype). Female: (I) vagina with refractive folds; arrow indicates a postvulvar lateral body pore (paratype). Scale 10 µm.

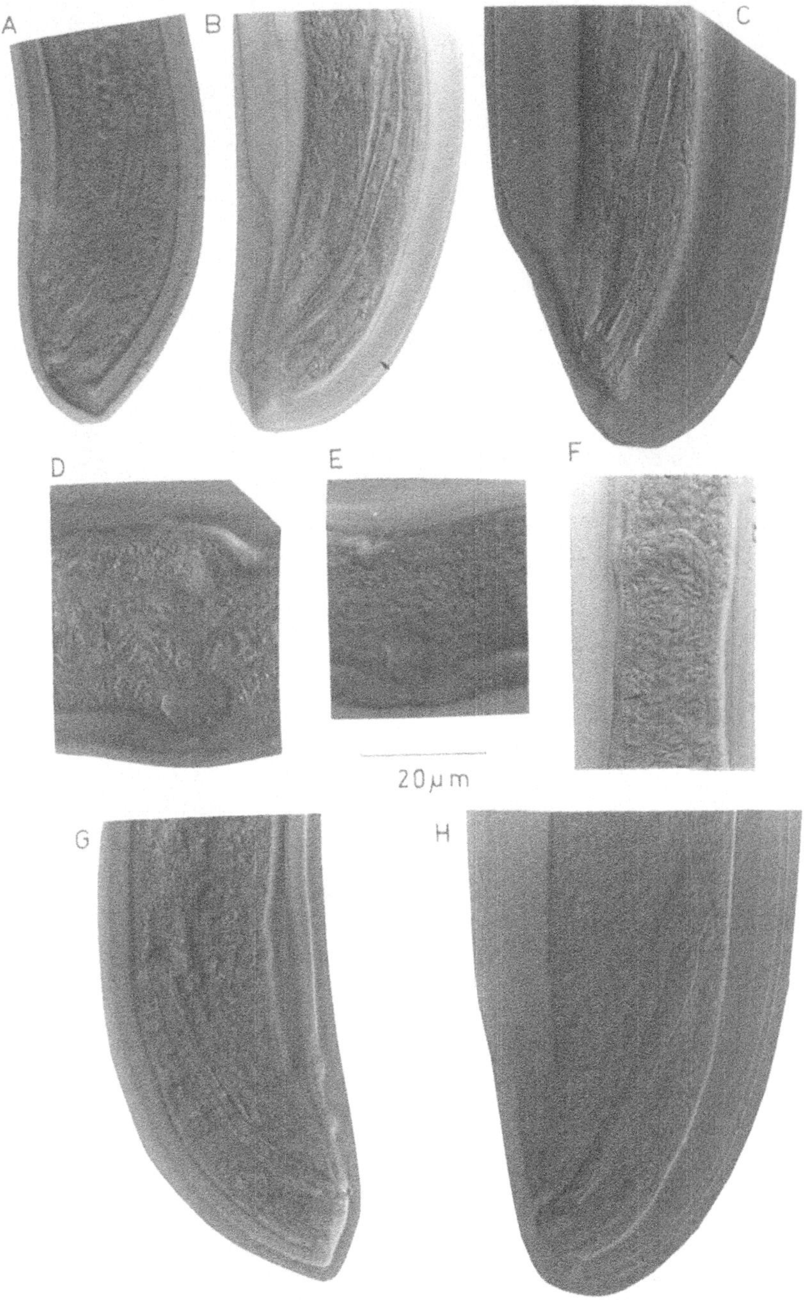

Fig. 4.28. *Paratrichodorus lobatus*. Males: (A-C, G-H) copulatory apparatus and tail, showing variability in swelling of the body cuticle (A-C Australian specimens, G-H specimens from South Africa), (F) part of testis with sperm (Australian specimen). Females: (D-E) vaginal region (Australian specimens).

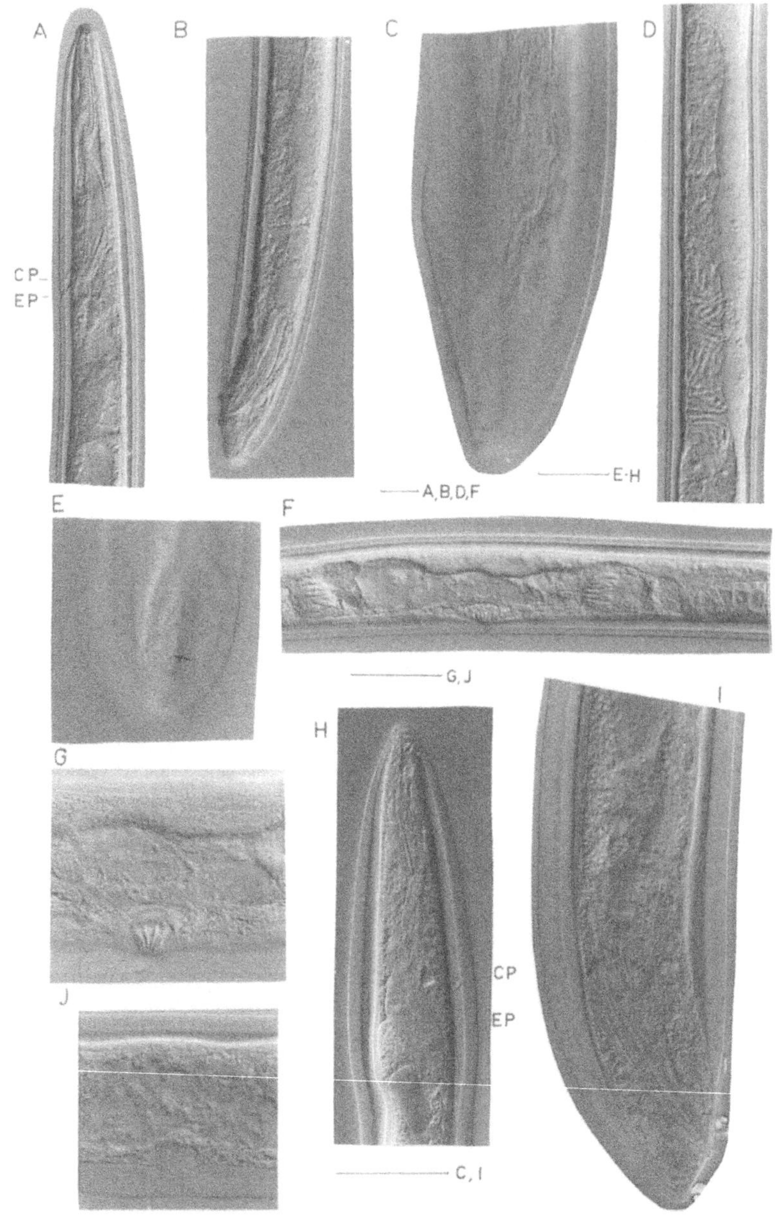

Fig. 4.29. *Paratrichodorus macrostylus*. Male paratypes: (A) anterior body region, (B) posterior body region, (C) tail region and caudal alae, (D) testis, (E) ventral view of cloacal region with bilobate anterior anal lip (arrow). Female paratype: (F) part of genital system, vagina region, uteri and spermathecae, (G) refractive folds of the vagina. *Paratrichodorus meyeri*. Males: (H) anterior body region (paratype), (I) posterior body region with copulatory apparatus (paratype). Female: (J) vaginal region (paratype) (scale= 20 µm).

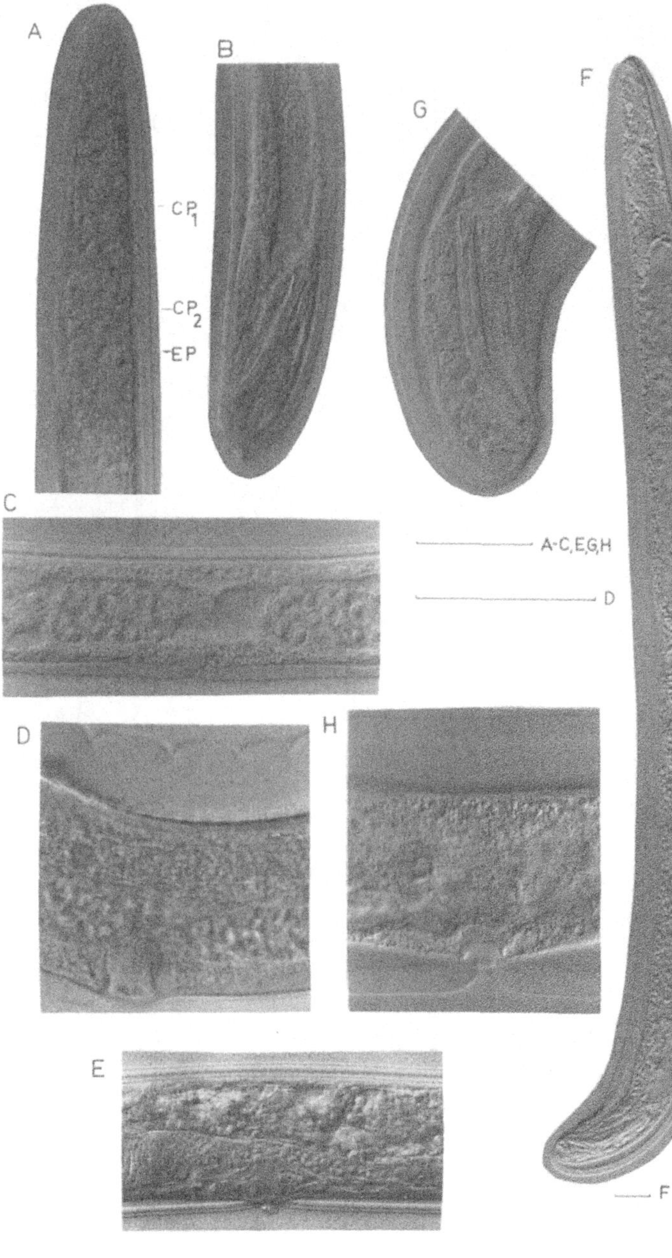

Fig. 4.30. *Paratrichodorus mirzai*. Males: (A) anterior body region, (B) copulatory apparatus and tail. Female: (C) vaginal region and part of uteri with sperm. *Paratrichodorus nanus*. Females: (D-E) vaginal region. *Paratrichodorus orrae*. Males: (F) total view (paratype), (G) copulatory apparatus and tail (paratype). Female: (H) vaginal region (paratype) (scale= 20 µm).

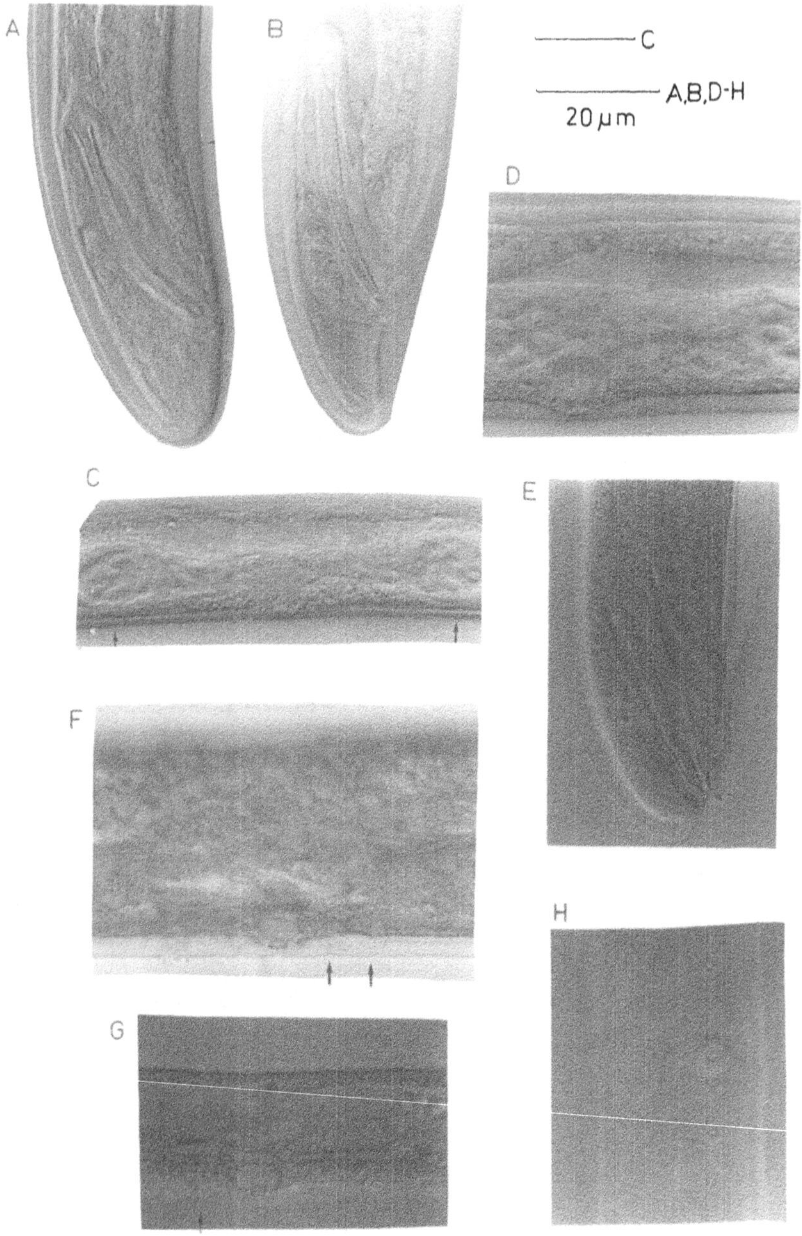

Fig. 4.31. *Paratrichodorus pachydermus*. Males: (A-B) copulatory apparatus and tail, respectively with accent on spicule and gubernaculum (Dutch specimens). Females: (C-D) vaginal region and part of uteri with sperm (Dutch specimen). *Paratrichodorus porosus*. Male: (E) copulatory apparatus and tail (paratype). Females: (F-G) vaginal region, (H) ventral view of vulva and ventromedian advulvar body pores (arrows).

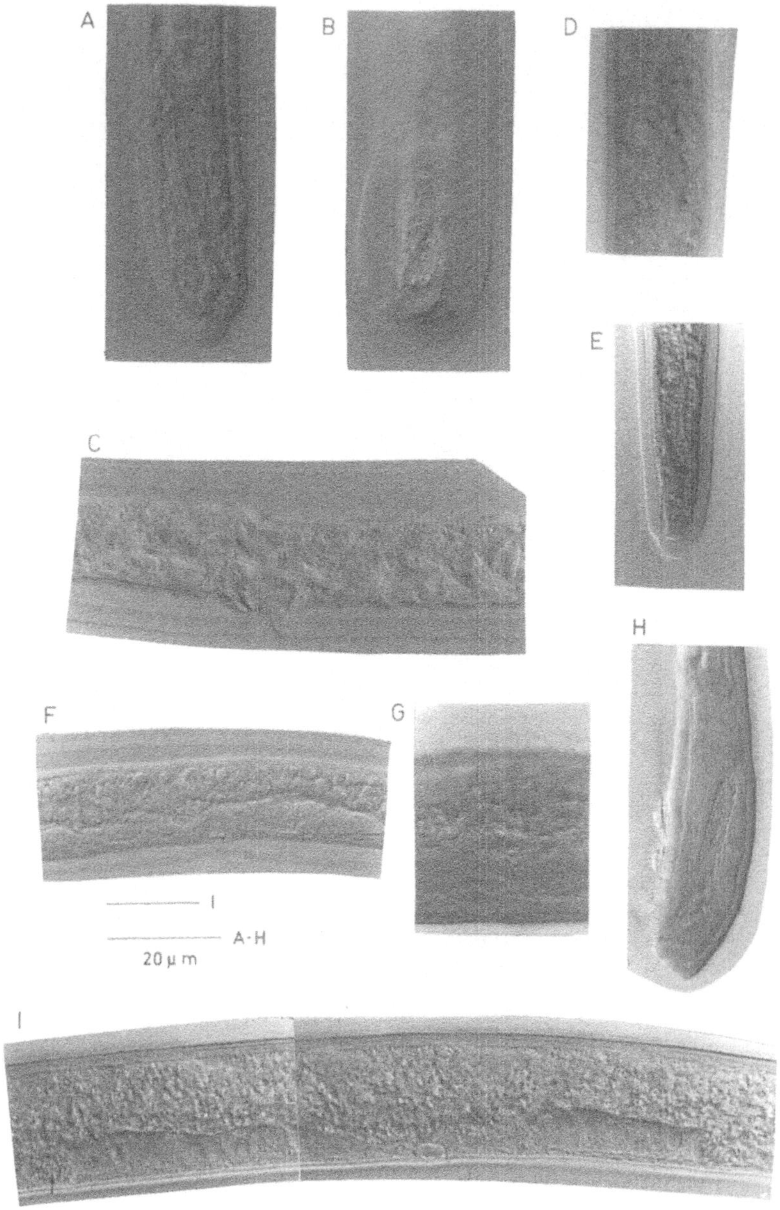

Fig. 4.32. *Paratrichodorus queenslandensis*. Male paratypes: (A) posterior body region, (B) ventral view of tail region with caudal alae. Female paratype: (C) vaginal region and part of uteri with sperm. *Paratrichodorus renifer*. Females: (D) ventral view of vulva (paratype), (E) tail region (paratype), (F) vaginal region and part of uteri with thread-like sperm. *Paratrichodorus rhodesiensis*. Females: (G) vaginal region, (I) genital system with unusual sperm location (arrow). Male: (H) copulatory apparatus and tail (paratype).

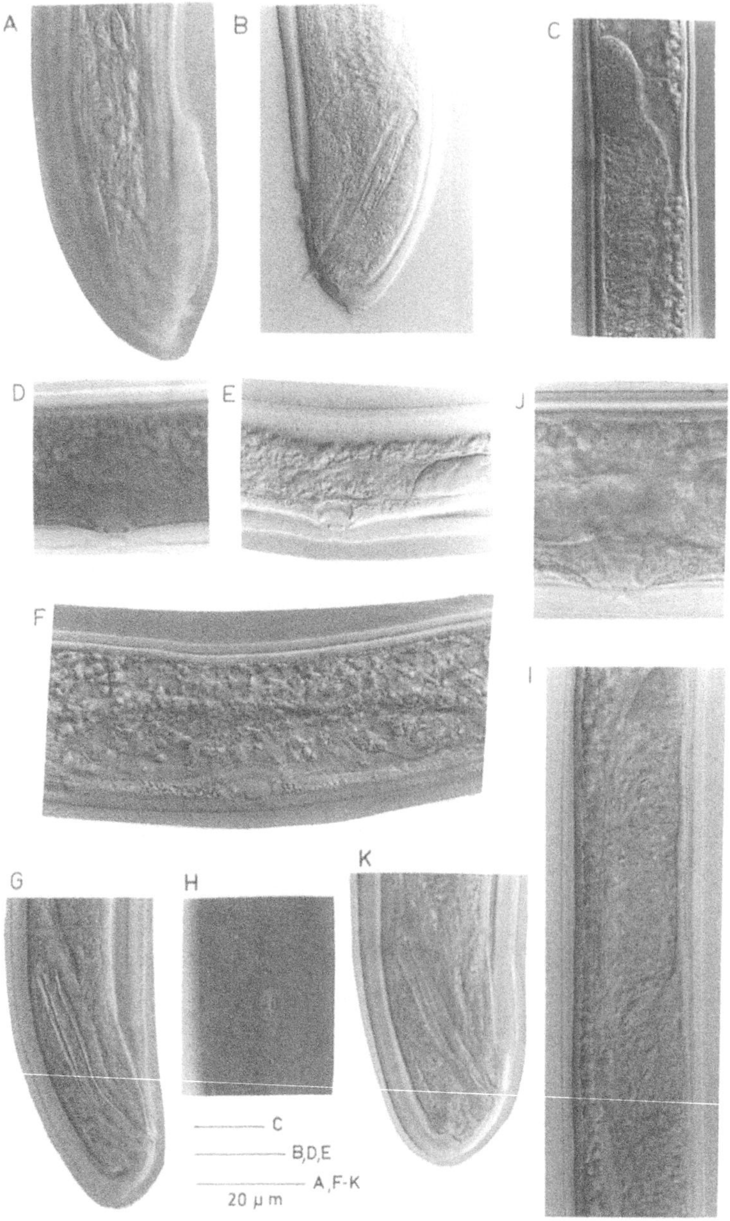

Fig. 4.33. *Paratrichodorus teres*. Males: (A-B) tail region respectively with accent on caudal alae and on copulatory apparatus, (C) testis with sperm. Females: (D-E) vaginal region (paratypes), (F) vaginal region and part of uteri. *Paratrichodorus tunisiensis*. Male: (G) copulatory apparatus and tail, (I) testis with sperm. Female: (H) vulva, ventral view. *Paratrichodorus weischeri*. Female: (J) vaginal region. Male: (K) copulatory apparatus and tail.

4.3. *Monotrichodorus*

Currently the genus *Monotrichodorus* Andrássy, 1976 is comprised of 7 species, three recently described by Siddiqi (1991). Apart from the type species, *M. monohystera* and the secondly described species, *M. vangundyi*, the species have only been recorded once i.e. by their type population.

Study of type material and other specimens, revealed that most *Monotrichodorus* species are very similar and differ only by one or two characteristics. Table 4.9. gives a review of the diagnostic characters used to differentiate the species (based on Siddiqi, 1991). Additionally, the swelling of the body cuticle, the position of the caudal pores (subterminal or terminal) in males and the position of the lateral advulvar body pores in females have been used by Siddiqi (1991) to differentiate species.

4.3.1. DISCUSSION

Females

1. The primary distinction between species of the genus is the ventral shape of the vulva: two of the nominal species (*M. muliebris*, *M. sacchari*) have a longitudinal slit, whilst the other four (*M. acuparvus*, *M. monohystera*, *M. parvus*, *M. vangundyi*) have a transverse slit.

2. In relation to the shape of the vulva, the shape of the vaginal sclerotizations in lateral view can be considered. Again, the same two groups can be distinguished: - *M. muliebris* and *M. sacchari* with small, more or less trapezoidal sclerotized pieces in lateral view, well separated whereas the other four species each have elongated, rod-like pieces, more or less oblique in orientation and with the tips close to one another. The differences reported for the vaginal sclerotizations (especially in the second group) appears to be a result of different degrees of (a generally weak) sclerotization; some pieces are entirely weakly sclerotized, others partly. However, independent of the different degrees of sclerotization, the outline of the pieces is about the same (=rod-like) and can best be observed with interference contrast (Fig. 3.18.).

3. The presence of lateral cervical pores in females as observed in *M. vangundyi* is an acceptable feature, but is difficult to observe as it is not always visible, even among paratype specimens.

Table 4.9. Review of the diagnostic characters to differentiate the *Monotrichodorus* species, based on the literature

	M. acuparvus	*M. monohystera*	*M. muliebris*	*M. parvus*	*M. proporifer*	*M. sacchari*	*M. vangundyi*
L	♂: 620-920 µm ♀: 610-830 µm	♂: 670-900 µm ♀: 610-940 µm	♂:- ♀: 1040-1180 µm	♂: 490-590 µm ♀: 480-590 µm	♂: 630-810 µm ♀: 600-830 µm	♂: 530-800 µm ♀: 530-800 µm	♂: 670-830 µm ♀: 650-810 µm
onchiostyle	♂: 41-47 µm ♀: 40-48 µm	♂: 43-48 µm ♀: 45-58 µm	♂:- ♀: 55-59 µm	♂: 46-51 µm ♀: 45-50 µm	♂: 55-62 µm ♀: 55-63 µm	♂: 52-61 µm ♀: 53-59 µm	♂: 49-56 µm ♀: 48-57 µm
spicule length	38-49 µm (44 µm)	51-59 µm (55 µm)	-	40-44 µm (42 µm)	46-50 µm (48.6 µm)	47-59 µm (52 µm)	50-57 µm (54 µm)
gubernaculum	10-18 µm	10-14.5 µm	-	10-14 µm	16-20 µm	12-18 µm	13-16 µm
V	75-80 %	77-83 %	76-77 %	77-81 %	79-85.5 %	79-84 %	82.3-85.6 %
vulva	transverse slit	transverse slit	longitudinal slit	transverse slit	transverse slit	longitudinal slit	transverse slit
shape vaginal sclerotization	dots, irregular dots, close, asymmetrical	clear, about triangular, oblique, close	weak, about trapezoid, separated	rod-like, close, oblique	rectangular rods, oblique, close	about triangular to trapezoid, separated	clear rod-like, close
position CP to S-E pore	just anterior to S-E pore	just anterior to S-E pore	-	just anterior to S-E pore	anterior to S-E pore	anterior to S-E pore	anterior to S-E pore
position CP to nerve ring	just posterior to nerve ring	just posterior to nerve ring	-	just posterior to nerve ring	from anterior to posterior to the nerve ring	at level of nerve ring	posterior to nerve ring
n° SP in spicule range	1	1	-	1	1	2	2
lateral cervical pore in female	none	none	none	none	none	none	one pair
geographic distribution	Columbia	Venezuela, San Salvador, Costa Rica, Peru, Brazil	Ecuador	Columbia	Brazil	French Guyana	Ecuador, Panama

Males

1. The distinction between species is less clear with males than with females. The shape of the spicules, the most important diagnostic character in trichodorid males, is similar for all nominal species of the genus.

The position of the second precloacal supplement in relation to the retracted spicules is the only other character, apart from morphometric data such as body length, length of onchiostyle and spicule length, which is considered of diagnostic importance. Based upon this character, two species (*M. sacchari* and *M. vangundyi*) can be distinguished from the other four by possessing two supplements within the region of retracted spicules *vs* one supplement in the other species (males are unknown for *M. muliebris*).

Other characters used to distinguish species are based on morphometric data such as differences in body length, onchiostyle length, length of the postvulvar uterine sac, the position of the vulva, the length of the vagina in female, and distance between the ventromedian cervical papilla and the S-E pore, minor differences in the position of the caudal pores (terminal *vs* subterminal) and the position of the ventromedian cervical papilla and S-E pore in relation to the nerve ring in male. For example, *M. parvus* is mainly differentiated from *M. acuparvus* by being smaller (480-590 µm *vs* 610-920 µm) and having a slightly longer onchium (23-26 µm *vs* 22-24 µm), a longer postuterine sac, caudal pores terminal *vs* subterminal and the ventromedian cervical papilla in males located somewhat more distant from the S-E pore, than *M. acuparvus*. For all former characters, intraspecific variability has been shown for species of the other genera of the Trichodoridae (Decraemer, 1989b), and has also been observed among the type specimens and other specimens of the seven species of *Monotrichodorus*. For example, the position of the ventromedian cervical papilla and S-E pore in relation to the nerve ring varies in *M. proporifer* from anterior to posterior to the nerve ring.

Decraemer (1986) and Andrássy (1989) questioned the validity of *M. vangundyi* being regarded as different from *M. monohystera*. Siddiqi (1991) considered it a valid species on the basis of the presence of lateral cervical pores near the amphids in the female and the position of the second supplement behind the head of retracted spicules in the male. These restricted differences are not very decisive for retaining both as separate species and hereby, *M. vangundyi* is considered as a subspecies of *M. monohystera*.

Furthermore, *M. acuparvus*, *M. parvus* and *M. proporifer* are very similar to *M. monohystera*, apart from some minor differences in morphometric data (body length, length of onchiostyle and spicule length) in males and females. They are considered as synonyms of *M. monohystera monohystera*.

M. muliebris is close to *M. sacchari* in the shape of their respective vulva (a longitudinal slit), but differs from it by having a longer postvulvar uterine sac (about one body width long *vs* insignificant to about half a body diameter long in *M. sacchari*), the longer body (1040-1180 µm *vs* 530-800 µm in *M. sacchari*), the more anterior position of the vulva (76-77% *vs* 79-84%) and the length and shape of the vagina (elongated, about cylindrical *vs* shorter pear-shaped in *M. sacchari*). Apparently, there is only minor difference in size and shape of the vaginal sclerotizations. Most of the former differences (e.g. shape, length of vagina) are probably induced by fixation (the only two type specimens of *M. muliebris* are described as having an abnormally swollen body cuticle and the head originally drawn retracted, both evidence of badly fixed specimens). Males of *M. muliebris* are unknown, but the female specimens have a distinct spermatheca filled with sperm cells; probably, *M. muliebris* is a bisexual species. Since type specimens of *M. muliebris* were not available, this species is still considered a valid species.

Monotrichodorus monohystera monohystera (Allen, 1957) Andrássy, 1976

Fig. 3.16.R; 3.18.A, C-G; 4.34.A-C, E-G, J; 4.35.A-C, F-J

syn. *Trichodorus monohystera* Allen, 1957
syn. *Monotrichodorus acuparvus* Siddiqi, 1991, n.syn.
syn. *Monotrichodorus parvus* Siddiqi, 1991, n.syn.
syn. *Monotrichodorus proporifer* Siddiqi, 1991, n.syn.

Measurements

Holotype male: L= 900 µm, a= 19, b= 5.0, onchiostyle= 47 µm, spicule length= 58 µm, gubernaculum= 12 µm
Males: L= 490-920 µm (785 µm*; 740 µm ± 83; 540 µm ± 31; 730 µm ± 52), a= 18-31, b= 3.9-6.2, T= 57-72%, onchiostyle= 41-62 µm (45.5 µm*: 44 µm ± 2; 47.5 µm ± 1.4; 58.5 µm ± 1.6), spicule length= 38-59 µm (55 µm*; 44

µm ± 2.2; 42 µm ± 1.4; 48.6 µm ± 1.1), gubernaculum= 10-20 µm; anterior end to EP= 74-96 µm; distance CP1-EP= 2-12 µm

Females: L= 480-940 µm (755 µm*; 770 µm; 870 µm; 730 µm ± 57; 520 µm ± 33; 750 µm ± 66), a= 15-31, b= 3.6-5.7, V= 75-85.5% (80%*; 78.5%; 81.5%*; 77.5% ± 1; 79.7% ± 0.96; 81.2% ± 1.6), onchiostyle= 40-63 µm (48.5 µm*; 52 µm; 50.5 µm; 44.7 µm ± 2; 47.4 µm ± 1.7; 59 µm ± 1.8); anterior end to EP= 78-98 µm (from Allen, 1957; Loof, 1964: ♀ only; Rodriguez-M *et al.*, 1978: ♀ only; Siddiqi, 1991:resp. *M. acuparvus*, *M. parvus*, *M. proporifer*).

Emended diagnosis

M. monohystera monohystera males are characterized by the position of the ventromedian precloacal supplements in relation to the retracted spicules (with only the posterior supplement within the region of retracted spicules, the second supplement varying from just anterior to the level of the proximal end of retracted spicules), the position of the single ventromedian cervical papilla (immediately anterior to the S-E pore and usually just posterior to the nerve ring), the position of the paired, lateral, cervical pores (slightly posterior to the S-E pore), the small sperm cells and the length of the spicules (38-59 µm). Females are distinguished by the shape of the vaginal sclerotizations in lateral view (rod-like with tips close to one another, oblique), one pair of lateral advulvar body pores and the vulva shape (a transverse slit).

Relationships

M. monohystera monohystera resembles *M. sacchari* in males, but females differ by the ventral vulva shape (a transverse slit *vs* a longitudinal slit) and by the shape of the vaginal sclerotizations (elongated, about rod-shape, oblique, close together *vs* small about trapezoid, well separated in *M. sacchari*). *M. m. monohystera* is close to *M. m. vangundyi* in males and females, but differs in males by the more anterior position of the second precloacal supplement and in females by the absence of lateral cervical pores.

Remarks

Allen (1957) described the females of the type population as having two lateral body pores on one side (one slightly posterior to the level of the vulva and one less than a body width anterior to the vulva). All other descriptions of type specimens and specimens from other populations described females with one lateral advulvar body pore on each side, varying from slightly anterior, at the level or slightly (less than one body width) posterior to the vulva (Loof, 1964;

Rodriguez-M *et al.*, 1978; Decraemer, 1986).

The body cuticle is usually not swollen upon fixation, but was described as swollen by Loof (1964); it was also observed to be swollen among other specimens (present study).

Rodriguez-M. *et al.* (1978) observed one male specimen without ventromedian cervical papilla.

All males and females have a pair of caudal pores, close together, not a single pore (Decraemer, 1986).

Distribution and host

M. monohystera monohystera is restricted to Central and South America. It was originally described from Venezuela; further records are from San Salvador, Costa Rica, Brazil and Peru. An additional record as a result of synonymization is Columbia.

M. monohystera is associated with the roots of forest trees, frailejon (*Espeletia* sp.), sugar-cane, banana, *Solanum* sp., *Allstroameria* sp., grass (*Andropogon gayanus*), primary rain forest, oil palm *(Elaeis guineensis*), citrus and *Ceiba pentandra* (Allen, 1957; Loof, 1964; Rodriguez-M *et al.*, 1978; Siddiqi, 1991).

Monotrichodorus monohystera vangundyi Rodriguez-M, Sher & Siddiqi, 1978: n.grad.
Fig. 3.18.B; 4.34.D, H; 4.35.F-G

syn. *M. vangundyi* Rodriguez-M, Sher & Siddiqi, 1978, n.syn.

Measurements

Holotype male: L= 780 µm, a= 21.6, b= 4.9, T= 57%, onchiostyle= 55 µm, spicule length= 53 µm, gubemaculum= 17 µm, anterior end to EP= 95 µm, distance CP-EP= 8 µm

Males: L= 670-830 µm (730 µm), a= 18-24, b= 4.3-5.7, T= 57-70%, onchiostyle= 49-56 µm (53 µm), spicule length= 50-57 µm (54 µm), gubernaculum= 13-16 µm, anterior end to EP= 94 µm ± 5, anterior end to CP= 89.3 µm ± 4.4 µm

Females: L= 650-810 µm (720 µm), a= 15.5-22.0, b= 3.5-5.8, V= 82.3-85.6% (83.8%), onchiostyle= 48-57 µm (53 µm).

Emended diagnosis

M. monohystera vangundyi males are characterized by the number and position of the ventromedian precloacal supplements (two supplements clearly within the region of the retracted spicule; the second supplement at a quarter length of the spicule head), the position of the ventromedian cervical papilla (shortly anterior to the S-E pore and posterior to the nerve ring), the position of the paired lateral cervical pores (immediately posterior to the nerve ring) and the length of the spicules (50-57 μm). Females by the presence of paired lateral cervical pores immediately behind the amphid openings, the shape of the vaginal sclerotizations in lateral view (rod-like, with tips close to one another), the shape of the vulva (a large transverse slit) and by the paired lateral body pores within one vulvar body width anterior to the vulva level.

Relationships

M. monohystera vangundyi is closely related to M. *monohystera monohystera* from which it is distinguished in males by the number and position of the precloacal supplements in relation to the retracted spicules (two supplements clearly within the region of retracted spicules *vs* one supplement), and in females by the presence of lateral cervical pores (absent in *M. m. monohystera*).

Remarks

Lateral cervical pores in females were not always visible, even among paratypes.

The ventromedian cervical papilla and the S-E pore in males are not always located posterior to the nerve ring, both may be at the level of the nerve ring. Large sperm cells have been observed in the spermathecae.

Distribution and host

M. monohystera vangundyi was described from soil around the roots of oil palm (*Elaeis guineensis*) near Rosa Zarate, Ecuador and from citrus, banana and native forest soils near the type locality. It was also found in Rio Corutu river bed soil, Puerto Armuelles and around roots of *Ceiba pentandra* Barro Colorado, Panama (Rodriguez-M, Sher & Siddiqi, 1978).

Monotrichodorus muliebris Andrássy, 1989
Fig. 3.18.J, K

Measurements

No separate data were given for the holotype female.

Holotype female and paratype female: L= 1040-1180 µm (1100 µm*), a= 20-21, b= 4.7-4.5, V= 76-77%, onchiostyle= 55-59 µm (57 µm*) (from Andrássy, 1989).

Emended diagnosis

M. muliebris female is characterized by a long body (1040-1180 µm), a longitudinal slit-like vulva, a pronounced postvulvar uterine sac, the length (32-36 µm) and shape of the vagina (elongated, about cylindrical), the size and shape of the vaginal sclerotizations (small, trapezoid, well separated), spermatheca filled with large globular sperm (5.5-6.5 µm), one pair of preadvulvar body pores (at 12-23 µm from vulva), pharyngeal bulb offset, S-E pore shortly anterior to mid-pharynx. Males unknown.

Relationships

M. muliebris females are close to *M. sacchari* having a longitudinal slit-like vulva, similar vaginal sclerotizations (more or less trapezoidal). They differ mainly in body length (1040-1180 µm *vs* 530-580 µm) and show also minor differences in size and shape of the vaginal sclerotizations (larger and not rounded triangular in *M. muliebris* (see also previous general discussion, p 234).

Remarks

Males are unknown in the population from Ecuador described by Andrassy (1989) as *M. muliebris* but the two female type specimens had a distinct spermatheca filled with 5.5-6.5 µm large globular spermatozoa. Furthermore, these specimens were badly fixed, with the body cuticle abnormally swollen (7-8 µm), the head region retracted (see original figure) and the vagina well elongated. Consequently, the morphometric data must be interpreted with caution.

Distribution and host

M. muliebris was described from Volcan Cotopaxi, Ecuador, where it was found associated with the roots of paramograss (Andrássy, 1989.)

Monotrichodorus sacchari Baujard & Germani, 1985

Fig. 3.18.H-K; 4.36.A-H

Measurements

Holotype female: L= 650 µm, a= 15, b= 4.8, V= 81%, onchiostyle= 55 µm

Females: L= 530-800 µm (650 µm ± 60), a= 14-21, b= 3.5-6.8, V= 79-84% (82% ± 1.2), onchiostyle= 53-59 µm (56 µm ± 1.7), anterior end to EP= 74-88 µm

Males: L= 530-800 µm (650 µm ± 60), a= 16-22, b= 3.5-7.7, onchiostyle= 52-61 µm (57 µm ± 2), spicule length= 47-59 µm (52 µm ± 2.6), gubernaculum= 12-18 µm, anterior end to EP= 74-88 µm, anterior end to ĊP1= 68-82 µm (from Baujard & Germani, 1985).

Emended diagnosis

M. sacchari is distinguished by the length of its onchiostyle (52-61 µm), and in males by the position of the three ventromedian precloacal supplements with the posterior two clearly within the region of the retracted spicules, the position of the single ventromedian cervical papilla (shortly anterior to the S-E pore, and at the level of the nerve ring), the pair of lateral cervical pores situated at the level of the S-E pore, two very close caudal pores and the slightly thickened subventral body cuticle giving the impression of rudimentary caudal alae. It is further characterized in females by the vulva shape (a longitudinal slit), one pair of lateral advulvar body pores and by the shape of the vaginal sclerotizations in lateral view (about rounded triangular to trapezoid, well separated).

Relationships

M. sacchari females are closest to *M. muliebris* having a longitudinal slit-like vulva and similar well separated vaginal sclerotizations (see general discussion, p 234).

M. sacchari females can be clearly distinguished from *M. monohystera* (with two subspecies) by the shape of the vaginal sclerotizations (round triangular to trapezoid *vs* rod-like) and by the vulva shape (a longitudinal slit *vs* a transverse slit) and from *M. monohystera vangundyi* by the lack of lateral cervical pores in females. Males, however, cannot be clearly differentiated from *M. monohystera vangundyi* , but differ from *M. monohystera monohystera* by having two precloacal supplements clearly within the region of retracted spicules *vs* one supplement.

Siddiqi (1991) further distinguished males of *M. sacchari* and *M. vangundyi* based upon different swelling reactions of the body cuticle when fixed (swollen in *M. sacchari vs* non-swollen in *M. vangundyi*), and the position of the S-E pore and ventromedian cervical papilla (near the nerve ring in *M. sacchari vs* well behind the nerve ring in *M. vangundyi).*

Remarks

Among type specimens of *M. vangundyi* males have been observed with the ventromedian cervical papilla and the S-E pore at the level of the nerve ring. Furthermore, the swelling of the body cuticle appeared to be influenced by fixation and was not species correlated. Consequently, the differentiation of males of *M. sacchari* and *M. vangundyi* are based on variable features.

Baujard & Germani (1985) described males and females with a single caudal pore but Decraemer (1986) observed two very closely opposed pores in males as well as in females.

Distribution and host

M. sacchari was described from French Guyana, from soil of the rhizosphere of *Saccharum officinarum.* (Baujard & Germani, 1985).

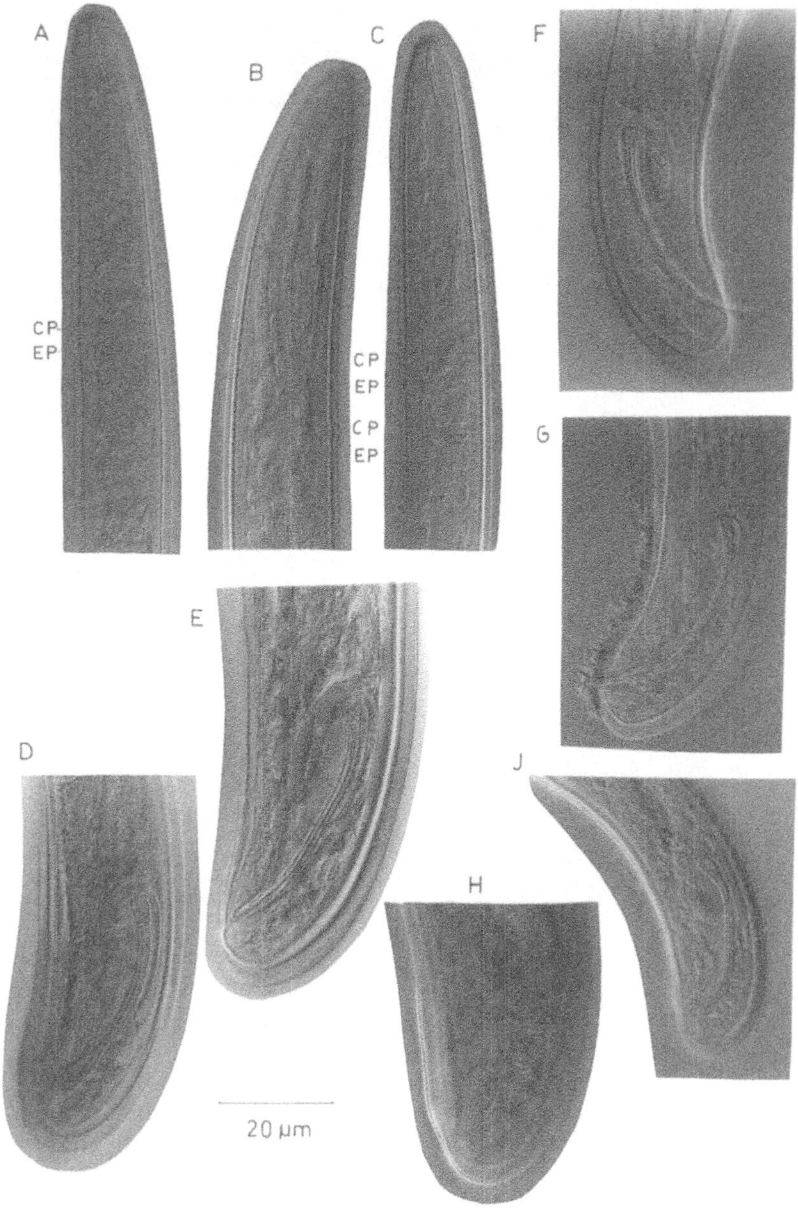

Fig. 4.34. *Monotrichodorus monohystera monohystera*. Males. (A-C) anterior body region (A: paratype *M. acuparvus* (= *M. m. monohystera)*; B-C: paratypes of *M. proporifer (= M. m. monohystera)*, (D-J) posterior body region (D,H: respectively paratype and male from Panama of *M. m. vangundyi*; E specimen of *M. m. monohystera* from Panama; F: paratype of *M. proporifer* (= *M. m. monohystera)*; G: paratype of *M. acuparvus* (=*M. m. monohystera)*.

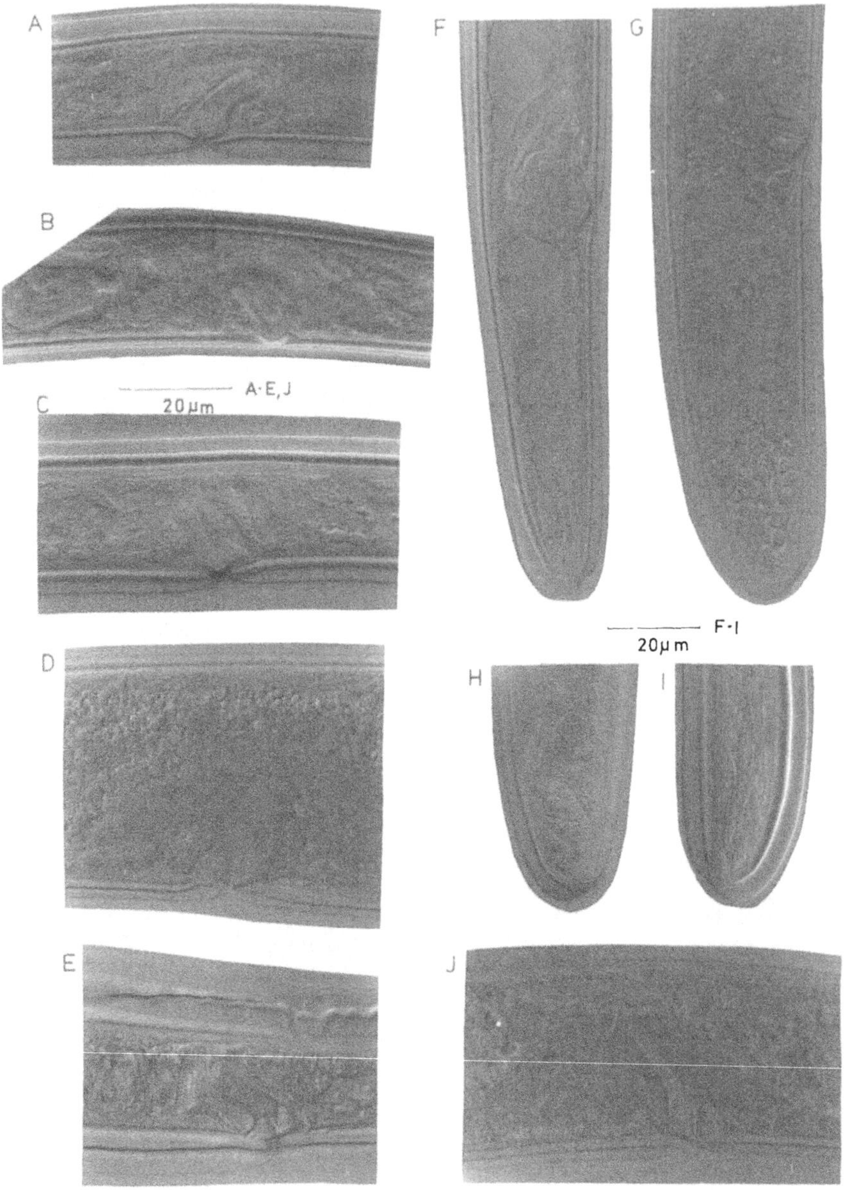

Fig. 4.35. *Monotrichodorus monohystera monohystera*. Females. Vaginal region. (A) paratype of *M. acuparvus* (=*M. m. monohystera*), (B) paratype of *M. parvus* (= *M. m. monohystera*), (C) paratype of *M. proporifer* (=*M. m. monohystera*), (D-E) *M. m. vangundyi*, specimens of San Salvador and Ecuador, (J) *M. m. monohystera*, specimen from San Salvador; posterior body region. (F-G) respectively of paratype of *M. m. vangundyi* and *M. m. monohystera*, specimen from Panama; tail region. (H-I) paratypes of *M. proporifer* (= *M. m. monohystera*).

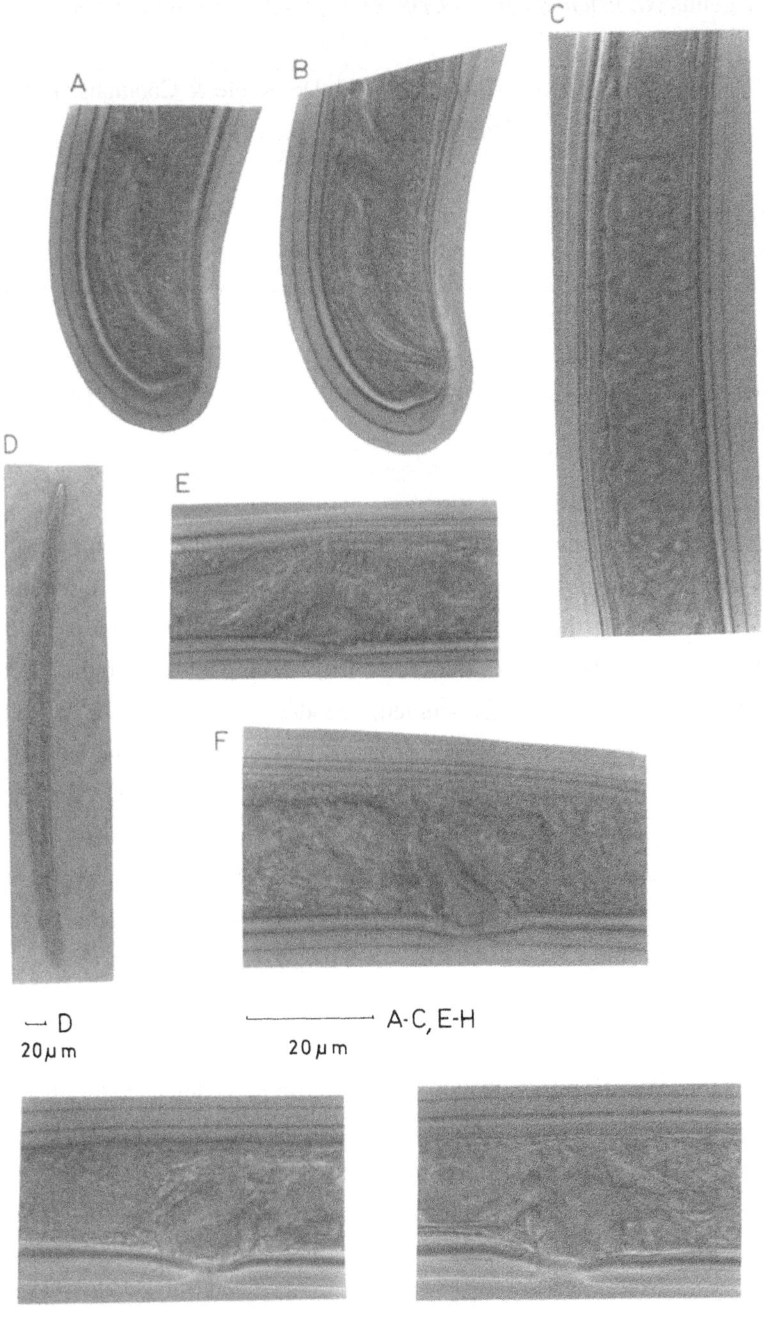

Fig. 4.36. *Monotrichodorus sacchari*. Male paratypes: (A-B) posterior body region, (C) part of testis. Female paratypes: (D) total view, (E-H) vaginal region.

4.4. The genus *Allotrichodorus* Rodriguez-M, Sher & Siddiqi, 1978

Allotrichodorus brasiliensis Rashid, De Waele & Coomans, 1986
Fig.3.19.D; 3.20.D; 4.37.A-D

Measurements

Holotype female: L= 700 µm, a= 18.0, b= 5.0, V= 88%, onchiostyle= 61 µm, anterior end to EP= 89 µm

Females: L= 640-670 µm, a= 19.0 -20.3, b= 4.3-4.6, V= 89%, onchiostyle= 60-62 µm, anterior end to EP= 85 µm

Males: L= 570-730 µm (668 µm; 730 µm), a= 17.5-20.3, b= 4.2-5.7, T= 72-126%, onchiostyle= 59-65 µm (61 µm; 63 µm), spicule length= 58-66 µm (61 µm; 65 µm), gubernaculum= 10-11 µm, anterior end to EP= 85-93 µm (from Rashid *et al.*, 1986).

Emended diagnosis

A. brasiliensis females are characterized by the shape of the vaginal sclerotizations in lateral view (conspicuous, consisting of two parts: a usual small differentiation with sclerotized granules at the vulva lips, and a much lighter sclerotization in the vaginal wall), the ventral vulva shape (a transverse slit). Males by the length (58-66 µm) and shape of the spicules (slightly curved, slender, shaft striated), the absence of ventromedian cervical papillae, a pair of lateral cervical pores at the level of the nerve ring, three ventromedian precloacal supplements (two clearly within the region of the retracted spicules, the third one from immediately anterior to just posterior to the retracted spicules) and caudal alae narrow but clear. The pharynx has an anteriodorsal overlap of the intestine in males and females and the S-E pore is situated about at the level of the nerve ring.

Relationships

Males of *A. brasiliensis* resemble those of *A. guttatus* in having no ventromedian cervical papilla but can be distinguished by their shorter onchiostyle (59-65 µm *vs* 65-72 µm in *A. guttatus*), shorter spicules (58-66 µm *vs* 65-76 µm in *A. guttatus*) and by the absence of a capitular extension of the spicules (present in *A. guttatus*) (Rashid *et al.*, 1986).

Remarks

One male out of five had an offset pharyngeal bulb and another had four ventromedian precloacal supplements.

The vaginal sclerotization in lateral view was described as being bipartite: (1) a small differentiation with sclerotized granules at the vulva lips and (2) a much lighter sclerotization in the vaginal wall. The small sclerotized pieces at the vulva lips are similar to the vaginal sclerotizations in other trichodorid species; the second much lighter sclerotization in the vaginal wall is a differentiation of the vaginal wall which was also marked in *M. monohystera* and in some *Trichodorus* species, but without some degree of sclerotization as in *A. brasiliensis*.

Distribution and host

A. brasiliensis is recorded only from Brazil, associated with the roots of cocoa *(Theobroma cacao)* (Rashid *et al.*, 1986).

Allotrichodorus campanullatus Rodriguez-M, Sher & Siddiqi, 1978
Fig. 3.19.G; 3.20.A; 4.37.E-H

Measurements

Holotype male: L= 690 µm, a= 16, b=4.4, T= 69%, onchiostyle= 53 µm, spicule length= 58 µm, gubernaculum= 16 µm, anterior end to EP= 90 µm, distance CP1-EP= 9 µm

Males: L= 530-780 µm (620 µm; 611 µm; 718 µm; 540 µm; 584 µm; 570 µm), a= 11-27.7, b= 3.6-6.2, T= 52-63%, onchiostyle= 45-64.5 µm (50.3 µm; 48 µm; 56 µm; 54.6 µm; 60 µm; 60 µm), spicule length= 46-66 µm (52 µm; 49 µm; 61 µm; 49.5 µm; 61.6 µm; 59 µm), gubernaculum= 8-17 µm, anterior end to EP= 75-98 µm, distance CP1-EP= 3-14 µm

Females: L=530-750 µm (670 µm; 570 µm; 720 µm; 640 µm; 600 µm), a= 12-20.6, b= 3.8-5.9, V= 83-86.4% (84.6%; 84%; 85%; 84.7%; 85%), onchiostyle= 45-65.5 µm (52.4 µm; 46 µm; 55 µm; 63.2; 60 µm), anterior end to EP= 72-94 µm (from Rodriguez-M *et al.*, 1978; Rashid *et al.*, 1986: 4 populations of ♀, 5 of ♂).

Emended diagnosis

A. campanullatus males are characterized by the length (46-66 µm) and shape of the spicules (slender, arcuate, proximally provided with a capitular extension with dorsal groove, shaft striated and with bristles), three ventromedian precloacal supplements (all within the region of the retracted spicules or third supplement anterior to the retracted spicules), caudal alae narrow but clear, a ventromedian cervical papilla at the level of the nerve ring and anterior to the S-E pore and a pair of lateral cervical pores at the level of the nerve ring. Females by the shape of the vaginal sclerotizations in lateral view (conspicuous, about triangular to trapezoid), the vulva shape (a transverse slit) and caudal pores terminal. S-E pore shortly posterior to nerve ring.

Relationships

A. campanullatus males are similar to those of *A. guttatus* in the structure of the spicules, both with a capitular extension. They differ by the presence of a ventromedian cervical papilla (absent in *A. guttatus),* by the length of the spicules (46-66 µm *vs* 65-76 µm in *A. guttatus*). The females differ in the shape of the vaginal sclerotization (large triangular to trapezoid *vs* smaller and with non-sclerotized additional differentiation of the vaginal wall in *A. guttatus*). Furthermore, the onchiostyle is somewhat shorter than in *A. guttatus* (45-65.5 µm *vs* 63-72 µm).

Remarks

Variable within and between populations is the kind of pharyngo-intestinal junction, either with an anteriodorsal intestinal overlap (rare or common) or with an offset pharyngeal bulb (Rashid *et al.*, 1986).

Among different populations some variability was observed in the following morphometric data: onchiostyle length, length of spicules, length of postvulvar uterine sac, and in the shape of vaginal sclerotizations in lateral view: rounded triangular to trapezoid (Rashid *et al.*, 1986).

Distribution and host

A. campanullatus is recorded from different localities in Brazil, often in heavy, loamy soils. It was found in the rhizosphere of cocoa (*Theobroma cacao),* coffee, banana, wild palm (palm-Mane-Veio), cloves (*Eugenia carophyllata*), lilac, *Hevea brasiliensis* and in native woodland (Rodriguez-M *et al.*, 1978; Rashid *et al.*, 1986).

Allotrichodorus guttatus Rodriguez-M, Sher & Siddiqi, 1978

Fig. 3.19.I-J

Measurements

Holotype male: L= 680 µm, a= 13.0, b= 4.3, T= 58%, onchiostyle= 67 µm, spicule length= 76 µm, gubernaculum= 15 µm, anterior end to EP= 97 µm

Males: L= 670-740 µm (710 µm), a= 12-16, b= 4.3-5.3, T= 58-74%, onchiostyle= 65-72 µm (70 µm), spicule length= 65-76 µm (70 µm), gubernaculum= 14-18 µm
Females: L= 650-720 µm (690 µm), a= 12-15, b= 4.1-5.2, V= 84-88% (86%), onchiostyle= 63-71 µm (69 µm) (from Rodriguez-M *et al.*, 1978).

Emended diagnosis

A. guttatus males are distinguished by the length and structure of the spicules (slender, elongated, nearly straight, with capitular extension, shaft striated), the absence of ventromedian cervical papilla, the position of paired lateral cervical pores (slightly posterior to the nerve ring, about at the level of the S-E pore, shortly posterior to nerve ring, three ventromedian precloacal papillae within the region of the retracted spicules and caudal alae inconspicuous. Females by the shape of the vaginal sclerotizations in lateral view (well developed, more or less triangular) and vulva shape (a transverse slit). Furthermore, males and females have an anteriodorsal intestinal overlap.

Relationships

A. guttatus and *A. campanullatus* are the only species of the genus in which male spicules have a capitular extension. They differ in males by the length of the spicules, the absence of ventromedian cervical papilla, present in *A. campanullatus*, and in females by the shape of the vaginal sclerotizations in lateral view.

Remarks

Uniquely for trichodorids, the S-E pore and a short posteriorly bent inner canal have been described associated with an inner reniform cell (Rodriguez-M *et al.*, 1978). Study of type material could not confirm this beyond doubt. In this region a small cell and nucleus may often be observed, also present in other species of the genus e.g. in *A. loofi*, however, similar cells (?epidermal) also occur more anteriorly. Therefore, its association with the S-E system can not be ascertained.

Distribution and host

A. guttatus is known from the type locality in Brazil, where it was found in soil around the roots of cocoa trees.

Allotrichodorus longispiculis Rashid, De Waele & Coomans, 1986
Fig. 3.19.C; 3.20.E; 4.38.A

Measurements

Holotype female: L= 530 µm, a= 17.5, b= 3.4, V= 82%, onchiostyle= 60 µm, anterior end to EP= 83 µm
Female: L= 700 µm, a= 23.0, b= 4.4, V= 83%, onchiostyle= 61 µm, anterior end to EP= 103 µm
Male: L= 530 µm, a= 17.5, b= 3.8, T= 60%, onchiostyle= 58 µm, spicule length= 69 µm, gubernaculum= 10 µm, anterior end to EP= 78 µm, distance CP1-EP= 5 µm (from Rashid *et al.*, 1986).

Emended diagnosis

A. longispiculis male is characterized by the length (69 µm) and shape of the spicules (elongated, slender, ventrally curved at distal end), a single ventromedian cervical papilla immediately anterior to the S-E pore and just anterior to the nerve ring, paired lateral cervical pores about at the level of the S-E pore i.e. at level of nerve ring and the posterior border of the nerve ring, three ventromedian precloacal supplements, all clearly within the region of the retracted spicules and caudal alae inconspicuous. Females by the shape of the vaginal sclerotizations in lateral view (distinct, but relatively small, rod-like, largely parallel to the longitudinal body axis).

Relationships

A. longispiculis females closely resemble females of *A. loofi* but have a slightly different shape of the vaginal sclerotizations in lateral view (Fig. 3.19.C *vs* Fig. 3.19.A-B). Males of *A. longispiculis* have one ventromedian cervical papilla as do males of *A. loofi*, but differ from the latter by the longer, slightly more slender spicules (69 µm *vs* 50 µm), and, to a lesser extent by the position of the third precloacal supplement which is clearly within the region of retracted spicules *vs* within or just outside the retracted spicules in *A. loofi*.

Remarks

The holotype female was described as having a slight anteriodorsal overlap of the intestine, but no overlaps in the paratype female. The male has the intestine overlapping the pharynx. The original illustrations of the holotype female and paratype male show an oblique pharyngeal base indicating, at most, a minute intestinal and pharyngeal overlap.

Distribution and host

A. longispiculis is known from the type locality from Brazil where it was found in soil around the roots of *Cecropia*.

Allotrichodorus loofi Rashid, De Waele & Coomans, 1986

Fig. 3.19.A-B; 3.20.B; 4.38.B-E

Measurements

Holotype female: L= 610 µm, a= 17.2, b= 4.9, V= 84%, onchiostyle= 54 µm, anterior end to EP= 89 µm

Females: L= 480-670 µm (584 µm), a= 14.2-20.0, b= 4.1-5.1, V= 82-85% (83%), onchiostyle= 51-63 µm (56 µm), anterior end to EP= 74-91 µm

Male: L= 600 µm, a= 17.3, b= 4.0, T= 60%, onchiostyle= 60 µm, spicule length= 50 µm, gubernaculum= 13 µm, anterior end to EP= 86 µm, distance between CP1-EP= 4.5 µm (from Rashid *et al.*, 1986).

Emended diagnosis

A. loofi male is characterized by the presence of a ventromedian cervical papilla immediately anterior to the S-E pore and at the level of the nerve ring, paired lateral cervical pores situated at the level of the nerve ring, length (50 µm) and shape of the spicules (slender, distally arcuate, shaft with striae and bristles), caudal alae rather inconspicuous and three ventromedian precloacal supplements with the anterior one just anterior or posterior to the head of retracted spicules. Females by the shape of the vaginal sclerotizations in lateral view (conspicuous, small, about rod-shaped to narrow triangular, largely parallel to the longitudinal body axis). Males and females have a short anteriodorsal intestinal overlap of the pharynx.

Relationships

A. loofi females are similar to those of *A. longispiculis* by the comparable shape of the vaginal sclerotizations in lateral view, and the male by possessing a ventromedian cervical papilla anterior to the S-E pore and spicules without capitular extension. The two species differ mainly by the somewhat larger and less sclerotized vaginal sclerotizations in females in *A. loofi*, and the shorter spicules (50 µm *vs* 69 µm) in males.

Remarks

From the original illustrations there appears to be some variability in the length and shape of the vaginal sclerotized pieces in lateral view.

Distribution and host

A. loofi is known from three localities in Brazil. It was found in soil around the roots of coffee.

Allotrichodorus sharmai Rashid, De Waele & Coomans, 1986

Fig. 3.19.E-F; 3.20.F; 4.38.F-G; 4.39.A-B

Measurements

Holotype female: L= 690 µm, a= 14.2, b= 4.5, V= 84%, onchiostyle= 64 µm, anterior end to EP= 100 µm

Females: L= 580-710 µm (670 µm; 640 µm), a= 12.5-20, b= 4.1-4.7, V= 81-85% (83%; 83%), onchiostyle= 54-65 µm (64 µm; 56 µm), anterior end to EP= 87-100 µm

Males: L= 620-800 µm (680 µm; 660 µm), a= 12.2-18.4, b= 3.8-4.7, T= 58-62%, onchiostyle= 60-66 µm, spicule length= 55-67 µm (62 µm; 59 µm), gubernaculum= 10-11 µm, anterior end to EP= 86-105 µm, distance CP1-EP= 6-10 µm (from Rashid *et al*., 1986).

Emended diagnosis

A. sharmai is characterized by the length of the onchiostyle (54-66 µm) and the pronounced anteriodorsal intestinal overlap. It is further distinguished in females by the shape of the vaginal sclerotizations in lateral view (conspicuous, large, mitten-shaped) and vulva shape (a transverse slit). The males by the presence of a ventromedian cervical papilla (at the level of the nerve ring and immediately anterior to the S-E pore), paired lateral cervical pores anterior to the nerve ring, length (55-67 µm) and shape of the spicules (slender, slightly curved, smooth), three ventromedian precloacal supplements, all within the region of the retracted spicules and distinct caudal alae in males.

Relationships

Females of *A. sharmai* can be distinguished from those of the other species in the genus by the shape and size of the vaginal sclerotizations. Males of *A. sharmai* closely resemble those of *A. brasiliensis* and *A. loofi* in spicule shape (without capitular extension) and of *A. brasiliensis* also in spicule length. They differ from *A. brasiliensis* by the presence of a ventromedian cervical papilla (absent in *A. brasiliensis*) and from

A. loofi by the length of the spicules (55-67 µm *vs* 50 µm in *A. loofi*).

Remarks
One male has been described with the anterior precloacal supplement outside the region of the retracted spicules (Rashid *et al.*, 1986).

Distribution and host
A. sharmai is recorded from the type locality in Bahia State, Brazil where it was found in light sandy soil around the roots of cocoa (*Theobroma cacao*).

Allotrichodorus westindicus (Rodriguez-M, Sher & Siddiqi, 1978) Rashid, De Waele & Coomans, 1986

Fig. 3.19.H; 4.39.C-G

syn. *Paratrichodorus (Nanidorus) westindicus* Rodriguez-M, Sher & Siddiqi, 1978;
syn. *Nanidorus westindicus* Rodriguez-M, Sher & Siddiqi, 1978

Measurements
Holotype female: L= 460 µm, a= 23, b= 4.5, V= 64%, onchiostyle= 34 µm, anterior end to EP= 98 µm
Females: L= 380-500 µm (460 µm; 390 µm; 426 µm), a= 16-25, b= 3.8-5.4, V= 60-66% (64.7%; 66%; 63%), onchiostyle= 30-37 µm (34.5 µm; 34 µm; 32 µm), anterior end to EP= 74-102 µm (from Rodriguez-M *et al.*, 1978; Baujard & Germani, 1985; Rashid *et al.*, 1986).

Emended diagnosis
A. westindicus is characterized by the relatively short body (380- 500 µm), the length of the onchiostyle (30-37 µm), the position of the S-E pore (usually at the level of the pharyngo-intestinal junction or just posterior), the usually offset pharyngeal bulb, the position of the vulva (60-65%), the shape of the vulva (a transverse slit), the shape of the vaginal sclerotizations in lateral view (minute, about triangular, close), the short vagina and by the small cuneate sperm cells.

Relationships
A. westindicus differs from the other species in the genus by the absence of males and

by the more anterior position of the vulva (60-66% *vs* 81-89%), the shorter and not anteriorly directed vagina (25-30% of corresponding body width *vs* 38% to more than 50%), the shorter body length (380-500 µm *vs* 480-750 µm) and shorter onchiostyle (30-37 µm *vs* 45-71 µm) in females. Also, it is the only species with populations recorded from outside Brazil.

A. westindicus has a somewhat intermediate position between the genera *Allotrichodorus* and *Paratrichodorus* (see discussion chapter 3).

Remarks

The S-E pore may also be observed more anteriorly than in previous descriptions, up to halfway along the pharyngeal bulb.

The pharyngeal bulb is usually offset, but may slightly overlap the intestine (Rodriguez-M *et al.*, 1978) or be shortly overlapped by the intestine (see illustration by Rashid *et al.*, 1986).

The tail end shows some variability in shape: from obtusely rounded to subdigitate (Rodriguez-M *et al.*, 1978).

The reproduction of this species could be hermaphroditic; sperm cells have been observed in females and males appear to be absent.

Distribution and host

A. westindicus has been recorded from Trinidad (West Indies), Martinique and Itacara, Brazil. It has been found in soil around the roots of cocoa (*Theobroma cacao*), sugar-cane and *Anthurium* sp. (Rodriguez-M *et al.*, 1978; Baujard & Germani, 1985; Rashid *et al.*, 1986).

Table 4.9. Review of the diagnostic features for *Allotrichodorus* species

	A. brasiliensis	*A. campanullatus*	*A. guttatus*	*A. longispiculis*	*A. loofi*	*A. sharmai*	*A. westindicus*
L	♂: 570-730μm ♀: 640-700μm	♂: 530-780μm ♀: 530-750μm	♂:670-740μm ♀:650-720μm	♂:530μm ♀:530-700μm	♂:600μm ♀:480-670μm	♂:620-800μm ♀:580-710μm	♂:- ♀:380-500μm
onchiostyle	♂: 59-65μm ♀: 60-62μm	♂:45-64.5μm ♀:45-65.5μm	♂:65-72μm ♀:63-71μm	♂:58μm ♀:60-61μm	♂:60μm ♀:51-63μm	♂:60-66μm ♀:54-65μm	♂:- ♀:30-37μm
spicule length	58-66μm	46-66μm	65-76μm	69μm	50μm	55-67μm	-
gubernaculum	10-11μm	8-17μm	14-18μm	10μm	13μm	10-11μm	-
V	88-89%	83-86.4%	84-88%	82-83%	82-85%	81-85%	60-66%
spicule shape	slender, shaft with striae	slender, with capitular extension, striae, bristles	slender, long, with capitular extension	slender,long, distally bent, with striae	slender, with striae, bristles	slender, smooth	-
shape vaginal sclerotization	conspicuous, bipartite	triangular to trapezoid	about triangular	small, rod-like	small, rod-like	conspicuous, mitten-shaped	minute, about triangular
vagina: length, orientation	> half mbd, directed anteriorly	> half body width, anteriorly directed	about half body width (mbd)	about half body width, directed anteriorly	40% of body width, directed anteriorly	38-43% mbd, anteriorly directed	short, 25-30% of body width
n° CP	0	1	0	1	1	1	-
geographic distribution	Brazil	Brazil	Brazil	Brazil	Brazil	Brazil	Trinidad Martinique, Brazil

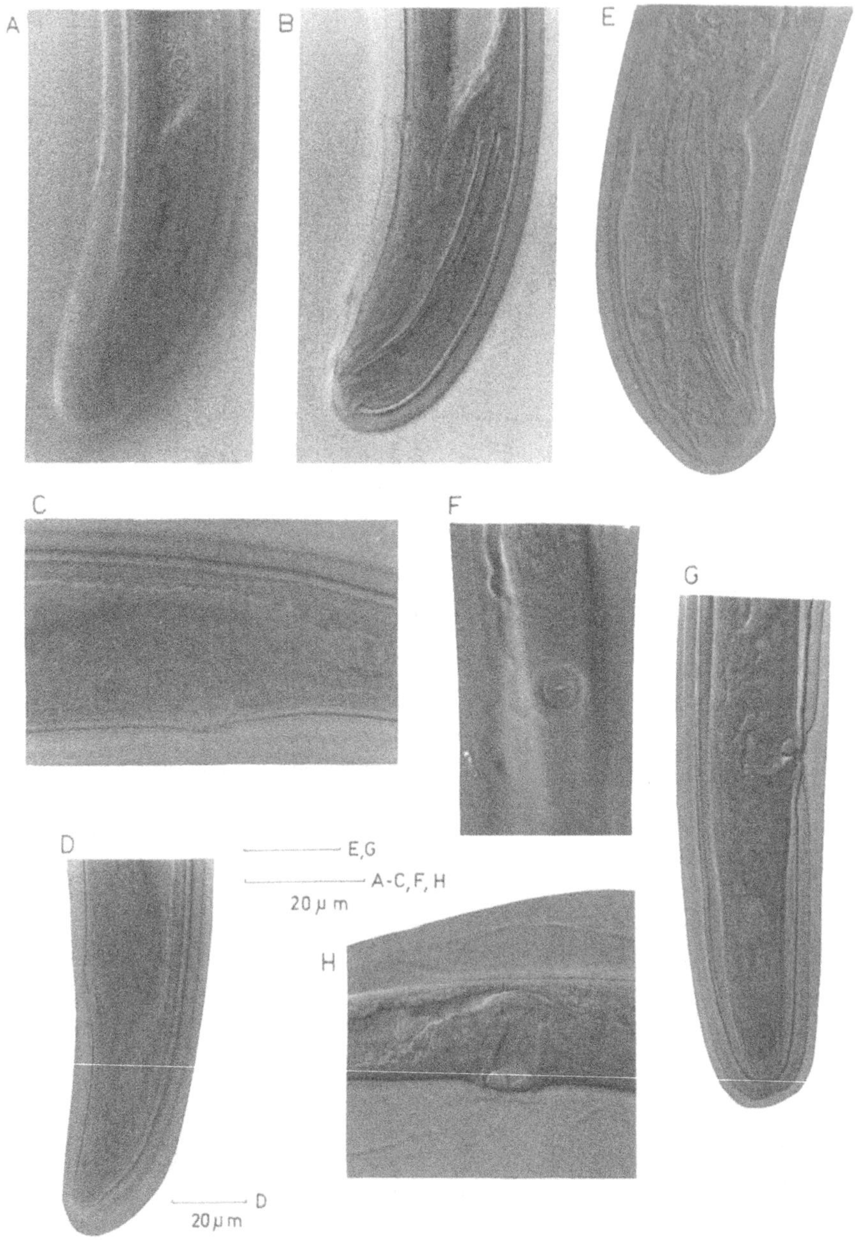

Fig. 4.37. ***Allotrichodorus brasiliensis***. Male paratype: (A-B) posterior body region, respectively with accent on caudal alae and on copulatory apparatus. Females: (C) vaginal region (holotype), (D) posterior body region (paratype). ***Allotrichodorus campanullatus***. Male: (E) copulatory apparatus and tail. Females: (F) vulva, ventral view, (G) posterior body region, (H) vaginal region.

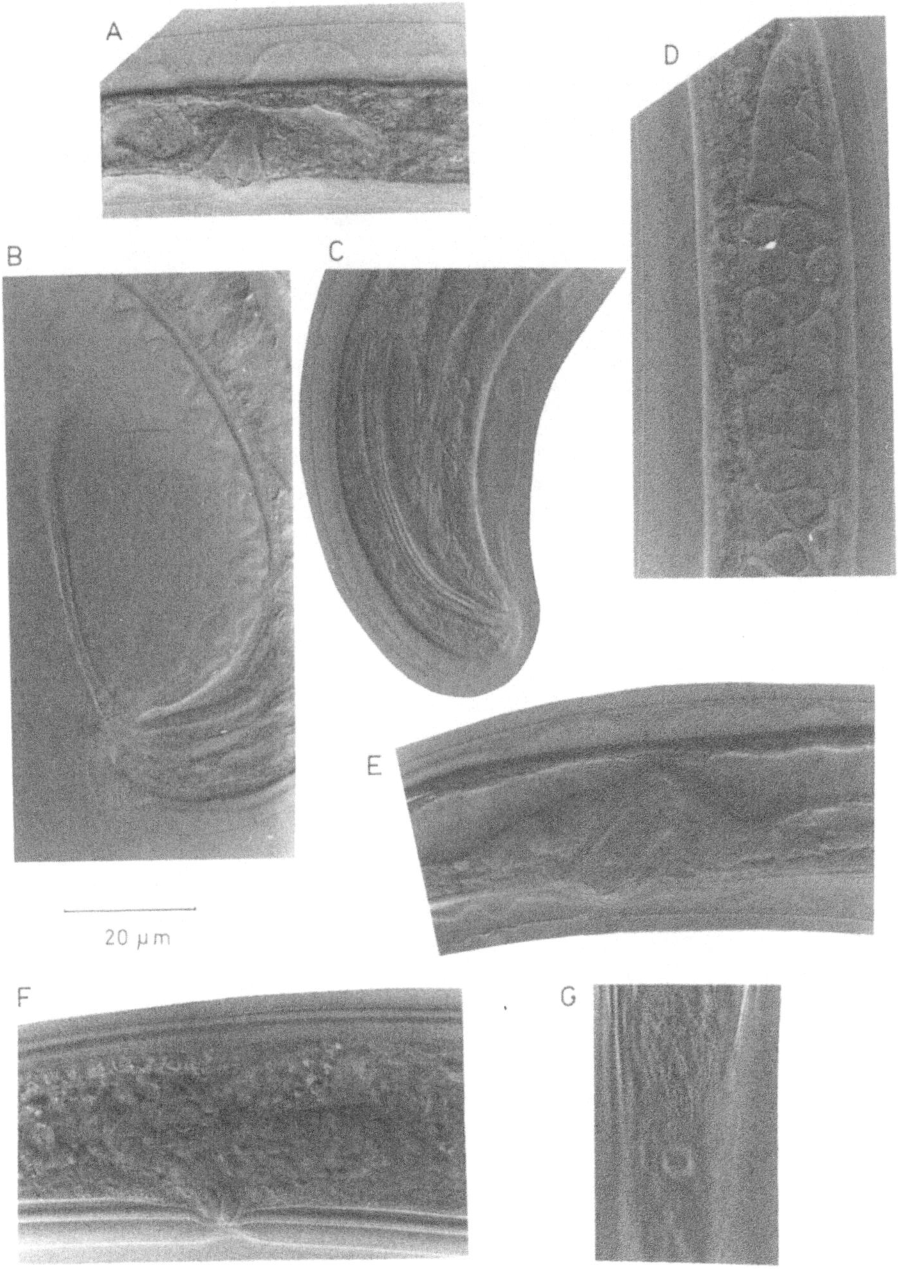

Fig. 4.38. *Allotrichodorus longispiculis*. Female holotype: (A) vaginal region. *Allotrichodorus loofi*. Male paratypes: (B) tail with protruding spicule, (C) copulatory apparatus and tail, (D) testis, (E) vaginal region. *Allotrichodorus sharmae*. Female paratypes: (F) vaginal region, (G) vulva, ventral view.

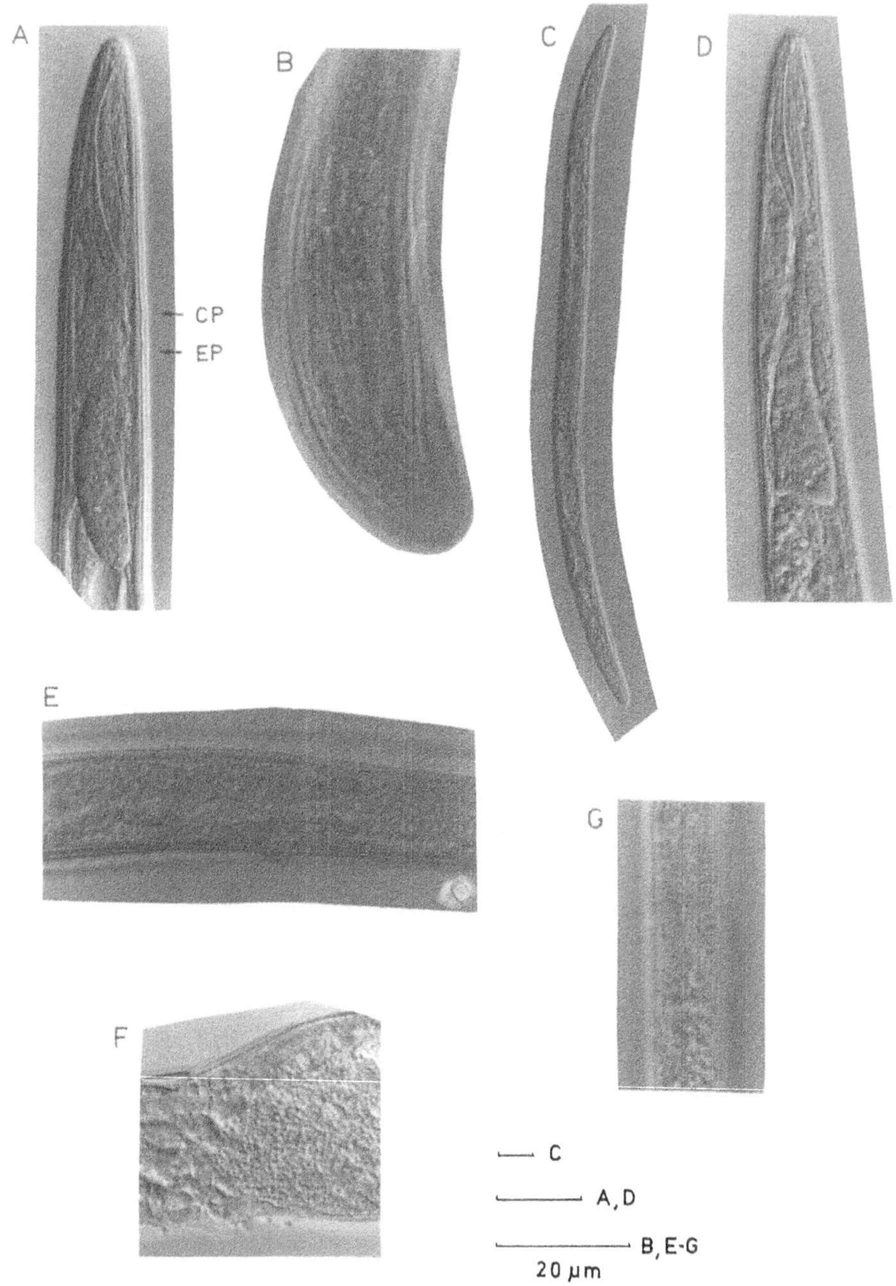

Fig. 4.39. *Allotrichodorus sharmae*. Male paratypes: (A) anterior body region, (B) posterior body region. *Allotrichodorus westindicus*. Females: (C) total view, (D) anterior body region, (E-F) vaginal region, (G) vulva, ventral view.

4.5. Position of the family Trichodoridae.

Up to 1973, the genus *Trichodorus* was the sole member of the subfamily Trichodorinae Thorne, 1935, which together with the Diphtherophorinae (containing the genera *Diphtherophora*, *Tylolaimophorus*, *Triplonchium*, *Brachynemella*) were assigned to the family Diphtherophoridae (Micoletzky, 1922) Thorne, 1935 of the superfamily Dorylaimoidea.

Thorne (1939, 1961) tentatively put Diphtherophoridae under Dorylaimoidea and Trichodorinae under Diphtherophoridae. Clark (1961) gave Diphtherophoridae a superfamily rank. Coomans & Loof (1970) proposed two new suborders Bathyodontina and Diphtherophorina under Dorylaimida which also included Dorylaimina, Mononchina and Trichosyringina. Andrassy (1976) also considered Mononchina and Diphtherophorina under Dorylaimida.

Detailed study of the structure of the reproductive system of the Dorylaimida (including Trichodoridae) and Enoplida by Geraert *et al.* (1980), revealed that the Trichodoridae show such important dissimilarities in the structure of this system that their position within the Dorylaimida became questionable.

Siddiqi (1980) removed the Diphtherophorina from the Dorylaimida to the Enoplida, and later he (1983) suppressed it in favour of Triplonchida (Triplonchia of Cobb, 1920). Lorenzen (1981) recognized under Enoplia three orders (Enoplida, Dorylaimida, Trefusiida) but did not recognize Alaimina and Diphtherophorina as suborders but assigned their families to the Dorylaimida. At present, the Trichodoridae are considered to belong to the Triplonchida (syn. Diphtherophorina Coomans & Loof, 1970, apud Siddiqi, 1983) classified under the Enoplia rather than to the Dorylaimida.

Triplonchida Cobb, 1920 is related to the order Tripylida Siddiqi, 1983 in possessing 1) a cuticle swelling upon fixation, 2) an evertible stoma which is mostly a pharyngostoma, 3) spermatheca usually present and 4) the protractors of the spicules are in the form of suspensors forming a capsule around each spicule and apparently are not attached to the body-wall; the latter character making both orders easily recognizable amongst the Enoplia. Triplonchida differs from Tripylida in having 1) the cephalic sensilla in one circlet (6 + 4), 2) amphids located at base of the cephalic region, 3) dorsal pharyngeal gland not opening in the stomatal region but at the beginning of the pharyngeal bulb, and 4) in possessing a protrusible stylet and its associated protractor muscles, lacking in the Tripylida (Siddiqi, 1983).

CHAPTER 5

Keys to the trichodorid species

Separate keys to species are provided for males and females for the genera *Trichodorus* and *Paratrichodorus*. The most important diagnostic features used in the keys are presented in Figs 5.1 -5.5. For figures of the respective species see chapter on taxonomy. Species with a 50% variability of a diagnostic feature are indicated with an asterisk and occur twice in the key; rare variability of the characters used are not included in the key. A coputerized polytomous key will be published separately in cooperation with Dr. P. Baujard.

Trichodoridae extracted by the centrifugal-flotation method (Gooris & D'Herde, 1972), using a colloidal solution of silica, Ludox LS, give the best recovery of good quality nematodes. Specimens extracted by centrifugal-flotation with $MgSO_4$-solution are poor in quality and may give problems with identification (Decraemer *et al.*, 1979). Further, it is recommended to fix the extracted trichodorid specimens in hot (60-70°C) 4% formaline = 100 ml formaldehyde 40% + 10 µm glycerine + 890 ml water (De Grisse, 1969).

5.1. Key to Trichodorus males

1.	- No ventromedian cervical papilla	*T. obscurus*
	- Ventromedian cervical papilla(e) present	2
2.	- One ventromedian cervical papilla (Fig. 5.1.F)	3
	- Two ventromedian cervical papillae (Fig. 5.1.D)	16
	- Three ventromedian cervical papillae (Fig. 5.1.A)	35

3. - Two ventromedian precloacal supplements within region of retracted spicules (second one may be near spicule head, just posterior or just anterior to it) 4
   - A single supplement within or just anterior to region of retracted spicules 5
4. - Both supplements clearly within region of retracted spicules (Fig. 5.2.G); spicule length 42-50 µm *T. intermedius*
   - Middle supplement opposite spicule head (Fig. 5.2.D); spicule length 48-55 µm *T. dilatatus*
5. - Spicule shaft with indentation (Fig. 5.3.A) 6
   - Spicule shaft without indentation (Fig. 5.3.F) 9
6. - S-E pore near pharyngo-intestinal junction *T. sanniae*
   - S-E pore positioned more anteriorly 7
7. - Spicules stout; spicule length 52-57 µm; onchiostyle 62.5-71 µm *T. magnus*
   - Spicules slenderer and shorter 8
8. - Spicules with septum in front of constriction (Fig. 5.3.B); spicule length 43-46.5 µm; onchiostyle 45-49 µm *T. kilianae*
   - Spicules without septum; spicule length 43-49 µm; onchiostyle 55-57 µm *T. vandenbergae*
9. - Very long onchiostyle (140-155 µm) *T. elegans*
   - Onchiostyle shorter than 100 µm 10
10. - Spicule shaft widened posteriorly (Fig. 5.3.M) ; spicule length 44-53 µm *T. cottieri*
    - Spicules different 11
11. - Spicules with anterior part curved ventrally, then shaft largely straight, distal tip straight or curved (Fig. 5.3.K) 12
    - Spicules curved ventrally or about straight (Fig. 5.3.J) 14
12. - Spicules slender, shaft finely striated, no bristles 13
    - Spicules stouter, shaft striated and largely provided with bristles; spicule length 43-70 µm *T. californicus**
13. - Onchiostyle 40-67.5 µm, gubernaculum 17-26 µm, spicule length 42-55 µm *T. eburneus*
    - Onchiostyle 27.5-30.5 µm, gubernaculum 10.5-15 µm, spicule length 39-45 µm *T. philipi*
14. - Spicule shaft striated, largely provided with bristles; spicule length 43-70 µm *T. californicus**

- Spicules smooth 15

15. - Spicules 43-65 μm *T. obtusus*
- Spicules shorter, 27-33 μm *T. borai*

16. - Two ventromedian precloacal supplements in region of retracted spicules *T. borneoensis*
- One precloacal supplement in region of retracted spicules 17
- No precloacal supplements in region of retracted spicules 34

17. - One ventromedian cervical papilla in region of onchiostyle (Fig. 5.1.C) or near base of onchiostyle 18
- No ventromedian cervical papilla in onchiostyle region 20

18. - Spicules with knob-like manubrium (Fig. 5.3.F,G) *T. variopapillatus**
- Spicule shape different 19

19. - Spicule shaft slender with sharp constriction, spicule length 34-43 μm *T. orientalis*
- Spicule shaft slender, its striation interrupted at narrower part near mid-corpus; spicule length 33-50 μm *T. elefjohnsoni*

20. - Spicules with knob-like manubrium 21
- Spicules different 23

21. - S-E pore between ventromedian cervical papillae; posterior precloacal supplement posterior to spicule head; spicule length 40-54μm; onchiostyle 50-66 μm *T. coomansi*
- Ventromedian cervical papillae anterior to S-E pore; posterior precloacal supplement opposite spicule head 22

22. - Spicule length 43 μm; onchiostyle 45 μm; *T. complexus*
- Spicules slenderer; spicule length 44-53 μm; onchiostyle 54-62 μm *T. hooperi**

23. - Spicule with a wide, long manubrium, more or less offset (Fig. 5.3.H,I) 24
- Spicules different 26

24. - Spicule head sharply offset from shaft by indentation (Fig. 5.3.H); spicule length 57-65 μm *T. taylori*
- Spicule head marked off differently; spicules longer 25

25. - Spicule length 63-71 μm, gubernaculum 30-37 μm; onchiostyle 53-62 μm; terminal tail cuticle usually slightly thickened (Fig. 5.2.E) *T. minzi*
- Spicule length 51-61 μm, gubernaculum 6-7 μm; onchiostyle 65-72 μm; terminal tail cuticle thickened *T. persicus*

26. - Proximal part of spicules curved ventrally, shaft largely straight 27
- Spicule shape different 30
27. - Spicules slender, anterior part clearly curved ventrally (Fig. 5.3.K); striation fine 28
- Spicules stouter, curvature less obvious 29
28. - Onchiostyle 67-73 µm, with inner onchium (Fig. 5.1.E); spicule length 62-74 µm *T. petrusalberti**
- No inner onchium observed; onchiostyle shorter, 41-64 µm; spicule length shorter, 40-55 µm; *T. rinae*
29. - Spicule manubrium short, marked; shaft usually without bristles; ventromedian cervical papillae both posterior to the nerve ring *T. nanjingensis*
- Manubrium wider, gradually tapering to a narrow shaft, usually with bristles (indicated by irregular outline); ventromedian cervical papillae positioned differently *T. sparsus**
30. - Spicule ventrally bent, shaft slender, clearly indented; S-E pore between ventromedian cervical papillae (Fig. 5.1.D) *T. parorientalis*
- Spicule shaft different; S-E pore posterior to ventromedian cervical papillae 31
31. - Spicules about equally wide, slightly ventrally bent 32
- Spicule manubrium wider, gradually tapering to a narrow shaft; spicules ventrally bent 33
- Spicule manubrium wider, gradually tapering to a straight narrow shaft, no bristles; spicule length 44-53 µm; onchiostyle 54-62 µm; terminal tail cuticle well thickened *T. hooperi**
32. - Spicule shaft smooth, without bristles; spicule length 35-47 µm; terminal tail cuticle not thickened *T. aequalis*
- Spicule shaft striated, striation interrupted at narrow mid-corpus with distinct bristles; spicule length 41-50 µm; terminal tail cuticle slightly thickened *T. paucisetosus*
33. - Posterior precloacal supplement opposite spicule head; terminal tail cuticle variously thickened; spicule shaft with a few bristles; spicule length 39.5-50 µm; onchiostyle 44.5-60 µm *T. giennensis*
- Posterior precloacal supplement opposite or posterior to spicule

head; terminal tail cuticle well thickened; spicule shaft without bristles; spicule length 43-65 μm; onchiostyle 44-73 μm — *T. sparsus**

- Posterior precloacal supplement posterior to spicule head; terminal tail cuticle not thickened; spicule length 62-74 μm — *T. petrusalberti**

34. - Spicule shaft enlarged posteriorly and provided with a velum (Fig. 5.3.D,E), spicule length 44-47 μm; onchiostyle 57-67 μm — *T. carlingi*
  - Spicule shaft with constriction, no velum, spicule length 29-38 μm; onchiostyle 33-59 μm — *T. lusitanicus**
35. - Two ventromedian cervical papillae in onchiostyle region (Fig. 5.1.A), second one may be just posterior to onchiostyle base — 36
  - One ventromedian cervical papilla in onchiostyle region — 39
  - No ventromedian cervical papilla in onchiostyle region — 44
36. - No precloacal supplements in region of retracted spicules — 37
  - One supplement in region of retracted spicules — 38
37. - Spicules rather stout, with long narrow zone at mid-spicule (Fig. 5.3.T), usually with marked bristles; spicule length 29-38 μm; onchiostyle 33-59 μm — *T. lusitanicus**
  - Spicules slenderer with a usually shorter indented zone at mid-spicule, bristles less obvious (Fig. 5.3.A); spicule length 22-37 μm; onchiostyle 32-53 μm — *T. viruliferus*
38. - Spicule shaft very slender in distal half (Fig. 5.3.P), constriction at mid-spicule may be obscure, bristles not always distinct; spicule length 32-54 μm — *T. primitivus*
  - Spicules wider anteriorly, shaft slender posteriorly, without constriction, no bristles; spicule length 31.5-41.5 μm — *T. aquitanensis*
39. - Spicules with a knob-like manubrium — 40
  - Spicules different — 41
40. - Spicules slender, shaft narrowing towards manubrium (Fig. 5.3.F); spicule length 40-53 μm; onchiostyle 50-56 μm — *T. variopapillatus**
  - Spicules stouter; spicule length 30-44 μm (Fig. 5.3.G); onchiostyle 35-50 μm — *T. similis*
41. - Posterior body region straight, caudal alae present (Fig. 5.2.A); spicule shaft enlarged posteriorly; spicule length 28-49 μm — *T. cylindricus*

- Posterior body region curved; no caudal alae; spicules different 42

42. - Spicules with a ventral flange or velum (Fig. 5.3.D); spicule length 31-37 µm *T. velatus*
- Spicules without velum 43

43. - Spicules with clearly indented shaft (Fig. 5.3.S), and distal end usually grooved; spicule length 27-34 µm; onchiostyle 32-54 µm *T. beirensis*
- Spicules with slightly indented shaft (Fig. 5.3.T) and distal end usually not grooved; spicule length 29-34 µm; onchiostyle longer, 55-66 µm *T. azorensis*

44. - Posterior body region straight; caudal alae present (Fig. 5.2.B); spicule length 38-47 µm *T. paracedarus*
- Posterior body region curved ventrally (Fig. 5.2.C); no caudal alae 45

45. - No precloacal supplements in region of retracted spicules 46
- One supplement in region of retracted spicules 47

46. - Spicules curved ventrally; manubrium long and wide; shaft with marked striae, no bristles; onchiostyle 57-82 µm *T. yokooi*
- Spicules almost straight, slenderer; manubrium short, striae less obvious, bristles present; onchiostyle 38.6-51 µm *T. tricaulatus*

47. - Spicules cephalated, anterior part of shaft curved ventrally; obvious striae interrupted about mid-spicule (Fig. 5.3.N); spicule length 46-58 µm; onchiostyle 39-48 µm *T. pakistanensis*
- Spicules cephalated, almost straight; fine striation of shaft continuous; spicule length 36-53 µm; onchiostyle 41-70 µm *T. cedarus*

5.2. Key to Trichodorus females

1. - Onchiostyle very long, 140-150 µm *T. elegans*
- Onchiostyle much shorter 2

2. - Vaginal sclerotizations rod-like pieces parallel to the vaginal lumen, well separated (Fig. 5.4.A) 3

- Vaginal sclerotizations different and not parallel to the vaginal lumen 4

3\. - Vaginal sclerotization rod-like; onchiostyle 28-57 µm *T. primitivus*
- Vaginal sclerotizations shorter, oval; onchiostyle 36-56 µm *T. viruliferus**

4\. - Vaginal sclerotizations parallel to the longitudinal body axis (Fig. 5.4.B) 5
- Vaginal sclerotizations differently orientated 7

5\. - Vaginal sclerotizations small, about triangular; vulva pore-like; onchiostyle 42-45 µm *T. complexus*
- Vaginal sclerotizations large, about rectangular; vulva a transverse slit 6

6\. - Vaginal sclerotizations roughly trapezoid or rectangular; vagina cylindrical; onchiostyle 60-72 µm; known from Virginia, U.S.A. *T. obscurus*
- Vaginal sclerotizations slightly smaller and rectangular; vagina wider and with different shape; onchiostyle 57-68 µm; known from New Zealand *T. cottieri*

7\. - Vaginal sclerotizations as large rounded quadrangular pieces close to one another (Fig. 5.4.C) 8
- Vaginal sclerotizations smaller and different 9

8\. - Vaginal sclerotizations large; vagina barrel-shaped (Fig. 5.4.N); vulva a transverse slit; one pre-advulvar and two postvulvar lateral body pores; onchiostyle 47-58 µm *T. aquitanensis*
- Vaginal sclerotizations large; vagina rhomboid (Fig. 5.4.M); vulva a transverse slit; one prevulvar and one postadvulvar lateral body pore; onchiostyle 50-54 µm *T. variopapillatus*
- Vaginal sclerotizations slightly smaller, slightly separated; vagina rhomboid; one postadvulvar lateral body pore; onchiostyle 55-63 µm *T. azorensis*

9\. - Onchiostyle small, less than 30 µm *T. philipi*
- Onchiostyle longer than 30 µm 10

10\. - Vaginal sclerotizations large, inner side concave (Fig. 5.4.E) 11
- Vaginal sclerotizations different 12

11\. - Vagina pear-shaped (Fig. 5.4.O); vulva a transverse slit; onchiostyle 60.5-69 µm *T. dilatatus*
- Vagina pear-shaped; vulva shape unknown; onchiostyle 55-80 µm *T. californicus*
- Vagina rhomboid; vaginal sclerotizations slightly larger; vulva

a transverse slit; onchiostyle 62-75.2 μm *T. intermedius*

The females of these three Californian species are difficult to differentiate, showing minor differences in shape and size of the vaginal sclerotizations.

12. - Vaginal sclerotizations large, rounded triangular, rarely oval (Fig. 5.4.D) 13

- Vaginal sclerotizations different 15

13. - Vagina cylindrical; vaginal sclerotizations large triangular with longest side parallel to lumen vagina, slightly separated; vulva a transverse slit; no lateral body pores; onchiostyle 39-56 μm *T. beirensis*

- Vagina rhomboid; vaginal sclerotizations slightly smaller, different 14

14. - Vulva pore-like; vaginal sclerotizations regular triangular, well to slightly separated; one postadvulvar lateral body pore or with an additional prevulvar body pore on each side; onchiostyle 35-58 μm *T. lusitanicus*

- Vulva a transverse slit; vaginal sclerotizations rounded triangular; close to one another; one postadvulvar lateral body pore; onchiostyle 36-52 μm *T. similis*

15. - Vaginal sclerotizations medium-sized triangular with tips pointing to the vulva, oblique (Fig. 5.4.F) 16

- Vaginal sclerotizations different 17

16. - Onchiostyle long, 62-77 μm; vagina pear-shaped; vaginal sclerotizations triangular *T. yokooi*

- Onchiostyle 42-50 μm; vagina rhomboid; vaginal sclerotizations rounded triangular *T. velatus*

- Onchiostyle 35-52 μm; vagina barrel-shaped; vaginal sclerotizations triangular *T. cylindricus*

17. - Vaginal sclerotizations large or medium-sized, rounded, well separated (Fig. 5.4.I) 18

- Vaginal sclerotizations smaller 22

18. - Vagina pear-shaped (Fig. 5.4.O) 19

- Vagina barrel-shaped (Fig. 5.4.N) 21

19. - Vaginal sclerotizations large, separated 20

- Vaginal sclerotizations smaller (medium-sized), well separated; onchiostyle 53-60 μm; one postadvulvar lateral body pore; vulva a transverse slit *T. hooperi*

20. - One postadvulvar lateral body pores; onchiostyle 60-72 µm *T. taylori*
- One prevulvar and one postadvulvar lateral body pore; onchiostyle 63-71 µm *T. persicus*
- One prevulvar and one postadvulvar lateral body pore; onchiostyle 53-64 µm; vaginal sclerotizations smaller *T. minzi*

21. - Sperm distributed along the uteri; onchiostyle 41-53 µm *T. nanjingensis**
- Sperm in spermathecae; onchiostyle 58-72 µm *T. paucisetosus*

22. - Vaginal sclerotizations minute dot-like or small rounded 23(1)
- Vaginal sclerotizations small or minute, narrowly triangular, rounded triangular, roughly ovoid or trapezoid 25(1)

23. - Vaginal sclerotizations dot-like; vulva a longitudinal slit; one postadvulvar lateral body pore; vagina short; onchiostyle 43 µm *T. sanniae*
- Vaginal sclerotizations small rounded; vulva different 24

24. - Vagina about pear-shaped, short; one prevulvar and one postadvulvar lateral body pore; vulva a pore; onchiostyle 38-44 µm *T. tricaulatus**(3)
- Vagina barrel-shaped, 1/3rd body width long; vulva a pore; one postadvulvar lateral body pore; onchiostyle 45-60 µm *T. elefjohnsoni*
- Vagina barrel-shaped, short to about 1/2 body width long; vulva a transverse slit; one postadvulvar lateral body pore; onchiostyle 38-51 µm *T. pakistanensis*

25. - Postadvulvar body pores subventral instead of lateral 26
- Postadvulvar body pores lateral, when present 27

26. - No prevulvar lateral body pores; onchiostyle 45-70 µm; vagina pear-shaped; geographic distribution Japan, China *T. cedarus*
- One pair of prevulvar lateral body pores; onchiostyle 42-60 µm; vagina usually barrel-shaped; geographic distribution W-Africa *T. eburneus* (4)
- One pair of prevulvar lateral body pores; onchiostyle 50-58 µm; known from South Africa *T. vandenbergae* (4)

27. - Mean onchiostyle longer than 62 µm 28
- Mean onchiostyle shorter 31

28. - Body length smaller than 1000 µm 29
- Body length longer than 1000 µm 30

29. - Vagina rhomboid; vaginal sclerotizations small triangular or

drop-like; onchiostyle 65-74 µm; one prevulvar and one postadvulvar lateral body pore *T. petrusalberti*

- Vagina pear-shaped; vaginal sclerotizations small triangular to oval; mean onchiostyle 62 µm; one pair of prevulvar and one pair of postadvulvar lateral body pores *T. aequalis**(2,3)

30. - Vagina long, narrow; vaginal sclerotizations small, rounded triangular to oval; onchiostyle 68-70 µm; one pair of postadvulvar lateral body pores; known from South Africa *T. magnus*

- Vagina short, wider; vaginal sclerotizations slightly smaller, triangular; one pair of postadvulvar lateral body pores with or without an additional prevulvar pair; mean onchiostyle 60.5 µm; known from U.S.A. *T. obtusus**

31. - 0 or one pair of postadvulvar lateral body pores 32

- Two pairs of lateral body pores: one prevulvar and one postadvulvar 39

32. - Sperm in spermathecae 33

- Sperm throughout the uteri; onchiostyle 41-53 µm *T. nanjingensis**

33. - Vagina pear-shaped; one pair of postadvulvar lateral body pores 34

- Vagina barrel-shaped 35

- Vagina rhomboid; onchiostyle 36-56 µm *T. viruliferus**

34. - Onchiostyle 53-57 µm; vulva a pore *T. borneoensis*

- Onchiostyle 39-45 µm; vulva a transverse slit *T. paracedarus*

35. - Body length shorter than 650 µm; vulva a pore; onchiostyle 46-51 µm vulva sclerotizations roughly ovoid, oblique; known from India *T. borai*

- Mean body length longer than 650 µm 36

36. - Vaginal sclerotizations fine, minute triangular, tip pointed to vulva 37

- Vaginal sclerotizations slightly larger, triangular to oval 38

37. - Vagina short; onchiostyle 41-62 µm; L 624-1083 µm; known from South Africa *T. rinae*

- Vagina slightly larger; onchiostyle 47-58 µm; L 651-941 µm; known from Middle-East *T. orientalis*

- Vagina narrowly elongated; onchiostyle 46-49 µm; L 878-962 µm; known from South Africa *T. kilianae*

38. - Mean onchiostyle 45 µm; body length 1100-1500 µm; known from USA *T. obtusus*

- Mean onchiostyle 53-57 µm; body length 700-1177 µm; known from Spain *T. giennensis*

39. - Vagina pear-shaped 40

- Vagina barrel-shaped or rhomboid 42

40. - Overlapping by ventrosublateral pharyngeal glands; onchiostyle 38-44 µm; body length 602-785 µm; vaginal sclerotizations small rounded (Fig. 5.4.J) to rounded triangular; known from Japan *T. tricaulatus**[(3)]

- Pharyngeal bulb offset 41

41. - Vaginal sclerotizations small oval (Fig. 5.4.K) ; onchiostyle 50-55 µm; body length 634-768 µm; known from South Africa *T. parorientalis*

- Vaginal sclerotizations slightly larger, triangular; mean onchiostyle 45 µm; body length 601-821 µm; known from U.S.A. and Japan *T. aequalis**[(2,3)]

42.- Body longer than 1100 µm; onchiostyle 49-74 µm; vaginal sclerotizations small triangular *T. obtusus**

- Body length shorter 43

43.- Vaginal sclerotizations small trapezoid or rectangular, oblique (Fig. 5.4.L); onchiostyle 56-65 µm, body length 711-886 µm; vagina barrel-shaped; known from Alaska *T. carlingi*[(2)]

- Vaginal sclerotizations small triangular; onchiostyle 45-67 µm; body length 530-1090 µm; vagina barrel-shaped; known from Europe *T. sparsus*[(2)]

- Vaginal sclerotizations small triangular; onchiostyle 50-65 µm; body length 795-1014 µm; vagina rhomboid or barrel-shaped; known from Kenya *T. coomansi*

Remarks

(1) Vaginal sclerotized pieces when small, appear rather similar in many species, showing small interspecific variations in size and shape which often fall within the range of intraspecific variability. These features are difficult for species differentiation in a key. The shape of the vagina is used as an additional diagnostic feature although its restricted value (common occurrence of variability in size and shape due to fixation and physiological condition of the specimens) (Decraemer, 1980a).

(2) According to Bernard (1992a) *T. carlingi*, *T. sparsus* and *T. aequalis* are very similar

and cannot be reliably separated in female; however, the vaginal sclerotizations show minor differences.

(3) *T. tricaulatus* and *T. aequalis* females differ only by the different type of pharyngo-intestinal junction (Shishida, 1979).

(4) *T. eburneus* and *T. vandenbergae* females are very similar, the vaginal sclerotizations showing only minor differences in size and shape. Currently, they differ by their geographic distribution.

5.3. Key to *Paratrichodorus* males

1. - Males with 1 ventromedian precloacal supplement; no ventromedian cervical papillae; inner onchium present — 2
   - Males with 2 to 4 ventromedian precloacal supplements, with or without ventromedian cervical papillae; inner onchium rare — 3
2. - S-E pore along anterior intestine; pharyngeal bulb usually offset; onchiostyle 21-26 µm; spicule length 42-50 µm; body length 430-550 µm — *P. nanus*
   - S-E pore more anteriorly (along posterior part of pharynx or at level of junction with intestine); ventrosublateral pharyngeal glands usually overlapping the intestine, pharyngeal bulb rarely offset; onchiostyle 32-39 µm; spicule length 48-73 µm; body length 540-678 µm — *P. minor*
3. - Two precloacal supplements, 1-2 within the region of retracted spicules — 4
   - Three or four precloacal supplements, respectively with 1-2 or 3 supplements within the region of the retracted spicules — 9
4. - Both precloacal supplements within the region of the retracted spicules — 5
   - Only one precloacal supplement within the region of the retracted spicules; pharyngeal bulb offset; spicules with marked manubrium; onchiostyle 31-35 µm; spicule length 30-34 µm; body length 410-520 µm — *P. delhiensis*
5. - One ventromedian cervical papilla; pharynx usually with well developed anteriodorsal intestinal overlap; well developed sperm with large sausage-shaped nucleus; onchiostyle 43-48 µm; spicule length 36-39 µm; body length 530-770 µm; males rather rare — *P. porosus*
   - No ventromedian cervical papilla; short subventral overlap of the pharyngeal glands; small sperm with small rounded nucleus; males rare

or common 6

6. - One pair of lateral cervical pores at base of amphids; onchiostyle long, 75-78 μm, anterior position of S-E pore (near base of onchiostyle); both precloacal supplements opposite distal third of retracted spicules; spicule length 67-68 μm; body length 750-847 μm *P. acaudatus*
   - No lateral cervical pores; onchiostyle shorter; S-E pore more posteriorly; both precloacal supplements more dispersed 7
7. - Spicules short, 30 μm; males rare; both precloacal supplements along retracted spicules; onchiostyle 42 μm; body length 590-680 μm *P. allius*
   - Spicules longer; males rare or common; both precloacal supplements opposite distal half of retracted spicules 8
8. - S-E pore opposite isthmus; bursa extending to halfway the retracted spicules; known from Congo Brazzaville *P. alleni*[(1)]
   - S-E pore opposite anterior end pharynx bulb; bursa extending to spicule head; spicules shorter in relation to anal body width; known from Australia, New Zealand and South Africa *P. lobatus*[(1)]
9. - Four precloacal supplements, two clearly within the region of the retracted spicules, one at spicule head; pharyngeal glands with slight ventral overlap; onchiostyle 39-41 μm; spicule length 28-36 μm; body length 570-690 μm *P. paraporosus*
   - Three precloacal supplements, one or two within the region of the retracted spicules 10
10. - Two ventromedian cervical papillae 11
   - One ventromedian cervical papilla 13
   - No ventromedian cervical papilla 26
11. - Two pairs of postcloacal papillae (Fig. 5.5.A); middle precloacal supplement just anterior to retracted spicules; medium-sized sperm with rounded nucleus (Fig. 5.5.H); onchiostyle 29-36 μm; spicule length 30-34 μm; body length 450-630 μm *P. mirzai*
   - One pair of postcloacal papillae; middle precloacal supplement at level of spicule head; sperm with rounded or sausage-shaped nucleus 12
12. - Onchiostyle 40-44 μm; spicule length 33-38 μm; body length 590-710 μm sperm with large sausage-shaped nucleus (Fig. 5.5.D) *P. paramirzai*
   - Onchiostyle 33-35 μm; spicule length 30-33.5 μm; body length

520-660 μm; medium-sized sperm with rounded nucleus *P. faisalabadensis* (2)
- Onchiostyle 30-32 μm; spicule length 30-31 μm; body length 430-510 μm; medium-sized sperm with rounded nucleus *P. psidii* (2)

13.- Onchiostyle long, 133-180 μm; long fusiform sperm cells (Fig. 5.5.G); ventromedian cervical papilla at base of onchiostyle; spicule length 67.5-87.5 μm; body length 1160-1630 μm *P. macrostylus*
- Onchiostyle shorter than 100 μm; sperm cells different 14

14.- Spicules long, 79-87 μm; two precloacal supplements within the distal third of retracted spicules; sperm with large sausage-shaped nucleus; onchiostyle 68-82 μm; body length 930-1200 μm *P. atlanticus*
- Spicules shorter than 65 μm 15

15.- The two posterior precloacal supplements both close to the cloacal opening; sperm large with large sausage-shaped nucleus 16
- The two posterior precloacal supplements further along the retracted spicules; sperm large or small 18

16.- Spicules slender with narrow, kinking anterior part (Fig. 5.5.B); onchiostyle longer than 45 μm 17
- Spicules stouter, differently shaped; onchiostyle 38-44 μm; pharynx with both posterioventral pharyngeal and anteriodorsal intestinal overlaps; spicule length 46-52 μm; body length 666-694 μm *P. sacchari*

17.- Ventral pharyngeal gland overlap; distance between posterior two precloacal supplements small; onchiostyle 47-52 μm; spicule length 46-53 μm; body length 710-900 μm *P. anemones*
- Anteriodorsal intestinal overlap; distance between posterior two precloacal supplements larger; onchiostyle 49-63.5 μm; spicule length 44-61.5 μm; body length 691-1103 μm *P. hispanus*

18.- Spicules short, 34-37 μm, with narrower kinking part posterior to spicule head; large sperm with sausage-shaped nucleus; anteriodorsal overlap of pharynx; onchiostyle 43-48 μm; body length 560-800 μm *P. weischeri*
- Spicules with different shape, usually longer 19

19.- Larger sperm with sausage-shaped nucleus 20
- Small sperm with small rounded nucleus 23

20.- Ventromedian cervical papilla at base of onchiostyle; onchiostyle 64-72 μm; anteriodorsal intestinal overlap; spicule length 39-43 μm;

body length 790-1040 µm *P. grandis*

- Ventromedian cervical papilla posterior to onchiostyle base; onchiostyle shorter 21

21.- Tail as long as anal body width; two postcloacal papillae on each side; usually with anteriodorsal intestinal overlap but all types of overlap and offset bulb possible; onchiostyle 45-60 µm; spicule length 40-54 µm; body length 610-990 µm *P. pachydermus*[(4)]

- Tail shorter; one pair of postcloacal papillae; anteriodorsal intestinal overlap or pharyngeal bulb offset 22

22.- Onchiostyle 49-60 µm; ventromedian cervical papilla close to the S-E pore; spicule length 36-43 µm; body length 440-705µm *P. queenslandensis*

- Onchiostyle 35-47 µm; ventromedian cervical papilla at larger distance from S-E pore; spicule length 35-47 µm; body length 470-680 µm *P. orrae*

23.- Ventromedian cervical papilla at or near base of onchiostyle 24

- Ventromedian cervical papilla posterior to nerve ring or at its level 25

24.- Onchiostyle 43 µm; body length 360 µm *P. rhodesiensis** (specimens from Ivory Coast)

- Onchiostyle 40-51 µm; body length 523-742 µm; spicule length 52-55 µm *P. catharinae**

25.- Anterior cloacal lip not protruding; ventral overlap pharyngeal glands; ventromedian cervical papilla at level nerve ring or posterior to it; onchiostyle 47-52 µm; spicule length 45-50 µm; body length 714-900µm *P. tunisiensis*

- Protruding anterior cloacal lip (Fig. 5.5.B); ventral pharyngeal gland overlap; ventromedian cervical papilla at level or anterior to nerve ring; onchiostyle 40-51 µm; spicule length 52-55 µm; body length 523-742 µm *P. catharinae**[(3)]

- Protruding cloacal lip; ventral and dorsal overlap of pharynx; ventromedian cervical papilla posterior to nerve ring; onchiostyle 44-54 µm; spicule length 48-58 µm; body length 510-865 µm; spicules slenderer *P. meyeri*[(3)]

26.- Second precloacal supplement near spicule head; small sperm with small oval nucleus (Fig. 5.5.E); slight ventral overlap pharyngeal glands; spicule length 42-44 µm; spicules slightly cephalated; onchiostyle

39-41 µm; body length 600-690 µm; ♂ less common than ♀ *P. rhodesiensis**

- Middle supplement opposite mid-spicule when spicule retracted; sperm with fibrillar appearance, small nucleus (Fig. 5.5.F); usually with marked ventral overlap of pharyngeal glands, but all types of overlap possible; spicule length 36-57.5 µm; onchiostyle 41-64 µm; body length 537-1029 µm; males rare *P. teres*

Remarks

No males are known for *P. acutus*, *P. anthurii* and *P. renifer*.

(1) *P. allius* (only known by ♂ holotype) and *P. lobatus* differ by minor differences, some of them within the range of intraspecific variability. *P. lobatus* males exceptionally with three ventromedian precloacal papillae.

(2) *P. faisalabadensis* and *P. psidii* could represent variations of the same species.

(3) *P. catharinae* and *P. meyeri* males are very close.

(4) One postcloacal papilla and a pore on each side according to Sturhan (1985).

5.4. Key to *Paratrichodorus* females

1. - Onchiostyle longer than 80 µm; S-E pore near base of onchiostyle 2
   - Onchiostyle shorter; S-E pore more posteriorly 3
2. - Onchiostyle 160-188 µm; vaginal sclerotizations rod-like, oblique; vagina wide, rounded trapezoid with refractive folds; sperm long fusiform *P. macrostylus*
   - Onchiostyle 82-83 µm; vaginal sclerotizations small quadrangular; vagina quadrangular; sperm shape unknown *P. acaudatus*
3. - Tail usually convex-conoid, than conoid with acute or slightly rounded terminus (Fig. 5.5.C); onchiostyle 22-27 µm; vaginal sclerotizations bold dots; uterus with thread-like bodies; males unknown *P. acutus*
   - Tail shape different; onchiostyle usually longer 4
4. - Body pores medioventral (advulvar) (Fig. 5.5.K) 5
   - Body pores absent or when present lateral-sublateral (Fig. 5.5.L) 6
5. - Medioventral body pores (4-5) restricted to post-vulvar region;

vulva a longitudinal slit; intestine with ventral overlap by pharyngeal glands; onchiostyle 39-42 µm; body length 620-750 µm; males common *P. paraporosus*

- Medioventral body pores prevulvar (1-3) and postvulvar (1-4); vulva a pore; usually with anteriodorsal intestinal overlap of pharynx; onchiostyle 39-58 µm; body length 420-867 µm; males rare *P. porosus*

6. - S-E pore along anterior intestine; pharyngeal bulb offset; onchiostyle 21-26.5 µm; unusual sperm location at tip of reflexed branches possible (Fig. 5.5.J) *P. nanus*
   - S-E pore in pharyngeal region or near pharyngo-intestinal junction; pharynx usually with overlaps 7
7. - Mean onchiostyle length shorter to equal to 30 µm 8
   - Mean onchiostyle longer 9
8. - Vaginal sclerotizations rod-like, parallel to longitudinal body axis; usually with ventral overlap of pharyngeal glands; sperm small or thread-like, dispersed throughout uteri; vulva a transverse slit; mean onchiostyle 28-29 µm, with inner onchium; males rare *P. minor**
   - Vaginal sclerotizations small triangular; pharyngeal bulb offset; vulva a pore; small oval sperm concentrated near oviduct; mean onchiostyle 30 µm; male unknown *P. anthurii*
9. - Onchiostyle longer than 64 µm 10
   - Onchiostyle shorter 11
10. - Large sperm with sausage-shaped nucleus dispersed throughout uteri; S-E pore along isthmus; usually with anteriodorsal intestinal overlap of pharynx; vulva a pore; vaginal sclerotizations small triangular; onchiostyle 65-72 µm; body length 900-1040 µm *P. grandis*
    - Large sperm with sausage-shaped nucleus in spermathecae (Fig. 5.5.I); S-E pore opposite mid-bulb; vulva a pore; onchiostyle 64-70 µm; body length 760-1130 µm *P. atlanticus*
11. - Sperm concentrated near oviduct, spermathecae more or less marked 12
    - Spermathecae absent, sperm when present dispersed throughout uteri (Fig. 5.5.J) 16
    - Unusual sperm location at tip of reflexed ovaries 29
12. - Sperm large with sausage-shaped nucleus 14

- Sperm smaller (medium-sized or small) with rounded nucleus 13

13.- Mean onchiostyle 32 μm; anteriodorsal overlap of pharynx; body length 400-520 μm *P. psidii*

- Mean onchiostyle 49 μm; usually with ventral overlap of pharyngeal glands; body length 707-1100 μm *P. tunisiensis**

14.- Vaginal sclerotizations narrow triangular, oblique; vulva a pore 15

- Vaginal sclerotizations small oval, parallel to lumen of vagina; vulva a transverse slit; onchiostyle 43-48 μm; body length 590-790 μm; anteriodorsal intestinal overlap *P. weischeri*

15.- Anteriodorsal intestinal overlap; one prevulvar lateral body pore; onchiostyle 43-63.5 μm; body length 660-1100 μm *P. hispanus*

- Generally with ventral overlap of pharyngeal glands; usually one prevulvar and one postvulvar body pore; onchiostyle 42-53 μm; body length 650-970 μm *P. anemones*

- Anteriodorsal or both anteriodorsal and ventral overlaps at pharyngo-intestinal junction; one pair of postvulvar lateral body pores; onchiostyle 46-48 μm; body length 589-733 μm *P. sacchari**

16.- With ventral overlapping of the ventrosublateral pharyngeal glands 17

- With anteriodorsal intestinal overlap of pharyngeal bulb 23

- Pharyngeal bulb offset 25

- With both anteriodorsal intestinal overlap and ventral pharyngeal overlap 30

17.- Sperm medium-sized with rounded nucleus; mean onchiostyle 30.5-35 μm; body length 420-640 μm; vulva a longitudinal slit; vaginal sclerotizations dot-like, well separated *P. mirzai**

- Sperm small with small oval nucleus; males common or rare 18

- Sperm small to thread-like/fibrillar; males rare 22

18.- Ventral pharyngeal overlap usually pronounced; vulva a longitudinal slit; vaginal sclerotizations rod-like, well separated, parallel to lumen vagina; onchiostyle 44.5-58 μm; body length 535-1051 μm; 1-2 lateral body pores (always one pore postvulvar) *P. lobatus*

- Ventral pharyngeal overlap small 19

19.- No postvulvar lateral body pores 20

- 1-2 postvulvar lateral body pores present 21

20.- Vaginal sclerotizations rod-like/oval, well separated; onchiostyle 35-53 μm;

body length 400-1044 μm *P. allius**(2)

- Vaginal sclerotizations small oval, 1.5-2 μm apart; onchiostyle 37.5-44 μm; body length 380-800μm *P. rhodesiensis**(2)

21.- Vaginal sclerotizations dot-like, well separated; vulva a pore; onchiostyle 34-50 μm; body length 546-766 μm; 1-2 lateral body pores *P. catharinae**

- Vaginal sclerotizations small oval, parallel to vaginal lumen; onchiostyle 48-51 μm; body length 707-1100 μm; 2 lateral body pores *P. tunisiensis**

22.- Vaginal sclerotizations rod-like, parallel to longitudinal body axis and their tips close to each other; vagina rounded quadrangular; onchiostyle 26-47 μm with inner onchium; vulva a transverse slit; body length 440-1530 μm; no lateral body pores *P. minor**

- Vaginal sclerotizations oval, well separated, oblique; vagina with characteristic shape (Fig. 3.13.L); onchiostyle 41-61 μm; vulva a longitudinal slit; body length 626-1325 μm; usually 1 lateral body pore on each side, but 0-6 pores in total possible *P. teres*

23.- Sperm large with sausage-shaped nucleus 24

- Sperm small with oval nucleus; vaginal sclerotizations dot-like, well separated; vulva a pore; onchiostyle 34-50 μm; body length 546-766 μm; 1-2 lateral body pores *P. catharinae**

- Sperm medium-sized with rounded nucleus; vaginal sclerotizations rod-like well separated; vulva a pore; onchiostyle 30-34 μm; body length 540-650 μm; *P. faisalabadensis*

24.- Vaginal sclerotizations round to oval, may be oblique; vulva a pore; onchiostyle 44-58 μm; body length 600-1020 μm; usually 2 (2-5) lateral body pores on each side *P. pachydermus*

- Vaginal sclerotizations narrow triangular, oblique; vulva a pore; onchiostyle 46-48 μm; body length 589-733 μm; one lateral body pore on each side *P. sacchari*

- Vaginal sclerotizations oval, oblique, well separated; vulva a longitudinal slit; onchiostyle 38-47 μm; body length 455-790 μm; one pair of postadvulvar lateral body pores *P. orrae*

25.- Mean onchiostyle smaller or equal to 35 μm 26

- Mean onchiostyle longer 27

26.- Vulva a transverse slit; uterus with thread-like structures; mean onchiostyle 31-32.5 µm; vaginal sclerotizations reniform; body length 420-560 µm; no lateral body pores *P. renifer*

- Vulva a longitudinal slit; uterus with medium-shaped sperm; vaginal sclerotizations dot-like, well separated; mean onchiostyle 30.5-35 µm; two lateral body pores; body length 420-640 µm *P. mirzai**

- Vulva a pore; mean onchiostyle 33.5 µm; body length 440-595 µm *P. delhiensis*

27.- Vaginal sclerotizations small oval 28

- Vaginal sclerotizations dot-like, well separated; onchiostyle 41 -44.5 µm; two postadvulvar lateral body pores on each side *P. paramirzai*

28.- Onchiostyle 38-47 µm; one postvulvar lateral body pore *P. orrae*

- Onchiostyle 46-55 µm; two postvulvar lateral body pores *P. queenslandensis*

29.- Pharyngeal bulb with short ventral overlap; onchiostyle 37.5-44 µm; shape of vagina rounded to oval *P. rhodesiensis**(2)

- Usually with overlapping pharyngeal glands or both pharyngeal and intestinal overlaps; onchiostyle 35-53 µm; vagina broad rounded *P. allius**(2)

30.- Sperm large with sausage-shaped nucleus; onchiostyle 46-48 µm; vulva pore-like *P. sacchari**

- Small oval sperm; onchiostyle 45-53 µm; body length 510-896 µm; vulva a longitudinal slit; 2-5 lateral body pores; vaginal sclerotization narrow triangular, separated *P. meyeri*

Remarks

(1) Females are unknown for *P. alleni*.

(2) Females of *P. allius* and *P. rhodesiensis* are difficult to differentiate.

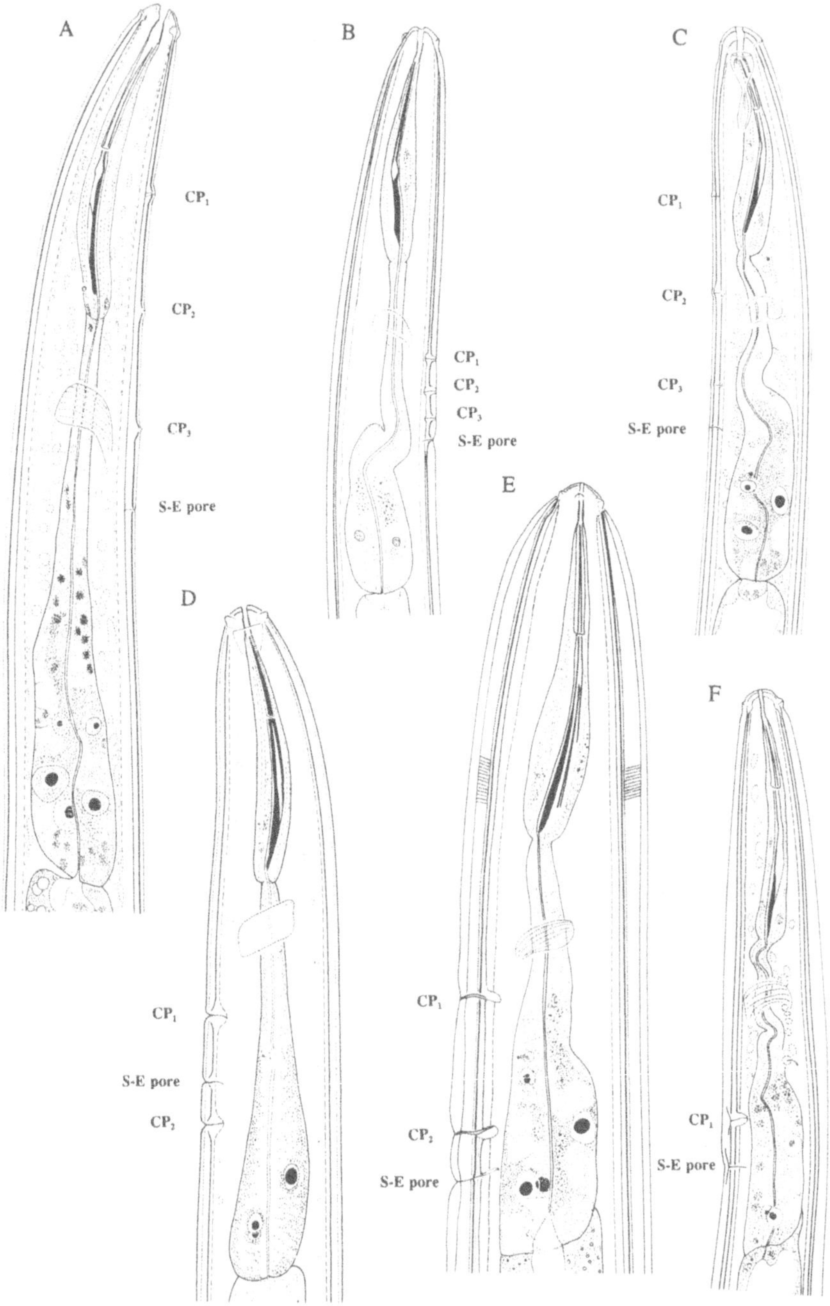

Fig. 5.1. Pharyngeal region in *Trichodorus* male with position of ventromedian cervical papillae (CP) and S-E pore: A. *T. viruliferus*; B. *T. paracedarus*; C. *T. similis*; D. *T. parorientalis*; E. *T. petrusalberti*; F. *T. kilianae*.

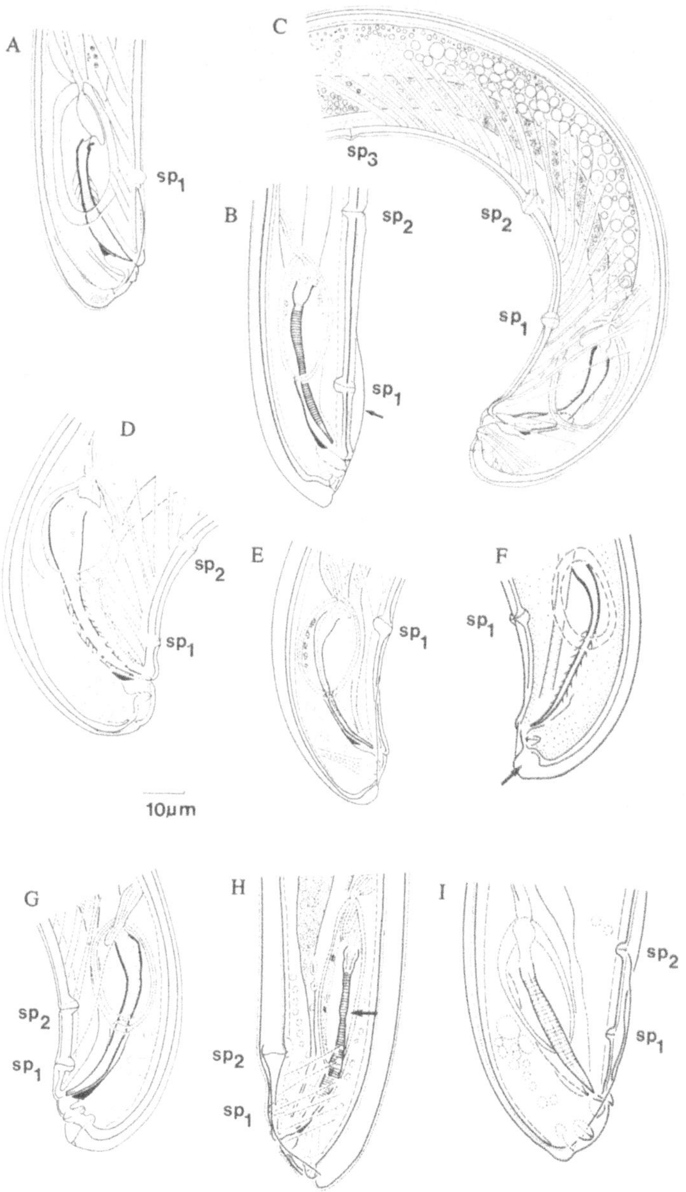

Fig. 5.2. Male tail region in *Trichodorus* (A-G) and *Paratrichodorus* (G-I) with position of the posterior precloacal supplements (SP1/SP2) in relation to the retracted spicule. A. *T. cylindricus* with bursa; B. *T. paracedarus* with bursa; C. *T. velatus*; D. *T. dilatatus*; E. *T. orientalis*; F. *T. sparsus* (after Loof, 1973; arrow points at swollen end cuticle); G. *T. intermedius*; H. *P. anemones* (arrow points at kinking part of the spicule); I. *P. allius*.

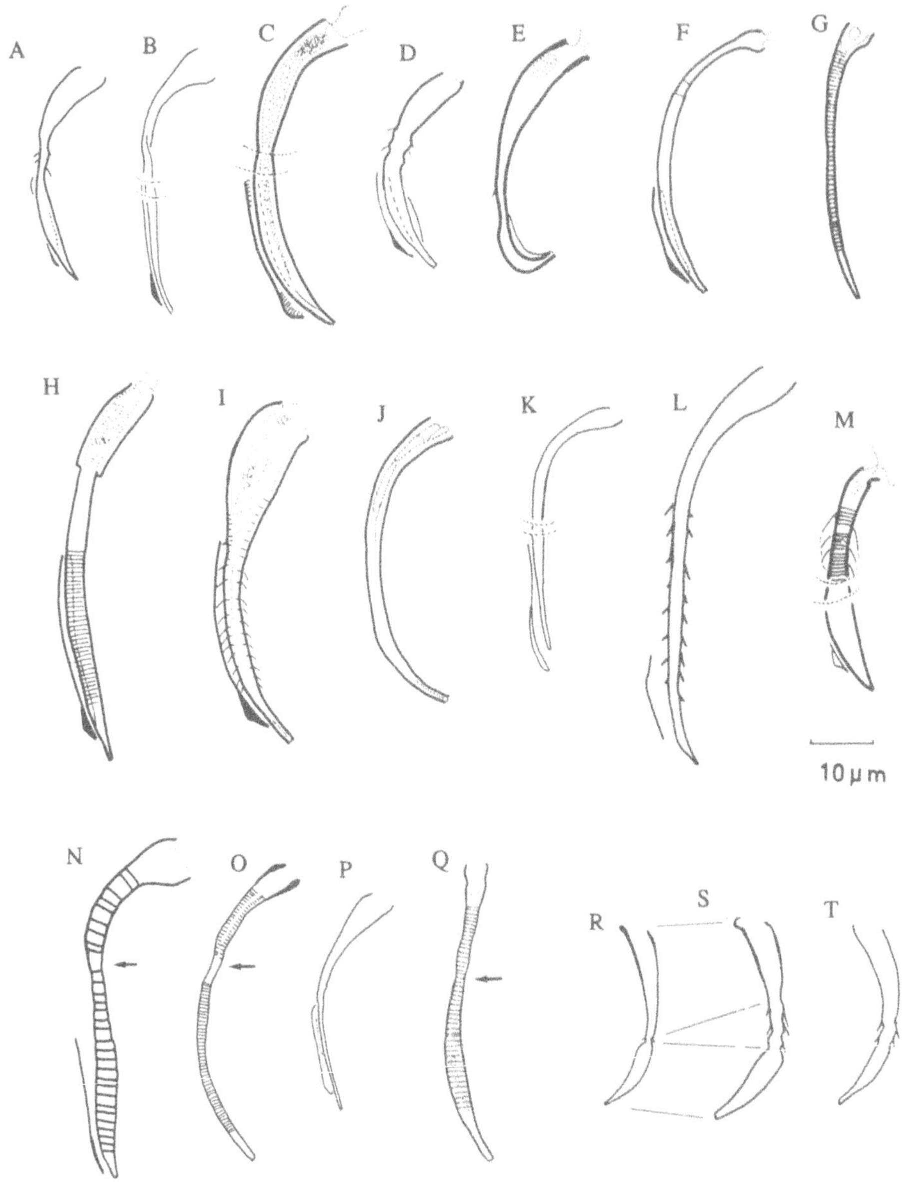

Fig. 5.3. Different types of spicules in *Trichodorus*: A. shaft indented, provided with spines/bristles; B. shaft indented, provided with septum; C. shaft indented; D-E. shaft provided with a velum; F-G. knob-like manubrium, respectively with curved and straight shaft; H. large, offset manubrium; I. long, wide manubrium, not offset; J. no marked manubrium, shaft equally wide and ventrally curved; K-L. anterior end of spicules curved, shaft respectively smooth and with bristles; M. shaft posteriorly widened; N-O. striation of shaft interrupted (see arrow); P. spicules anteriorly slightly wider, continuing into a fine shaft; Q. shaft with irregular outline (= kinking part); R-T. spicule shaft indented at different level and degree.

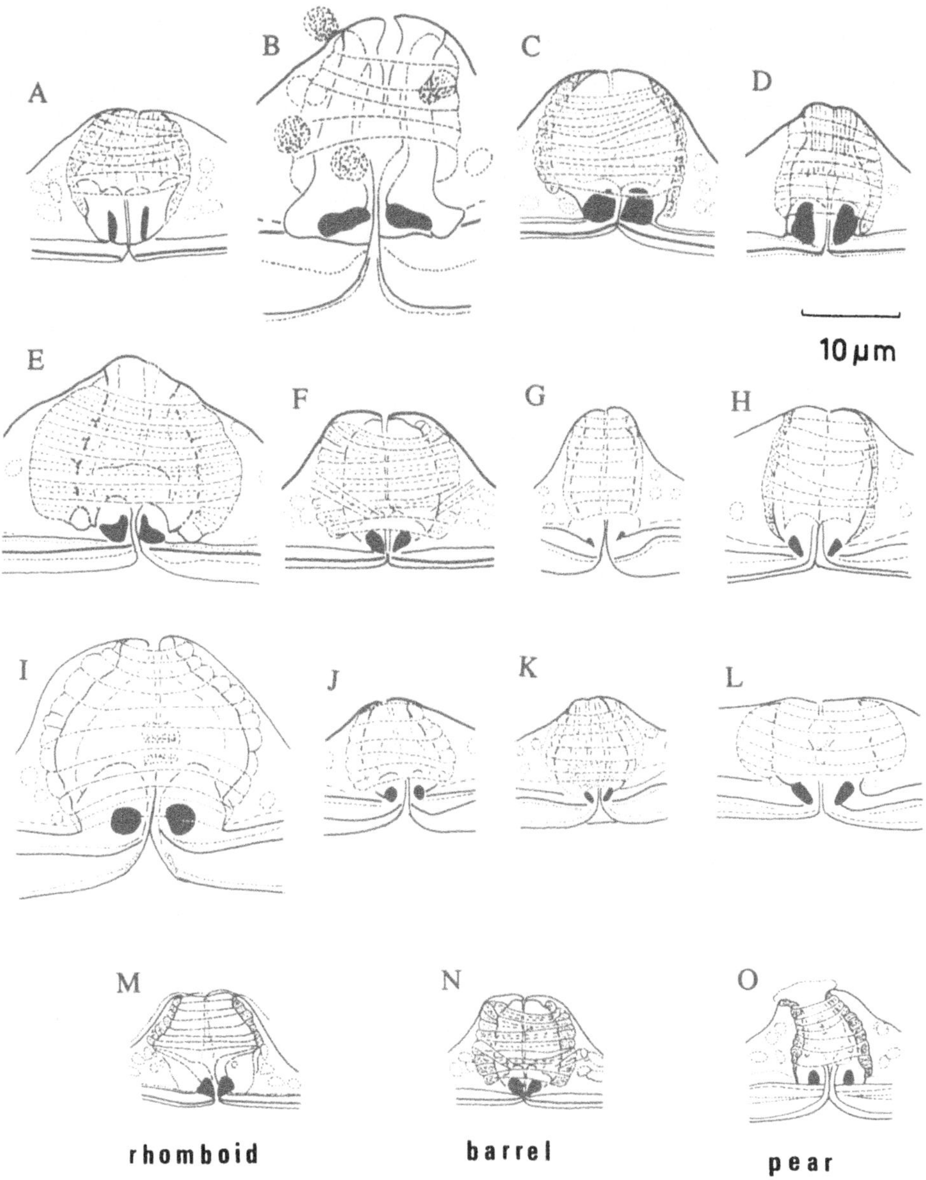

Fig. 5.4. Different shapes of vaginal sclerotizations in *Trichodorus* female: A. rod-like, parallel to vaginal lumen (*T. primitivus*); B. parallel to longitudinal body axis (*T. cottieri*); C. large, rounded quadrangular (*T. variopapillatus*); D. large, rounded triangular (*T. beirensis*); E. medium-sized, roughly triangular with inner side concave (*T. dilatatus*); F. medium-sized, triangular with tips pointing toward vulva (*T. cylindricus*); G-H. respectively minute and small triangular (*T. orientalis*, *T. sparsus*); I. large rounded, well separated (*T. taylori*); J. small, rounded (*T. tricaulatus*); K. small, oval (*T. parorientalis*); L. small, trapezoid (*T. carlingi*); M-O: different shapes of vagina: M: rhomboid-, N: barrel-; O: pear-shaped).

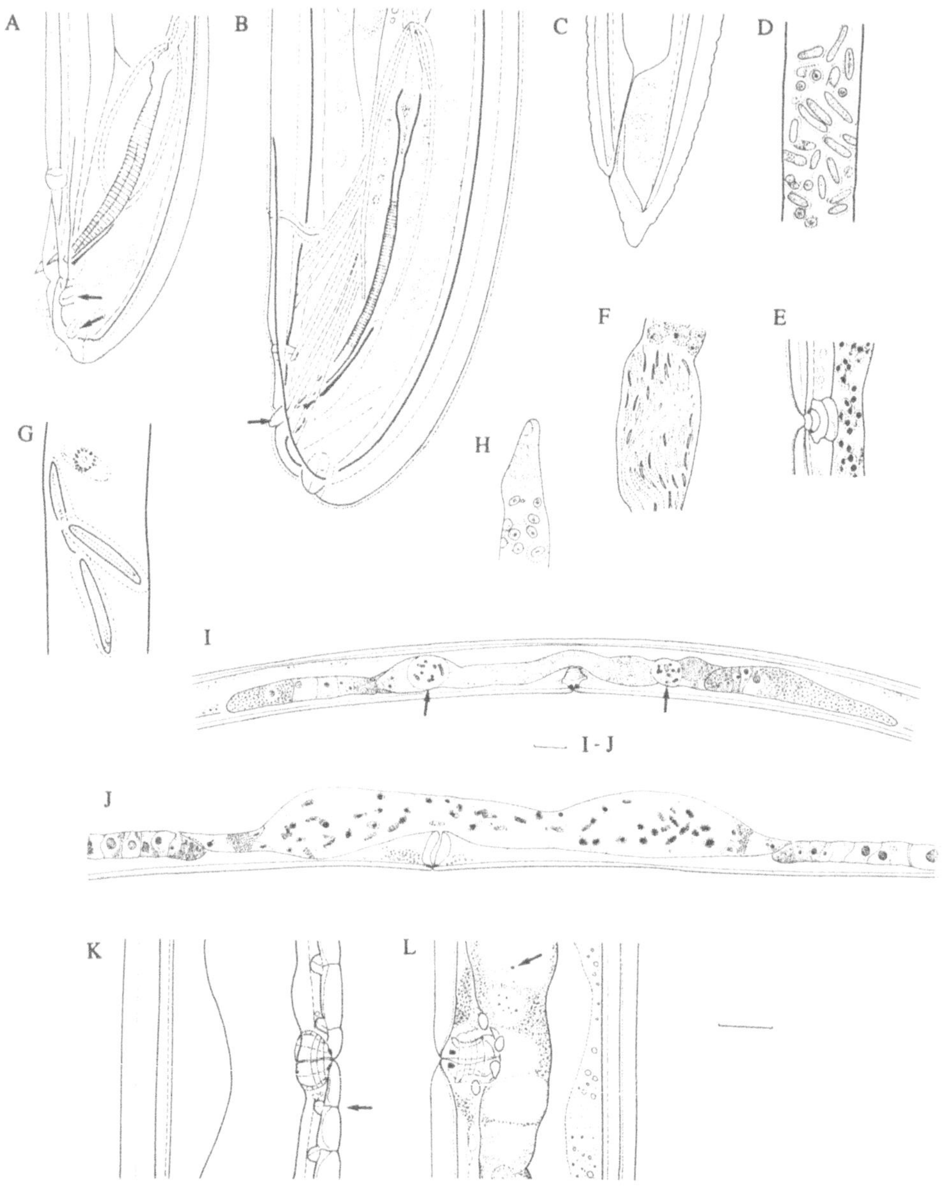

Fig. 5.5. A. Male, tail with two postcloacal papillae (arrows); B. male with protruding cloacal lip (arrow); C. female tail end conoid with acute tip; D. large sperm with sausage-shaped nucleus (♂); E. small sperm with small oval nucleus (♀); F. small sperm, thread-like (♂); G. large fusiform sperm (♂); H. medium-sized rounded sperm (♂); I. female reproductive system with indication of spermathecae; J. female reproductive system with sperm dispersed throughout uteri; K. vulvar region with indication of ventromedian advulvar body pore; L. vulvar region with indication of lateral advulvar body pore (scale 10 µm).

CHAPTER 6

ECONOMICALLY IMPORTANT SPECIES

The descriptions and data on distribution and bionomics of economically important species from Decraemer (1991) have been updated.

The morphometric data include mean values from different populations, for body length, onchiostyle length, spicule length and % V. These data are presented in parenthesis, beginning with the type population, and in the same order as given for the authors. The standard deviation is included when available; the mid-range (*) is calculated as an approximation of the mean value of each measurement in populations for which only the range of the morphometric character is available.

6.1. *Paratrichodorus minor* (Colbran, 1956) Siddiqi, 1974 (Figs 3.12.C.D.G; 3.14.Y; 3.15.N; 3.17.L.R.l.r; 4.27.E-F; 6.1. A-I)

syn. *Trichodorus minor* Colbran, 1956

syn. *Paratrichodorus (Nanidorus) minor* (Colbran, 1956) Siddiqi, 1974

syn. *Nanidorus minor* (Colbran, 1956) Siddiqi, 1974

syn. *Trichodorus christiei* Allen, 1957

syn. *Paratrichodorus (Nanidorus) christiei* (Allen, 1957) Siddiqi, 1974

syn. *Nanidorus christiei* (Allen, 1957) Siddiqi, 1974

syn. *Trichodorus obesus* Razjivin & Penton, 1975

syn. *Paratrichodorus (Nanidorus) obesus* (Razjivin & Penton, 1975) Rodriguez-M & Bell, 1978

Measurements

Females: L= 440-1530 µm (625 µm*; 585 µm*; 530 µm; 500 µm; 550 µm; 530 µm; 658 µm; -; 634 µm; 640 µm; 634 µm; 701 µm; 595 µm; 602 µm; 604 µm; 616 µm; 624 µm; 636 µm; 521 µm; 649 µm; 488 µm), onchiostyle= 26-47 µm (31.5 µm*; 40 µm*; 29 µm; 28.2 µm; 28 µm; 31 µm; 32.5 µm; -; 36 µm; 35 µm; 31 µm; 34.5 µm; 33 µm; 31.5 µm; 33 µm; 32 µm; 28.5 µm; 31.5 µm; 33 µm; 29 µm; 33.7 µm), a= 15-36, b= 3-6.8,

Fig. 6.1. *Paratrichodorus minor*. Female: (A) total view (paratype *P. christiei*), (B) reproductive system, (C) ventral view of reproductive system, (D) anterior body region (paratype *P. christiei*), (H) total view. Male: (E) anterior body region, (F) copulatory apparatus and tail, (G) ventral view of copulatory apparatus (after Heyns, 1975), (I) testis and part of vas deferens (After Decraemer, 1991. Courtesy Marcel Dekker Inc.)

V= 50-64% (53.5%*; 53%*; 54%; 55%; 54%; 57%; 52.1%; -; 53.5%; 50%; 53%; 53.5%; 54.5%; 52.5%; 53%; 55%; 54%; 53%; 52%; 55%; 54.1%)

Males: L= 540-678 µm (-; 635 µm*; -; -; 570 µm; -; -; 595 µm*; -; -; 628.5 µm; 654.5 µm; -;-;-), onchiostyle= 32-39 µm (-; 37 µm*; -;-; 28 µm; -; -; 34.5 µm*; -; -; 36 µm; 33.5µm; -; -; -), spicule length= 48-73 µm (-; 62.5 µm*;- ;-; 50 µm; -; -; 52.5 µm; -; -; -; 72.5 µm; -; -; -), a= 16-25, b= 3.6-6.5, c= 39-48 (from Colbran, 1956; Allen, 1957; Hooper, 1962; Siddiqi, 1962, 1963; Mamiya, 1967; Bird & Mai, 1968;Heyns, 1975; Shishida, 1979; Norton *et al.*, 1982; Vermeulen & Heyns, 1983: 8 populations ♀, 2 populations with ♂; Roca & Lamberti, 1984; Rashid *et al.*, 1986; Razjivin & Penton, 1975).

No separate data available for the holotype female.

Description

A rather short onchiostyle with inner onchium in adults. S-E pore usually near pharynx base. Pharyngeal glands usually overlapping the intestine ventrally and subventrally, pharyngeal bulb rarely offset.

Females without lateral body pores. No spermatheca and sperm when present, dispersed throughout uterus. Sperm cells very small, rounded to thread-like. Vaginal sclerotization rod-shaped, parallel to longitudinal body axis, close. Vagina quadrangular in side view. Vulva a short (2-2.5 µm) transverse slit. Caudal pores subterminal, obscure.

Males rare. No ventromedian cervical papillae nor lateral cervical pores present. Spicule capitulum marked slightly, shaft transversely striated, except at both extremities. Gubernaculum with small dorsal keel. Single ventromedian precloacal supplement present, shortly anterior to the cloacal opening. Caudal alae extending from just anterior to precloacal supplement to posteriorly to postcloacal papillae.

Remarks

The S-E pore varying from along posterior part of pharyngeal bulb to just posterior to the pharyngo-intestinal junction. No anteriodorsal overlap of the intestine except for the drawing given by Shishida (1979), showing both type of overlaps.

Stylet length of *P. minor* variable between populations and even within a population (Shishida, 1979).

Of five samples from different localities in Santa Fe, Argentina, four had populations with males and females. In two of them a group of males and females had globular sperm cells with a 1.5-2 µm nucleus, another group had males and females with thread-like sperm (different stages of sperm cells?) (Fig. 6.1.)(Decraemer & Chaves, 1988). Shishida (1979) observed a female specimen with extremely asymmetrical gonads, with the posterior branch being poorly developed. Loof (1975) synonymized *P. christiei* with *P. minor* upon a detailed study of various populations including paratype specimens of *P. christiei.*

Previously, several authors (Allen, 1957, 1960; Hooper, 1962; Siddiqi, 1962; Heyns, 1975) regarded *P. christiei* already as a possible synonym of *P. minor*. Allen (1960) remarked that *P. christiei* and *P. minor* could be geographic forms of the same species. Siddiqi (1963) described females and a male specimen from Nicaragua (L= 570 µm, onchiostyle length = 28 µm, spicule length= 50 µm) and suggested that *P. christiei* could be a synonym of *P. minor* but with a shorter spear and spicules.

In a key to species of *Paratrichodorus* Siddiqi (1974) differentiated both species upon onchiostyle length (*P. minor*: 27-34 µm; *P. christiei*: 33-47 µm).

Bird & Mai (1968) studied the morphometric variations of *P. christiei* based on 1,371 specimens (22 populations) of eight widely distributed areas of the United States and observed a range of 30-42 µm for the onchiostyle length.

Loof (1975) measured 76 females from all over the world and obtained a range for the onchiostyle length of 27-33 µm. He also remeasured paratype specimens of *P. christiei* and obtained an onchiostyle length between 27 and 33 µm. He calculated the spicule length from Allen's illustration was 57 µm. Former data reduce the differences (only based on morphometric data) between both species to intraspecific variations.

Although former convincing arguments for synonymizing *P. christiei* with *P. minor*, some authors still mention them as separate species (see Jairajpuri & Ahmad, 1992; Rhoades, 1985; Esser, 1990b; several authors in Evans *et al*, 1993).

Type host and locality

No type host was indicated in the original description; Moggill, Queensland, Australia.

Relationships

P. minor males resemble those of *P. nanus* in having a single ventromedian precloacal supplement, slender spicules (more than 2.5 times the anal body width) and absence of ventromedian cervical papillae and lateral cervical pores. The females with similar shape of vulva (a transverse slit) and vagina (about rectangular to quadrangular), small sperm (however not thread-like in *P. nanus*). Further *P. minor* and *P. nanus* have a rather short onchiostyle with an inner onchium in adult. *P. minor* differs from *P. nanus* by the more anterior position of the S-E pore (along posterior part of pharynx or at level of junction with intestine *vs* along anterior intestine), type of pharyngo-intestinal junction (usually with overlapping pharyngeal glands *vs* bulb usually offset), slightly longer onchiostyle (26-47 µm *vs* 21-26.5 µm); the males by somewhat longer spicules (48-73 µm *vs* 42-50 µm); the females by the shape of the vaginal sclerotizations in lateral view (rod-like, parallel to longitudinal body axis *vs* rounded triangular).

Distribution and Host

P. minor is cosmopolitan but may have partially been distributed by man. It is widespread in the USA, Australia and South Africa, and occurs in Japan, Java, Fiji, New Zealand, Philippines, Venezuela, Nicaragua, Brazil, Cuba, Argentina, Puerto Rico, Israel, Egypt, Senegal, Mauritania, Upper Volta, Ivory Coast, Canary Islands, India, Afghanistan, Taiwan, and USSR. Over 100 species of plants (economic crops, grasses, broad-leaf plants, weeds) are known hosts of *P. minor*. Some economically important crops that are common hosts are tomato, avocado, alfalfa, sugar-cane, wine grape, potato, *Phoenix canariensis*, red beet, sugar beet, endive, lettuce, cabbage, cauliflower, brussels sprouts, broccoli, mustard, radish, muskmelon, barley, millet, sweet corn, peanut, soybean, onion, okra, eggplant, celery, red clover, sweet pepper, boysenberry, peach, persimmon, walnut, wheat, blueberry, cranberry, chayote, lima bean, grapefruit, cowpea, carrot, castor bean, cotton, and azalea (Colbran, 1956, 1964; Mamiya, 1967; Hooper, 1977; Heyns, 1975; Allen, 1957; Rohde & Jenkins, 1957; Esser, 1990b). The nematode has also been associated with many other species of plants. According to Wyss (1970) *P. minor* is not endemic in Europe. Occasionally it has been found in Italy: Sicily, orange trees (Roca & Lamberti, 1984), Belgium: azalea, rhododendron (De Maeseneer, 1962), Sweden: *Gardenia*, *Saintpaulia ionantha* (Loof, 1975), Switzerland: *Anemone* sp. (Loof, 1975), Portugal: *Triticum aestivum* (Craveiro & Santos, 1984), and West Germany (Sturhan in Wyss, 1970). The record from The Netherlands (Loof, 1975) probably concerns *P. renifer* (Bongers, 1988).

Biology

P. minor is common in sandy soils but has also been found in peat and muck soils (Christie, 1959; Noffsinger, 1984). Brodie (1976) found the highest population densities of *P. minor* at 30 cm depth in soil with an 83% sand content. Above and below this depth (to 60 cm), significant reduction in density occurred. Soil temperature, texture, and moisture exert a marked influence on the vertical distribution of this species. Highest population densities occurred from December to March, when soil temperature at a 30 cm depth was 11-17°C and soil moisture was 18-23% by volume (Brodie, 1976). In cranberry bogs in the northern United States, highest populations of *P. minor* occurred in November and December, whilst in southeastern USA (Georgia) the highest numbers occurred in June (Noffsinger, 1984).

In field populations males are usually rare. Bird and Mai (1968) found only 12 males among thousands of females but Chaves (1984) found a population from Santa Fe, Mexico with a male-to-female ratio of 1:3. *Paratrichodorus minor* is parthenogenetic. On tomato, the life cycle was completed in 21-22 days at 22°C and in 16-17 days at 30°C (Rohde and Jenkins, 1957) which after 60 days gives a 10-fold or more increase in nematode numbers. Under controlled conditions, survival and reproduction on tomato seedlings were greatest at 10% soil moisture.

Host-Parasite Relationship

P. minor causes direct damage to the meristemic tissues by its feeding, resulting in an overall reduction in root development. Above-ground symptoms of infected plants are retarded growth, wilted foliage, and sensitivity to drought (Christie, 1959; Heyns, 1975). Chimaera symptoms, involving several kinds of partial chlorophyll defects, were found by Hafez *et al.* (1981) to be associated with the feeding of *P. minor.* Bird and Mai (1965) found morphometric and allometric variation among specimens from a single population of *P. minor* maintained on different plant species. Physiological races have also been shown to occur. *Paratrichodorus minor* is a vector of the Wisconsin isolate of TRV, which causes stem mottle and tuber spraing, or corky ringspot, in potato (Walkinshaw *et al.*, 1961; Ayala & Allen, 1968) and also transmits TRV from roots of aster plants (Komuro *et al.*, 1970). Liu & Ayala (1970) found a positive interaction between *Fusarium moniliforme* and *P. minor* on root growth but not on top growth of five sugar cane varieties. It is also a vector of pepper ringspot virus in South America (Salomao, 1973).

Control

Although *P. minor* is polyphagous in some circumstances crop rotation may be advantageous as populations increase only slightly in cropping systems involving *Crotalaria* sp. or under fallow conditions. Brodie *et al.* (1969) reported that "coastal" Bermuda grass suppressed populations of *P. minor.* Populations of *P. minor* did not increase on beggarweed or marigold. McGlochon *et al.*, (1961) reported that on cocksfoot grass, damage caused by *P. minor* was reduced 30-40% by applying dry foliage to the soil. Christie (1959) reported that fallow and dry tillage may give effective control of the nematode, whereas flooding does not. *Asparagus officinalis* is known to be antagonistic to many nematode species, including *P. minor* (Rohde & Jenkins, 1957). *P. minor* increased rapidly on corn and cotton, but the population densities were suppressed by peanut and soybean. Peanut was the most effective monocrop system for suppressing most nematode species, including *P. minor* (Johnson *et al.*, 1975).

Perry (1953) reported that after fumigation *P. minor* re-established itself much more quickly than any of the other numerous plant parasitic species present. Rhoades (1968) also reported higher populations of *P. minor* in cabbage plots 4 months after fumigation with dichloropropane-dichloropropene or ethylene dibromide than in unfumigated plots. Benson and Barker (1977) reported that control of *P. minor* on Japanese and Chinese hollies with aldicarb or dibromochloropropane was effective for 8 months, but after 12 months the nematode densities began to increase. Non-volatile nematicides such as aldicarb and carbofuran controlled *P. minor* better than ethylene dibromide, dichloropropane-dichloropropene and 1,2-dibromo-3-chloropropane (Johnson & Chalfaut, 1972). Rhoades (1985) reported in field trials on Myakka fine sand in Florida fenamiphos applied at 2.24 kg ai/ha decreased nematode numbers resulting in increased early growth and subsequent grain yield of field corn (*Zea mays*). Population re-establishment was slower and onion yields higher after treatment with fensulfothion, carbofuran, prophos, aldicarb, thionazin, and phorate than with dichloropropane-dichloropropene in plots with *P. minor* (Rhoades, 1969).

6.2. *Paratrichodorus pachydermus* (Seinhorst, 1954) Siddiqi, 1974 (Figs 3.12.R-T; 3.14.N; 3.16.D; 3.17.C.c; 4.31.A-D; 6.2.)

syn. *Trichodorus pachydermus* Seinhorst, 1954

syn. *Paratrichodorus (Atlantadorus) pachydermus* (Seinhorst, 1954) Siddiqi, 1974

syn. *Atlantadorus pachydermus* (Seinhorst, 1959) Siddiqi, 1974

Measurements

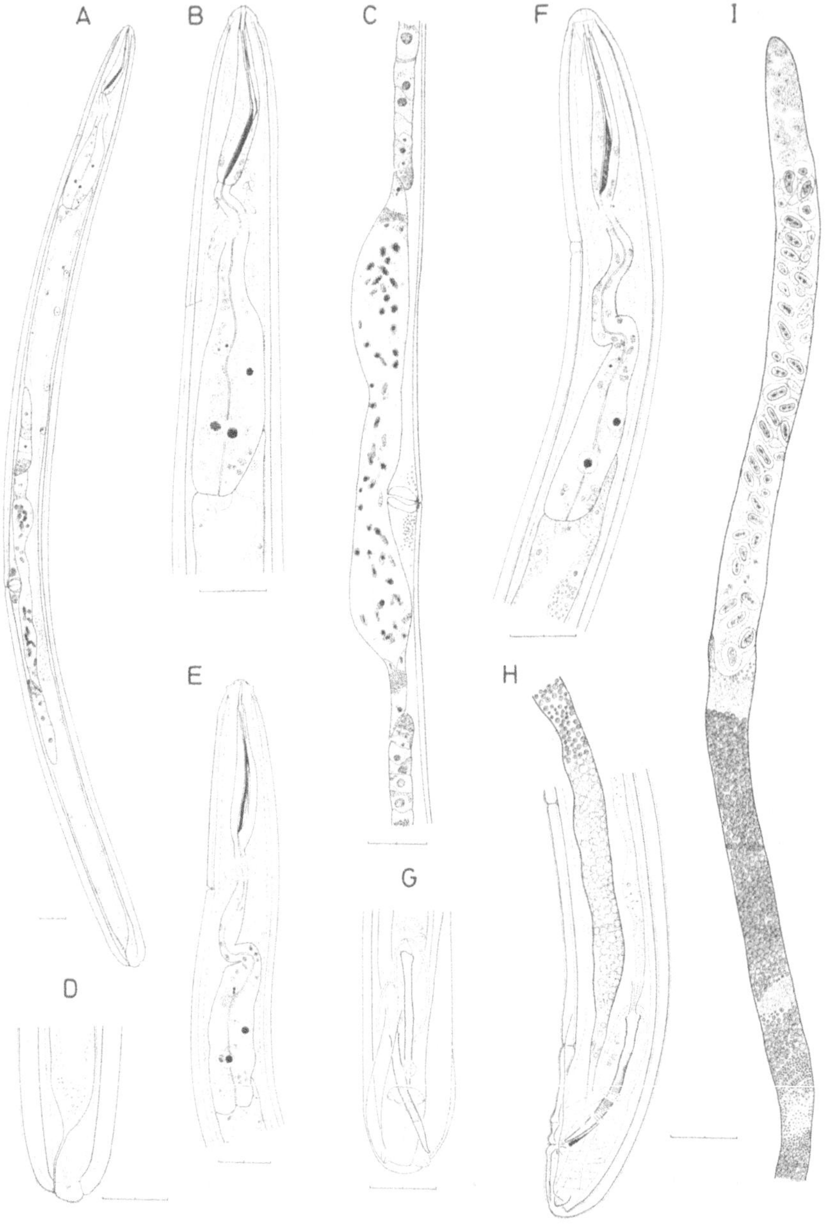

Fig. 6.2. *Paratrichodorus pachydermus*. Female: (A) total view, (B) anterior body region, (C) reproductive system (after Geraert *et al.*, 1981), (D) tail. Juvenile fourth stage: (E) anterior body region (after Decraemer, 1979). Male: (F) anterior body region, (G) ventral view of copulatory apparatus and tail, (H,I) reproductive system and tail (scale= 20 µm). (After Decraemer, 1991. Courtesy Marcel Dekker Inc.)

Females: L= 600-1020 µm (875 µm*; 825 µm*; -; 760 µm*; 830 µm; 810 µm), a= 13.9-32, b= 4.1.-7.7, V= 51-63% (57%*; 56%*; -; 56%*; 61%; 56.1%), onchiostyle= 44-58 µm (-; 49 µm*; -; 48.5 µm*; 52.5 µm*; 44.5 µm), anterior end to S-E pore= 70-103 µm.Males: L= 610-990 µm (800 µm*; 825 µm*; 810 µm; 780 µm*; 820 µm; -), a= 17.2-29, b= 4.1.-8.1., T= 50-70%, onchiostyle= 45-60 µm (-; 49.5 µm*; 52.3 µm; 45 µm*; 53.5 µm; -), spicule length= 40-54 µm (-; 49 µm*; 48.7 µm; 36.5 µm*; 50.0 µm; -), gubernaculum= 12-19 µm, anterior end to S-E pore= 77-98 (from Seinhorst, 1954; Allen, 1957; Coomans, 1962; Hooper & Southey, 1978; Baujard, 1980; Peneva, 1988).

No separate data are available for the holotype female.

Description

Males with a single ventromedian cervical papilla just anterior to the S-E pore. One pair of lateral cervical pores about level of ventromedian cervical papilla or just anterior. Sperm cells more or less oval-shaped, sausage-shaped nucleus 8.5-9.5 µm long. Spicules with distal part slightly ventrally curved, shaft transversely striated except at extremities. Gubernaculum parallel to spicules, slightly thickened at distal end. Caudal alae narrow. Three ventromedian precloacal supplements present (the posterior two within the region of the retracted spicules and well separated, the anterior supplement about two body widths anterior to the middle supplement). Two pairs of ventrosubmedian postcloacal papillae present (anterior pair minute, just posterior to anus, usually obscure; posterior pair large, subterminal near caudal pores). Tail about anal body diameter long.

Females with a pore-like vulva. Vaginal sclerotizations small, rounded triangular to oval slightly separated. No spermathecae; uterus functions as sperm reservoir, sperm throughout uterus, exceptionally with concentration near oviduct. Two to five lateral body pores on each side posterior to the vulva. One or two pairs of caudal pores; if one pair, positioned subterminally; if two pairs, positioned lateroterminally and subterminally. Pharyngeal bulb usually with a distinct anteriorly directed intestinal overlap of pharynx dorsally and dorsolaterally. Some specimens without distinct overlap; or posterior ventrosublateral pharyngeal glands overlap intestine; or both intestinal and pharyngeal overlaps present. S-E pore opening at anterior end of pharyngeal bulb, or just anterior.

Remarks

According to Sturhan (1985) the anterior pair of postcloacal papillae in male cannot be considered as papillae but at most as pores. Small pores are often present just posterior to the cloacal opening.

Males of a *P. pachydermus* population found in association with *T. aquitanensis* in the south of France showed a much larger variability of the distance between middle precloacal supplement (SP2) and the cloacal opening (53-83% of spicule length) than indicated in Loof (1975), overlapping the range known for *P. anemones* (38-64% of spicule length). According to Baujard (1980) this feature can not be used to distinguish *P. pachydermus* and *P. anemones.*

Sturhan (1985) described the presence of two pairs of caudal pores in female as in *P. weischeri*, a rare observation within the genus.

De Waele *et al.* (1985) described for the first time the presence of a pair of lateral cervical pores on both side in females. Those pores are less obvious than the more posterior lateral body pores in the region of the vulva and are rarely observed or described.

Specimens from Poland either with an offset bulb or a small intestinal overlap (Brzeski & Szcygiel, 1974).

Peneva (1988) observed one female from Bulgaria with four lateral body pores (one pore anterior and three pores posterior to the vulva).

Type host and locality

Soil around the roots of *Prunus serotina*, and other shrubs, Ede, the Netherlands.

Differential Diagnosis

P. pachydermus is closely related to *P. anemones*. They have similar morphometric data for body length, length onchiostyle and spicule length, well developed sperm with sausage-shaped nucleus; males with one ventromedian cervical papilla, three ventromedian precloacal supplements; females with a pore-like vulva and usually two lateral body pores postvulvar. *P. pachydermus* males differ from *P. anemones* in spicule shape (spicule shaft stout and about equally wide *vs* shaft slenderer, with narrow kinking part in anterior half), tail (long tail *vs* short tail less than one anal body diameter), presence of an additional pair of postcloacal papillae, absent in *P. anemones*. Females differ in shape of vaginal sclerotizations (rounded to oval *vs* fine triangular), the range of variability of the number of postvulvar lateral body pores (between 0 and 2 *vs* 2 and 5) and location of sperm (usually throughout the uterus, rarely concentrated near oviduct *vs* usually in spermathecae, rarely spread over uteri). Both species also differ in the most common type of pharyngo-intestinal junction (usually with anteriodorsal overlap of intestine *vs* usually with ventral overlap of pharyngeal glands).

Distribution and Host

P. pachydermus has been reported only from temperate regions, mainly Europe, where it is one of the most commonly occurring Trichodoridae species. Occasionally it has been found in Ontario, Canada, and the USA (Florida, Maryland, Michigan, New Jersey, Virginia) localities to which it has probably been introduced with planting material (Noffsinger, 1984). *P. pachydermus* is polyphagous and occurs in a wide range of habitats e.g. arable land, grassland, and woodland. Known hosts are tobacco, grape vine, potato, mulberry, strawberry, sugar beet, red clover, onion, beech, tulip (van Hoof, 1973; Cooke, 1976); *Gladiolus* sp. (Cremer & Schenk, 1967); oak, *Prunus serotina* (Seinhorst, 1954). Other plant associations are barley, rye, rye grass, lettuce, maize, fruit plantations, meadows,hop, maple, alder, birch, oak, *Sambucus* sp., *Castanea* sp., *Liquidambar* sp., *Prunus* sp., *Salix* sp., *Ulnus* sp., *Populus* sp., *Geranium molle*, *Mercurialis annua*, common chickweed, shepherds purse, *Pinus pinaster*, *P. sylvestris*, *P. contorta*, *Picea sitchensis*, *P. abies*, *Larix* sp., European beach grass, highbush blueberry, dahlia, juniper, white pine, white cedar, smoke tree, buckeye, *Rosa* sp. and *Cupressus* sp., *Crataegus* sp., *Clematis vitalba*, *Ligustrum sinense*, *Solarium nigrum*, *Syringa vulgaris*, *Tropaeolum majus* (in De Waele *et al.*, 1985); mimosa, *Elephantopus nudatus* (Esser, 1990a, 1993) and *Fraxinus caroliniana*, soybean and cotton (McGowan, 1977, 1987; Miller, 1982).

Biology

P. pachydermus is common in sandy or light sandy-loam soils, but occasionally occurs in loamy or peat soil. It is more common in slightly acid soils, usually at 15 and 70 cm depths (De Waele *et al.* 1985). Boag (1981) recorded no correlation between the vertical distribution of *P. pachydermus* and root distribution of the host plant, *Sitka* spruce. Also, *P. pachydermus* withstand both drought and waterlogged conditions. Males and females occur in equal numbers. Under controlled conditions, populations increased 30-fold in 4 months (Reepmeyer, 1973).

Host-Parasite Relationship

Through its feeding *P. pachydermus* causes serious damage and decreases yield of pine and onion (Whitehead & Hooper, 1970; Wasilewska, 1971; Banck, 1977). *P. pachydermus* is a vector of TRV (Sol & Seinhorst, 1961) and PEBV (van Hoof, 1962). TRV transmitted by *P. pachydermus* can infect a wide range of wild and cultivated plants; in potato it produces spraing and stem mottle (De Waele et al., 1985). *P.*

pachydermus can retain TRV for long periods (van Hoof, 1964) and it is specifically associated with TRV isolates belonging to the PRN serotypes (Ploeg *et al.*, 1995).

Control

P. pachydermus has a wide host range therefore crop rotation is not an effective means of control. The best level of control was achieved by soil fumigation with dichloropropane-dichloropropene or chloropicrin at 371 liters/ha (Whitehead *et al.*, 1970). More recently, systematic nematicides have been used to control these virus-vector nematodes. Alphey *et al.* (1975) reported that dichloropropane-dichloropropene decreased spraing by killing the nematode vectors. However, the oxime carbamates oxamyl and aldicarb controlled spraing caused by TRV not by decreasing the numbers of nematodes but by altering their behavior and ability to transmit the virus (De Waele *et al.*, 1985).

6.3. *Paratrichodorus porosus* (Allen, 1957) Siddiqi, 1974 (Figs 3.12.U; 3.14.J; 3.16.G.H; 3.17.F; 4.31.E-H, 6.3.)

syn. *Trichodorus porosus* Allen, 1957

syn. *Paratrichodorus (Nanidorus) porosus* (Allen, 1957) Siddiqi, 1974

syn. *Nanidorus porosus* (Allen, 1957) Siddiqi, 1974

syn. *Trichodorus bucrius* Lordello & Zamith, 1958

Measurements

Holotype male: L= 620 µm, a= 19, b= 4.6, T= 64%, onchiostyle= 48 µm, spicule length= 38 µm, gubernaculum= 12 µm.

Males: L= 530-770 µm (650 µm*), a= 15-25, b= 4.0-6.0, T= 55-74 %, onchiostyle= 43-48 µm (45.5 µm*), spicule length= 36-39 µm (37.5 µm*), gubernaculum= 12-13 µm (only from type population).

Females: L= 420-867 µm (615 µm*; 867 µm; 500 µm; 730 µm; 594 µm ± 77; 589 µm; 616 µm; 656 µm; 802 µm; 480 µm), a= 13.6-29, b= 3.1-6.2, onchiostyle= 39-58.1 µm (46.5 µm*; 49.0 µm; 42.0 µm; 50.0 µm; 50.4 ± 3.8; 48.4 µm; 42 µm; 52 µm; 47 µm*; 49.5 µm*), V= 51-59% (55.5%*; 53.1%; 56.2%; 56.0%; 55.9% ± 1.1; 55.1%; 51%; 52%; 47%*; 49.5%*), anterior end to S-E pore= 59-98 µm (one population) (from Allen, 1957; Lordello & Zamith, 1958; Siddiqi, 1962; Mamiya, 1967; Shishida, 1979; Vermeulen & Heyns, 1983: 4 populations; Jana & Baqri, 1984).

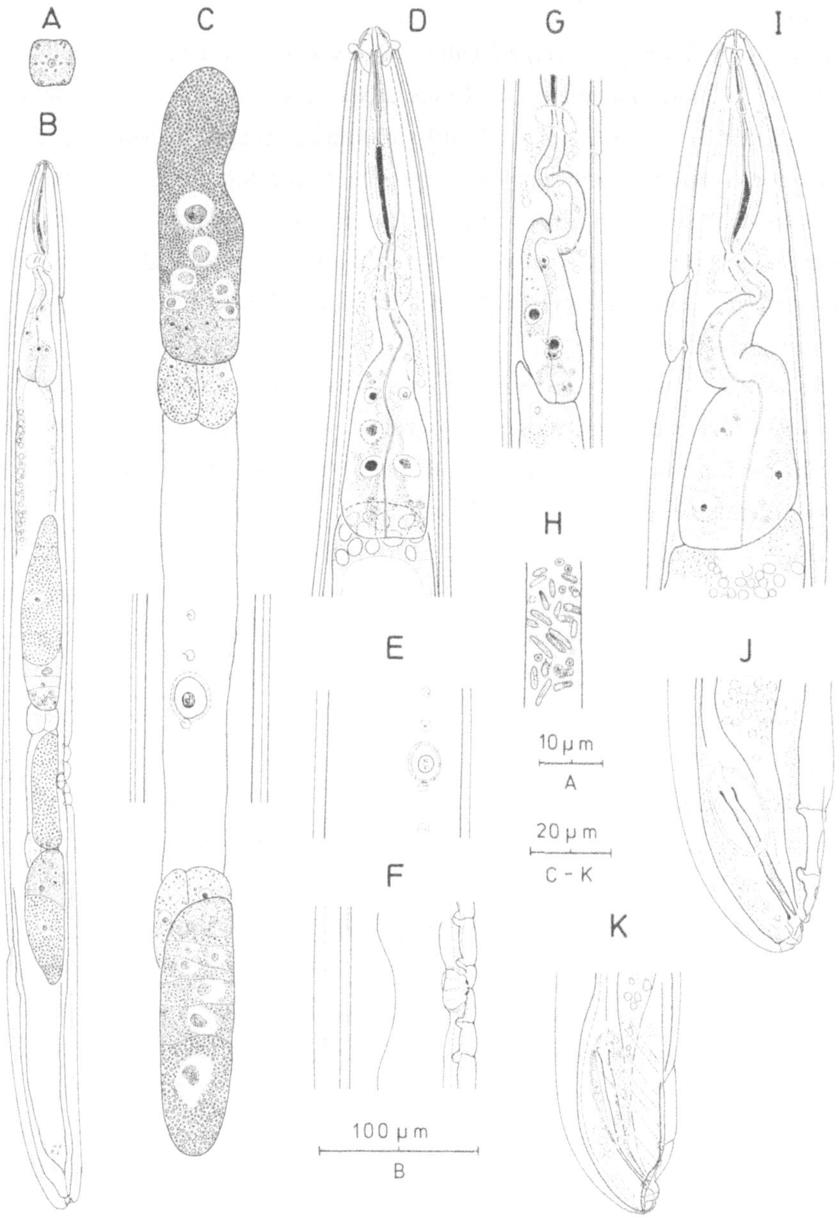

Fig. 6.3. *Paratrichodorus porosus*. Female: (A) *en face* view of head (after Siddiqi, 1962), (B) total view, (C) ventral view of reproductive system, (D) ventral view of anterior body region, (E) ventral view of vulva, (F) vaginal region. Male: (G) pharyngeal region (paratype), (H) sperm (paratype), (I) anterior body region (paratype), (J,K) copulatory apparatus and tail (paratypes). (After Decraemer, 1991. Courtesy Marcel Dekker Inc.)

Description

Males rare, only known from type locality. One ventromedian cervical papilla just anterior to S-E pore. Lateral cervical pore at the level of the ventromedian cervical papilla. Sperm cells large (3.5 x 10 μm) with sausage-shaped nucleus (8.5 μm long). Two ventromedian precloacal supplements present, both within the region of the retracted spicules and well separated. One pair of postcloacal subventral papillae. Spicules 36-39 μm long, corpus straight and transversely striated (except at both ends) and manubrium hardly marked. Gubernaculum parallel to spicules, distal end slightly enlarged. Caudal alae inconspicuous.

Females with ventromedian body pores located prevulvar and postvulvar (usually one to two anterior and one to two posterior to vulva). Lateral body pores absent. Rarely sperm in uteri; no differentiated spermathecae. Vulva pore-like. Vagina short, wide rounded rectangular to trapezoid, rarely barrel-shaped. Vaginal sclerotizations in lateral view small, clearly observable as slightly separated oval pieces.

Males and females with pharyngeal bulb usually with a well developed anteriodorsal intestinal overlap; bulb rarely offset. S-E pore between nerve ring and anterior end of pharyngeal bulb.

Remarks

The specimens described by Vermeulen & Heyns (1983) as *Paratrichodorus* ex. aff. *porosus* are considered as belonging to *P. porosus*, the differences in number of ventromedian body pores (from three to eight pores i.e. with one to four anterior or posterior to the vulva) being considered as intraspecific variability.

Japanese specimens usually with four ventromedian body pores, in some populations frequently with three advulvar pores (two anterior, one posterior; or one anterior, two posterior), exceptionally two or five pores. One Californian specimen with five pores (three anterior, two posterior to vulva).

Sperm cells in females were observed only in a few paratype specimens, a specimen from South-Africa and one from Brazil; when present, sperm distributed throughout the uteri.

In contradiction to all other authors, Jana & Baqri (1984) described the vulva as a transverse slit. All specimens examined had a pore-like vulva, which immediately beneath the surface opens into a longitudinal slit-like section of the vagina. Vagina short but variable in shape due to fixation. Dorsal intestinal overlap of pharynx usually pronounced, rarely short or absent.

Differential Diagnosis

P. porosus females most closely resemble *P. paraporosus* by possessing ventromedian advulvar body pores instead of lateral body pores as in the other species of the genus, but differ from *P. paraporosus* by the location of the ventromedian body pores: prevulvar and postvulvar *vs* postvulvar in *P. paraporosus*, the pore-like vulva *vs* vulva a longitudinal slit and rarity of sperm and absence of spermathecae *vs* presence of sperm located in spermathecae. *P. porosus* further differs from *P. paraporosus* in males, with males rare *vs* common in *P. paraporosus*, number and location of precloacal supplements: two precloacal supplements present, both within the region of the retracted spicules *vs* four supplements present, with two within the region of the spicules, one opposite spicule head and one well anterior to the spicule region.

P. porosus differs from the other species of the genus with rare males as *P. teres*, *P. minor*, *P. allius*, *P. nanus*, and *P. lobatus* by possessing well developed sperm cells with sausage-shaped nucleus instead of degenerated sperm or small sperm cells, by the shape and length of the spicules, the presence of a single ventromedian cervical papilla and by the number of ventromedian precloacal supplements (except *P. allius*).

Type host and locality

Soil around the roots of *Musa* sp.; Corona, California.

Distribution and Host

P. porosus is widespread in Australia, Japan, South Africa, and the USA (Alabama, Arkansas, California, Florida, Michigan, New Jersey, Tennessee, North Carolina, and South Carolina). It also occurs in Hawaii, Brazil, India, and Korea and has also been recorded from Azores, Madeira. In the USA *P. porosus* was reported in association with *Brassica oleracea* var. *capitata*, tobasco pepper, carrot, upland cotton, California walnut, lettuce, tomato, *Lycopersicon peruvianum*, apple, alfalfa, *Nicotiana glutinosa*, garden pea, peach, wild plum, mahaleb, pear, baby rose, spinach, wine grape, *Vigna sinensis* and corn (*Zea mays* var. *saccharata*) (Ayala and Allen, 1968; Ayala *et al*., 1970). *P. porosus* has also been found associated with many other plants such as *Bambusa tuldoides* (bamboo), *Butia* sp., *Crape myrtle*, *Camellia* sp., *Casuarina equisetifolia*, *Citrus* sp., *Citrus x paradisi*, *C. limon*, *C. sinensis*, *Ficus carica*, *Gardenia* sp., soybean, holly, *Ilex crenata rotundifolia*, walnut, English walnut, California privet, lupine, American crab apple, mulberry, banana, dwarf banana, common olive, avocado, philodendron, *Pittosporum tobira*, prune, *Prunus amygdalus*, *Rhododendron* sp., *Rosa* sp., *Rubus ursnus* var. *loganbaccus* and *Vigna sinensis* (Noffsinger, 1984; McGowan, 1972, 1988a, b, 1989b; Mead, 1979); sugar-cane in Madeira, Australia and Hawaii; *Carica papaya*, *Cypressus sempervirens* var. *stricta*, avocado and wine grape in Australia; *Camellia sinensis* in India and South Africa; pear and pumpkin plants in South Africa and cabbage in Hawaii. In Japan *P. porosus* has been found associated with five species of conifers (*Gingo biloba*, *Taxus cuspidata*, *Pinus thunbergii*, *P. densiflora*, *P. pentaphylla*) and 26 species of broad-leaf trees: *Juniperus* sp., *Carpinus tschonoskii*, *Quercus acutissima*, *Q. serrata*, *Q. salicina*, *Castanea crenata*, *Castanopsis cuspidata*, *Celtis sinensis*, *Aphanathe aspera*, *Cercidiphyllum japonicum*, *Magnolia obovata*, *Cinnamomum camphora*, *Neolitsea sericea*, *Daphniphyllum macropodum*, *Ilex integra*, *Acer palmatum*, *Ternstroemia japonica*, *Cleyera japonica*, *Eurya japonica*, *Idesia polycarpa*, *Aucuba japonica*, *Fatsia japonica*, *Rhododendron indicum*, *Callicarpa japonica*, *Viburnum awabuki* and *Cryptomeria japonica* (Mamiya, 1967; Shishida, 1979).

Biology

P. porosus occurs in sandy and sandy-loam soils. Under sugar-cane plantings, populations were concentrated in the upper 60 cm with highest numbers at 45 cm. From 60 to 105 cm depth, populations were low and numbers remained practically constant (Carneiro *et al.*, 1982). Under controlled conditions, reproduction on maize occurred at temperatures between 18 and 35°C, with the optimum being 24°C. At 29°C populations decreased suddenly, and the lowest reproduction occurred at 12 and 35°C (Ayala *et al.*, 1970). In North Carolina *P. porosus* reproduce at high rates with populations increasing from 750 to 17,730 individuals in 7 months on *Camellia* (Barriga, 1965).

Host-Parasite Relationships

P. porosus causes extensive damage to the root system of *Camellia* (Barriga, 1965) and under controlled conditions, *P. porosus* damaged maize and sorghum (*Sorghum vulgare*) (Chèvres-Roman *et al.*, 1971). Nishizawa (1973) suspected *P. porosus* to be the causal agent for black-rot disease of Chinese yam. Ayala and Allen (1968) reported that under controlled conditions *P. porosus* transmitted the Californian isolate of TRV after feeding on virus-infected Glurk tobacco and *Nicotiana glutinosa*.

Control

No detailed studies have been reported on the control of *P. porosus* but soil treatments with nematicidal fumigants, fallow (Wallace, 1971), or oxime carbamates probably would give control as these treatments are known to be effective against other trichodorids.

6.4. *Paratrichodorus teres* (Hooper, 1962) Siddiqi, 1974 (Figs 3.13.L.M; 3.14.M; 3.17.Q.q; 4.33. A-F; 6.4.)

syn. *Trichodorus teres* Hooper, 1962

syn. *Paratrichodorus (Paratrichodorus) teres* (Hooper, 1962) Siddiqi, 1974

syn. *Trichodorus flevensis* Kuiper & Loof, 1962

Measurements

Holotype female: L= 870µm, a= 23.4, b= 5.4, V= 55%, onchiostyle= 46 µm.

Females: L= 616-1325 µm (860 µm; 1020 µm*; 1015 µm*; 865 µm; 1020 µm*; 770 µm; 920 µm; 860 µm; 713 µm*; 650 µm; 694 µm ± 54; 1096 µm ± 149.9; 763.5 µm ± 111.4) a= 10.7-33.1, b= 4.0-9.3, V= 45.5-68.0% (54%; 56.6%*; 56.9%*; 55%*; 54%; 56.5%*; 61%; 55%; 55%; 55.7%; 51.4% ± 1.5; 53.5 ± 4.4), onchiostyle= 41-61 µm (45 µm; 47.5 µm*; 45 µm*; 47 µm*; 44.5 µm*; 45 µm; 47.5 µm; 44 µm; 43 µm; 56 µm; 45.7 µm; 47 µm ± 1; 48.5 µm ± 3.9; 45.0 µm ± 2.9), anterior end to S-E pore= 66.0-137.0 µm.

Males: L= 537-1029 µm (-; 790 µm*; 537 µm; 622 µm; 875.5 µm ± 71.6; 754 µm ± 82.3), a= 11.9-27.5, b= 4.3-7.5, T= 52.0-69.5%, onchiostyle= 41-64 µm (50.5 µm*; 46 µm; 49 µm; 47.5 µm ± 3.2; 50.5 µm ± 3.0), spicule length= 36-57.5 µm (48.5 µm*; 50 µm; 50 µm; 46.5 µm ± 4.5; 48.0 ± 3.5), gubernaculum= 10.5-17.0 µm, anterior end to S-E pore= 74-117 µm.

Juveniles 2nd stage: L= 475-503 µm, genital system= 14.0-14.8 µm; juveniles 3rd stage: L= 563-571 µm, genital system= 22.4-23.6 µm; juveniles 4th stage: L= 674-684 µm, genital system= 55.6-55.8 µm (from Hooper, 1962; Kuiper & Loof, 1962; Kuiper, 1977: only data of juveniles; Hooper & Southey, 1978; Decraemer & De Waele, 1981: 5 populations for ♀, 2 for ♂; Vermeulen & Heyns, 1983; Coiro *et al.*, 1989; De Waele & Kilian, 1992; Decraemer *et al.*, 1993: 2 populations).

Description

Males rare in most populations. No ventromedian cervical papillae nor lateral cervical pores. Small sperm cells with small rounded nucleus, usually with fibrillar appearance. Spicules cephalated, spicule shaft slightly tapering toward extremities, transversely

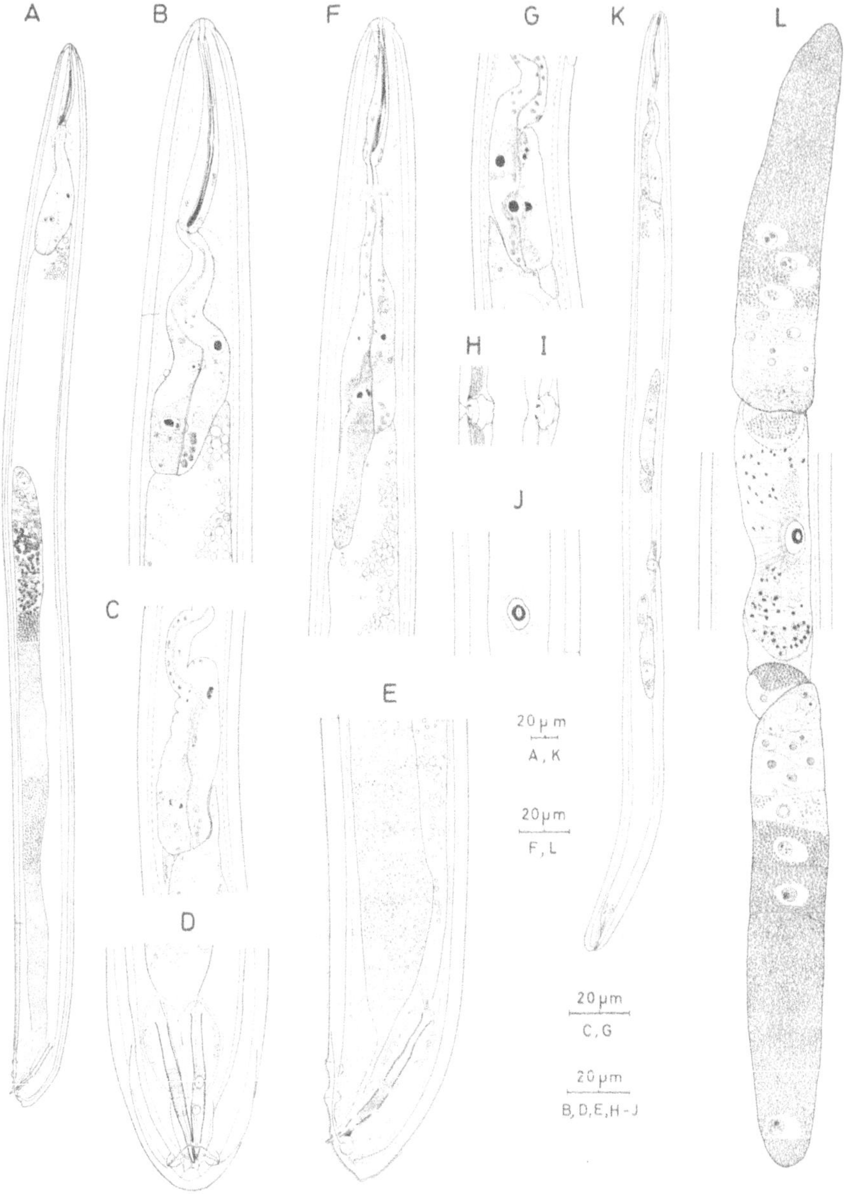

Fig. 6.4. *Paratrichodorus teres*. Male: (A) total view, (B) anterior body region, (C) pharyngeal region (after Decraemer & De Waele, 1981), (D) ventral view of copulatory apparatus and tail, (E) posterior body region. Female: (F) anterior body region, (G) pharyngeal region (F,G after Decraemer & De Waele, 1981), (H,I) vaginal region (paratype), (J) ventral view of vulva, (K) total view (paratype), (L) reproductive system (After Decraemer, 1991. Courtesy Marcel Dekker Inc.)

striated, except at both ends. Gubernaculum wider at distal end which is slightly bifurcated. Three ventromedian precloacal supplements (posterior two supplements clearly within the region of the retracted spicules, middle supplement opposite mid-spicule, anterior supplement less developed). Caudal alae narrow. One pair of large subventral postcloacal papillae. Cloacal opening with small lip-like extension of the anterior border.

Females with vaginal sclerotizations minute, oval-shaped, well separated in lateral optical section. Vulva a short longitudinal slit. Sperm usually absent, when present distributed along the uteri. Usually one lateral body pore posterior to the vulva, but number of lateral body pores variable.

Males and females with S-E pore opposite anterior end of pharyngeal bulb. Usually with overlapping ventrosublateral pharyngeal glands; an anteriodorsal intestinal overlap or both type of overlaps occur together; pharyngeal bulb rarely offset.

Remarks

The small populations of *P. teres* from Sierra Morena, Spain are bisexual, with an equal number of males and females, or even a higher number of males than females. However, no well developed sperm was observed in the Spanish specimens (Decraemer *et al.*, 1993). Anteriorly directed intestinal overlap of pharynx varies from slight to pronounced, but usually less obvious than the ventral/subventral pharyngeal gland overlap which varies from minute to very long; some specimens with offset bulb (Decraemer & De Waele, 1981). Spanish specimens with a pronounced overlap of the pharyngeal glands similar as in specimens from Poland (Decraemer & De Waele, 1981; Decraemer *et al.*, 1993).

Number of lateral body pores variable: from 0-2 on each side (= 1 to 4 pores in total) in Dutch populations, rarely up to six pores in total (Kuiper & Loof, 1962). Specimens from Margate, South Africa with a total of six to eight lateral body pores (two to three prevulvar, three, five or six postvulvar) (De Waele & Kilian, 1992), while females from Venda, South Africa with only one pair of lateral body pores posterior to the vulva (Vermeulen & Heyns, 1983).

Female specimens from Margate, South Africa have one pair of lateral cervical pores, a feature also observed in *P. pachydermus* females (De Waele & Kilian, 1992).

Type host and locality
Light sandy soil after lettuce at Hellesdon, Norwich, Norfolk, England.

Differential Diagnosis
P. teres females most closely resemble *P. lobatus* by similar vaginal sclerotizations (minute oval, well separated in lateral view), absence of spermathecae and small sperm; the males by similar shape of spicules, small sperm cells, a small lip-like extension of the anterior border of the cloacal opening and absence of ventromedian cervical papillae; both sexes by a pharyngeal gland overlap which may be very long. *P. teres* differs from *P. lobatus* by different sex ratio (males rare in *P. teres* vs common in *P. lobatus*); the males by the number of precloacal supplements (three *vs* two), slightly shorter spicules and slightly longer onchiostyle; the females by sperm usually absent in genital system *vs* usually numerous throughout the uteri and with a more distinct oval-shaped nucleus in *P. lobatus*.

P. teres also resembles *P. minor* by the rarity of males, minute thread-like sperm and absence of ventromedian cervical papilla. It differs from it in males by shape of the spicules (short stout spicules *vs* long slender spicules in *P. minor*), the number and position of the precloacal supplements (three supplements with posterior two in region of retracted spicules *vs* one supplement near cloacal opening in *P. minor*), and in females by the shape of vagina (Fig. 3.13.L-M *vs* Fig. 3.13.E-F) and vaginal sclerotizations (small oval-shaped, well separated and oblique *vs* small rod-like, closely located pieces, parallel to longitudinal body axis in *P. minor*).

Distribution and Host
P. teres is known mainly from temperate regions in Europe (Great Britain, The Netherlands, Belgium, West Germany, Poland, Italy, Spain and France). It also occurs in South Africa, and the U.S.A. (Florida and throughout the intermountain region in Oregon). The Australian specimens previously identified as *P. teres* belong to *P. lobatus* (Decraemer & Reay, 1991).
It has been found in arable soils and grassland and is polyphagous. Known hosts are sugar beet, potato, and onion. Other plant associations are wheat, carrots, lettuce, *Gladiolus*, rye grass, cole seed, barley, rye, red clover, Chinese holly, saltbush, sweet broom, wild cherry, Washington palm, soybean, tobacco, *Alnus glutinosa*, *Citrus* sp., *Fraxinus angustifolia*, *Dalhia*, *Tagetes*, cocksfoot grass, *Trollius* sp., *Apodites dimidiata*,

ferns and dune vegetation (Kuiper, 1977; Noffsinger, 1984; De Waele & Kilian, 1992; Decraemer *et al.*, 1993; McGowan, 1979; Mead, 1986a; Anonymous, 1987a, b, 1988d, 1989b).

Biology

P. teres is common in marine sandy or sandy-loam soils and with a low amount of silt or organic matter. It mainly occurs in the upper soil layers and aggregates around the roots of host plants at depths of 5-10 cm, and between 30-50 cm (Kuiper, 1977). It is sensitive to soil disturbance especially in dry soil (Bor & Kuiper, 1966). Males are rare, but do occur in some populations (Kuiper & Loof, 1964; Loof, 1965; De Waele & Kilian, 1992; Decraemer *et al.*, 1993). According to Wyss (1970) reproduction is mainly parthenogenetic. Under controlled conditions this nematode will increase 75-fold in 4 months (Reepmeyer, 1973).

Host-Parasite Relationship

P. teres can cause serious damage through its feeding (e.g. Docking disorder) and also is a vector of TRV. A relationship was found between the degree of virus infection and type of preceding crop (Kuiper, 1977). After the growth of a number of agricultural crops (spring wheat, potatoes, sugar beet, winter oilseed) and following fallow, a general transmission of tobacco rattle virus could be demonstrated from soil samples.

Control

Crop rotation does not provide a practical means for preventing damage by *P. teres*. Kuiper (1977) reported that over 80% of populations of *P. teres* may be killed after two rotavations. However, this treatment carries the risk of soil structure deterioration.
Reduction of nematode populations is obtained with relatively low doses of nematicides (dichloropropane-dichloropropene: 1-3 litres/acre) but a larger quantity of nematicide was necessary to prevent virus transmission. The great majority of the UK sugar beet crops which are at risk from Docking disorder receive row applications of a granular pesticide (usually aldicarb) at drilling (Evans *et al.*, 1993). When farmyard manure is used, infection of plants with rattle virus can be suppressed but nematode numbers are not reduced (Kuiper, 1977).

6.5. *Trichodorus primitivus* (de Man, 1880) Micoletzky, 1922 (Figs 3.2.Z; 3.4 (7c); 3.5.M; 3.9 (7c); 3.11.M; 4.19.F-J; 6.5.)

syn. *Dorylaimus primitivus* de Man, 1880

syn. *Trichodorus castellanensis* Arias Delgado, Jimenez Millan & Lopez Pedregal, 1965

syn. *Trichodorus mirabilis* Ivanova, 1977, n.syn.

Measurements

Males: L= 500-1000 µm (675 µm*; 780 µm*;720 µm; 760 µm; 765 µm*; 650 µm; 880 µm; 915 µm*; 750 µm*; 822 µm ± 55; 805 µm; 807 µm; 801 µm; 763 µm; 749 µm; 740 µm), onchiostyle= 47-58.5 µm (-; 45 µm*; 42.5 µm*; 56 µm; 47 µm; 52.5 µm*; 52 µm; -; 50 µm; 52 µm ± 3.1; 49 µm; 45 µm; 48 µm; 46 µm; 49 µm; 45.8 µm), spicule length= 32-54 µm (-; 36.5 µm*; 39 µm; 38 µm; 38 µm; 37.5 µm*; 36 µm; -; -; 45 µm ± 5.3; 39 µm; 35 µm; 37 µm; 41 µm; 42 µm; 34.6 µm), gubernaculum= 14.7-19, a= 18.6-38.9, b= 4.0-6.5, c= 40-65.

Females: L= 500-950 µm (600 µm*; 760 µm*; 735 µm*; 740 µm; 730 µm; -; 795 µm; 795 µm*; 750 µm*; 769 µm ± 33; 780 µm; 774 µm; 652 µm; 800 µm; 712 µm; 735 µm*), onchiostyle= 28-57 µm (-; 48.5 µm*; 42.2 µm*; 52 µm; 46 µm; -; 54.5 µm; -; 49.4 µm; 55 µm ± 2.3; 49 µm; 50 µm; 46 µm; 51 µm; 46 µm; 46 µm), a= 18-28, b= 4.4-5.8, V= 54-59.2% (-; 56%*; 54.5%*; 59%; 57%; -; 58%; 58.5%*; -; 57.5% ± 0.6; 56.6%; 57.7%; 57.5%; 59%; 53.9%; 56%*) (from de Man, 1880; Allen, 1957; Hooper, 1962: English and Dutch population; Seinhorst, 1963; Loof, 1963; Ivanova, 1977; Nesterov, 1979; Decraemer, 1979; Arias & Roca, 1986; Almeida *et al.*, 1989: 5 populations; Coiro et al., 1989).

Juveniles 2nd stage: L= 330-550 µm, onchiostyle= 34-42 µm, length genital system= 10-22µm; juveniles 3rd stage: L= 390-660 µm, onchiostyle= 39-50 µm, length genital system= 22-60 µm; juveniles 4th stage: L= 550-660 µm, onchiostyle= 43-47 µm, length genital system= 60-110 µm (from Decraemer, 1979).

de Man's male specimen: L= 630 µm, onchiostyle= 51 µm, spicule length= 40 µm, a= 19, b= 4.0 (data from Seinhorst, 1963).

Seinhorst (1963) considered de Man's specimen as the holotype. Loof (1961) advised to consider the primary types (1♂, 1♀) as destroyed since few diagnostic details are still visible and to accept the neotype of Allen (1957) although it does not fulfill the requirement that the neotype should be collected as near as possible to the original type locality.

Description

Males with three ventromedian cervical papillae anterior to S-E pore; the anterior papilla about mid-onchiostyle region, the middle papilla at base of onchiostyle. One lateral

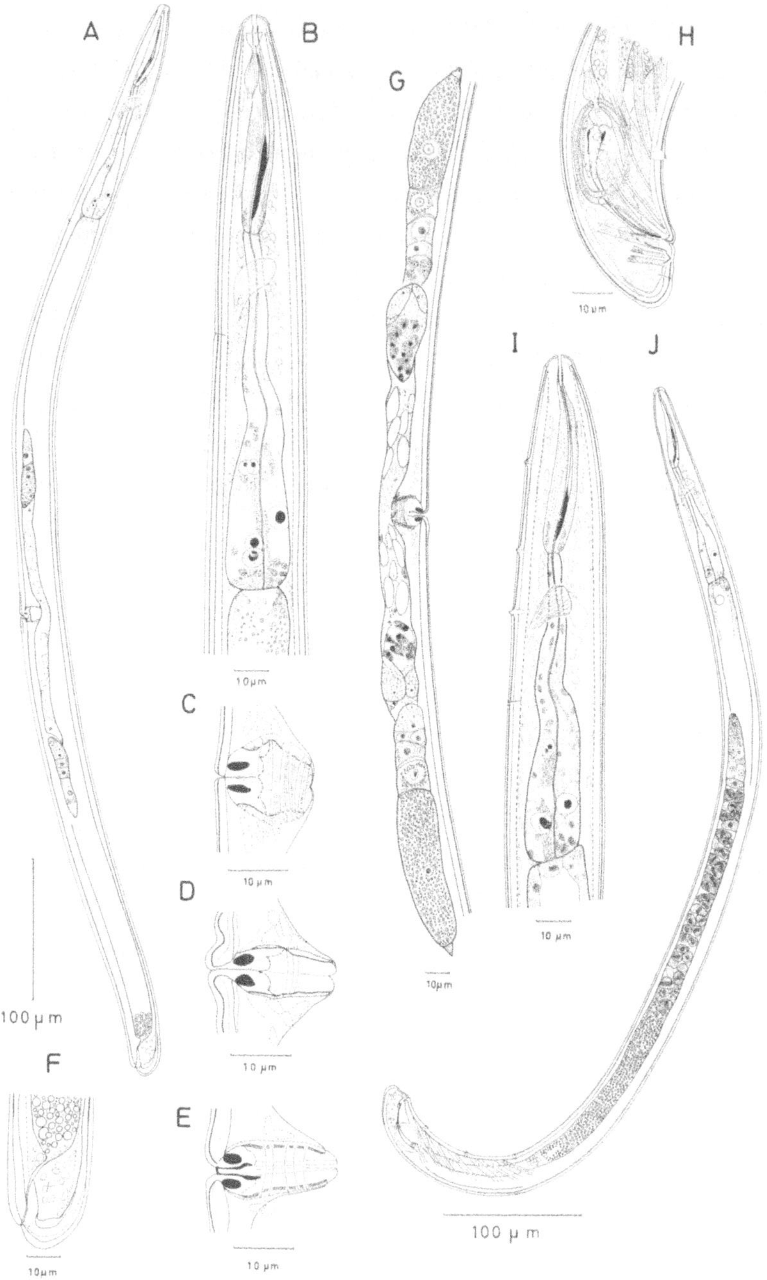

Fig. 6.5. *Trichodorus primitivus*. Female (A) total view, (B) anterior body region, (C-E) vaginal region, (F) tail, (G) genital system. Male: (H) copulatory apparatus and tail, (I) anterior body region, (J) total view. (After Decraemer, 1991. Courtesy Marcel Dekker Inc.)

cervical pore opposite posterior third of onchiostyle. Sperm cells large, striated; nucleus sausage-shaped, length 9 µm. Spicules, proximal end 3-4 µm wide, shaft gradually tapering to narrow (1 µm) distal half; shaft smooth except for a few fine bristles about mid-spicule (obscure by retracted spicules, position indicated by serrated outline of spicules). Gubernaculum convex anteriorly, usually positioned between the spicules, sometimes positioned ventrally. Three ventromedian precloacal supplements present, the posterior supplement opposite head of retracted spicules.

Females with one pair of lateral advulvar body pores just posterior to vulva; two pairs of lateral pores anterior to vulva (one pair three body widths anterior to vulva, one pair five to seven body width anterior to vulva). Vaginal sclerotizations rod-shaped, parallel to vaginal lumen. Vulva a transverse slit. Vagina rhomboid to elongated. Pharyngeal bulb usually offset, rarely with a minute ventral or a minute dorsal overlap.

Type host and locality

Clay soil covered with grass (Seinhorst, 1963), near Leiden, the Netherlands.

Remarks

No data on the shape of the vulva were available for *T. primitivus* except for *T. mirabilis* (syn. of *T. primitivus*) where the vulva was described as pore-like. A study of several specimens in ventral view revealed it to be a minute transverse slit. The illustration of *T. mirabilis* apparently is from a lower (=more inner) level with a section of the vaginal lumen, vagina and sclerotized ring.

Differential Diagnosis

T. primitivus males resemble *T. aquitanensis* and *T. hooperi* in spicule shape (proximally wide, tapering distally to a slender shaft). Further, they agree with *T. aquitanensis* males by three ventromedian cervical papillae, two in region of onchiostyle (only two ventromedian cervical papillae in *T. hooperi*). *T. primitivus* females are similar to those of *T. aquitanensis* by the transverse slit-like vulva and further to *T. viruliferus* (partim) by the orientation of the vaginal sclerotizations (parallel to the lumen of the vagina). *T. primitivus* differs from *T. aquitanensis* in male in detailed spicule shape and in female by the different shape of the vaginal sclerotizations (rod-like *vs* quadrangular). Further from *T. viruliferus* in male by a different spicule shape and in female by different vaginal sclerotizations (rod-shaped and parallel to vaginal lumen *vs* usually oval and parallel or oblique) and vulva (transverse lit *vs* pore).

Distribution and Host

T. primitivus has been reported mainly from temperate regions, mostly from Europe. It is widely distributed in Great Britain, Belgium, West Germany, Poland, southern Sweden, and occurs at various sites in France, Italy, the Netherlands, Norway, Denmark, Bulgaria, Portugal, and Rumania. It has not been found in Yugoslavia. Occasionally it has been reported from U.S.A. (Oregon, California, Arkansas, Virginia, Maryland, Florida) and New Zealand. Known hosts are sugar beet, *Brassica oleracea* var. *acephala* (cabbage) and *Trifolium pratense* (Whitehead & Hooper, 1970); mimosa, *Apium graveolens* var. *dulce* (celery),corn, alfalfa, cucumber and tobacco (Christie *et al.*, 1950; Gibbs & Harrison, 1964). Other plant associations are *Avena sativa*, *Buxus* sp., *Camellia japonica*, *Gossypium hirsutum* (cotton), butternut, *Lespedeza* sp., *Mimosa* sp., *Pinus* sp. and weat (Noffsinger, 1984); garden pea, potato, common chickweed, apple, prune, *Anemone coronaria*, *Fragaria ananassa*, *Brassica oleracea* var. *capitata*, *Prunus* sp., *Ribes nigrum* var. *europaeum*, tobacco, in vineyards and carrots (Hooper & Siddiqi, 1972); *Picea sitchensis*, *P. abies*, *Pinus sylvestris*, *P. concorta*, *Larynx* sp., and other coniferous and deciduous trees in Scotland (Boag, 1978); *Poa trivalis*, *Trifolium pratense*, *Mentha* sp., *Rumex* sp. in Portugal and comoon olive, buckhorn, *Ulmus procera*, *Citrus* sp., potatoes in the Azores (Almeida *et al.*, 1989). It was also found in association with chrysanthemum in Leningrad (Ivanova, 1977).

Biology

T. primitivus most frequently occurs in sand and sandy-loam soils but occasionally is found in heavy soils (Sturhan, 1967; Cooper, 1971). It can occur at a wide range of depths: from 0 to 70 cm with the highest populations occurring at 20-40 cm (De Pelsmaeker *et al.*, 1985) and from 60 to 120 cm under clover (Pietler, 1977). *T. primitivus* has a higher tolerance for copper in the soil than the other *Trichodorus* sp. and shows a relatively high tolerance for soil acidity (pH 5.1-6.5 in sand; 6.3-6.9 in sandy loam) (Cooper, 1971; De Pelsmaeker *et al.*, 1985). *Trichodorus primitivus* reproduces bisexually and has a life cycle similar to that of *T. viruliferus*, about 45 days at 15-20°C (Pitcher & McNamara, 1970).

Host-Parasite Relationship

T. primitivus causes direct damage to the plant roots by its feeding (Whitehead & Hooper, 1970). It is a vector of TRV, causing spraing in potato tubers (Hooper & Siddiqi, 1972; De Pelsmaeker *et al.*, 1985). It also transmits PEBV, which damages pea crops and infects other plants such as lucerne, cucumber, tobacco, and sugar beet (Gibbs & Harrison, 1964; Harrison, 1964). It is also responsible for docking disorder in sugar beet.

Control

Good control of docking disorder in sugar beet was obtained with dichloropropane-dichloropropene or 1,3-dichloropropene (Whitehead *et al.*, 1970). Spraing disease in potatoes was controlled using aldicarb or oxamyl (Brown & Sykes, 1973) or oxamyl (Alphey, 1978). The latter systematic nematicide does not greatly decrease the numbers of trichodorids but does decrease the incidence of the nematode-transmitted TRV (Alphey, 1978).

6.6. *Trichodorus similis* Seinhorst, 1963 (Figs 3.1.H; 3.4 (8d); 3.6.F; 3.9 (8d); 3.11.N; 4.20 F-I; 6.6.)

Measurements

Holotype male: L= 870 µm, onchiostyle= 42 µm, spicule length= 40 µm, a= 27, b= 5.6, T= 43.

Males: L= 550-1059 µm (815 µm*; 810 µm; 775 µm*; 887 µm; 973 µm; 750 µm*; 678 µm ± 58; 666 µm ± 62), onchiostyle= 35-50 µm (40 µm*; 40 µm; 40.7 µm; 43 µm; 45.5 µm; 37.7 µm*; 39.2 µm ± 3.1; 45.4 µm ± 3.8), spicule length= 30-44 µm (38 µm; 37 µm; -; 42 µm; 41.6 µm; 32.5 µm*; 34.2 ± 1.7; 32.6 µm ± 0.9), gubernaculum= 12-16 µm, a= 19.5-32.7, b= 4.3-11.3, c= 54-120.

Females: L= 600-1000 µm (805 µm*; 820 µm; 800 µm*; 855 µm; 952 µm; 735 µm*; 692 µm ± 56.4; 664 µm ± 43.9), onchiostyle= 36-52 µm (42 µm; 40 µm; 40.6 µm; 43 µm; 45 µm; 37 µm; 42.4 µm ± 1.4; 46.5 µm ± 2.3), V= 43-60% (54.5; 56; -; 56.7; 56.6; 52; 54.7 ± 4.2; 55.6 ± 2.4), a= 17.9-30, b= 4.6-11 (Seinhorst, 1963; Wyss, 1974; Decraemer, 1979; Roca & Lamberti, 1981: 2 populations; Peneva, 1988: 3 populations).

Description

Males with three ventromedian cervical papillae anterior to S-E pore; anterior papilla usually opposite posterior fourth of onchiostyle region, rarely at base onchiostyle or just posterior. One lateral cervical pore on both sides near base onchiostyle. Sperm cells

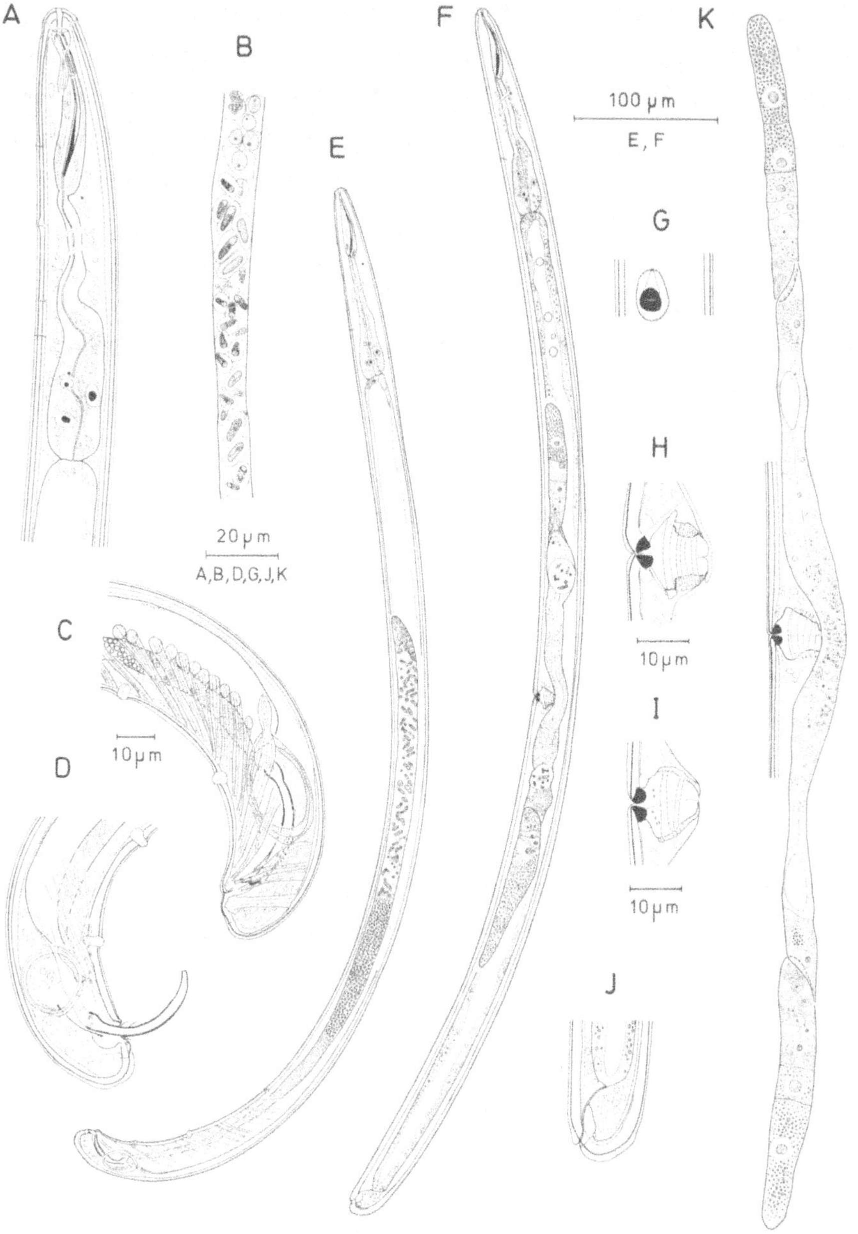

Fig. 6.6. *Trichodorus similis*. Male: (A) anterior body region, (B) sperm, (C,D) posterior body region (after Decraemer, 1980), (E) total view. Female: (F) total view, (G) ventral view of vulva, (H,I) vaginal region (after Decraemer, 1980), (J) tail, (K) reproductive system. (After Decraemer, 1991. Courtesy Marcel Dekker Inc.)

large; nucleus elongated, 2 x 5.5.-6.5 μm. Spicules, head offset, shaft mid-region widened, tapered distally to a blunt open end; retracted spicules usually with smooth appearance, protruding spicules with a few visible bristles in narrower calamus region. Gubernaculum with a characteristic proximal hook and distal knob. Three ventromedian precloacal supplements present, the posterior supplement opposite or just anterior to proximal end of retracted spicules.

Females with one pair of lateral advulvar body pores, just posterior to vulva. Vulva a transverse slit. Vagina round to rhomboid shaped, only proximal end surrounded by constrictor muscles. Vaginal sclerotization large, rounded triangular shape.

Pharyngeal bulb offset, rarely with a minute ventral overlap of pharyngeal glands.

Type host and locality

Peaty soil under grass, Vinkeveen, The Netherlands.

Remarks

Loof (1973) observed one male specimen out of 151 with four precloacal supplements. Peneva (1988) found two males with only two ventromedian cervical papillae and in some male specimens the anterior ventromedian precloacal supplement was not observed.

Differential Diagnosis

T. similis most closely resembles *T. variopapillatus* in males by the shape of the spicules (ventrally bent, with knob-like offset manubrium), three ventromedian cervical papillae (one at level of onchiostyle), three ventromedian precloacal supplements (posterior one at level spicule head); in females by vulva shape (a transverse slit). It differs from *T. variopapillatus* in males by shorter, stouter spicules with bristles *vs* long slender and smooth spicules and in females by shape of vagina (rhomboid *vs* barrel-shape), shape vaginal sclerotizations (medium-sized triangular *vs* large rounded triangular to trapezoid) and one pair of lateral body pores *vs* two pairs. *T. similis* males further show some similarity with *T. complexus* and *T. coomansi* in general spicule shape but differ in number of ventromedian cervical papillae and detailed spicule shape.

Distribution and Host

T. similis has been recorded mainly from the temperate regions in Europe and rarely in the USA (Florida, Michigan). It is widely distributed in Belgium, West Germany,

Bulgaria, and the Netherlands, but occurs less frequently in Great Britain, France, Poland, southern Norway, southern Sweden, southern Italy, and Sicily. It is found in various habitats such as woodlands, meadows and arable land (Wyss, 1974; De Waele & Coomans, 1983; Roca & Lamberti, 1984; Knobloch & Bird, 1981; Peneva, 1988). Plant associations are *Cypressus sempervirens*, *Sorghum* sp., *Pyrus domestica*, *Gladiolus* sp., *Picea* sp., *Hordeum vulgare*, *Brassica rapa* var. *silvestris*, *Nicotiana tabacum*, *Fragaria vesca* var. *semperflorens*, *Chenopodium quinoa*, *Vaccinium elliottii*, peach, English walnut, European plum, tobacco, common chickweed, carrot, alflafa, annual bleu grass, perennial rye grass, white clover, barley, onion, sugar beet, pepper, pea and potato (Wyss, 1974; Peneva, 1988; Anonymous, 1988a).

Biology

T. similis most frequently occurs in sandy or light sandy-loam soils but is occasionally found in medium loams (Rau, 1975; Wyss, 1974). It commonly occurs between 0 and 40 cm, with the highest populations between 20 and 30 cm; rarely found below 40 cm (i.e., 40-60cm) (De Pelsmaeker *et al.*, 1985; Steudel, 1969; Wyss, 1974). It prefers a slightly acid soil (pH 5.1.-6.1. in sand, pH 6.8-6.9 in sandy loam) (De Pelsmaeker *et al.*, 1985). *T. similis* reproduces bisexually and under optimum conditions the life cycle is probably completed in about 45 days, which is similar to *T. viruliferus* (Wyss, 1974; Pitcher & McNamara, 1970).

Host-Parasite Relationships

T. similis causes damage by its feeding resulting in stubby root symptoms. *T. similis* is a vector of TRV and can retain TRV for long periods (van Hoof, 1970). In potato TVR causes spraing and notched leaf disease in *Gladiolus* (Cremer & Kooistra, 1964; van Hoof, 1968, 1970; Eriksson, 1974).

Control

Satisfactory control on sugar beet growing in infested sandy soils was obtained by fumigating with small doses of dichloropropane-dichloropropene or 1,3-dichloropropene (Whitehead & Hooper, 1970). Best control of potato spraing disease caused by TRV in sandy soils was achieved by injecting dichloropropane-dichloropropene before planting (Cooper & Thomas, 1971). Fenamiphos and oxamyl controlled the virus spread more effectively than *Trichodorus* (Reepmeyer, 1973; Alphey, 1978). However, aldicarb greatly reduced spraing and numbers of *Trichodorus* (Steudel, 1969).

6.7. *Trichodorus viruliferus* Hooper 1963 (Figs 3.2.AA, aa; 3.3 (1e); 3.5.C; 3.8 (1e); 3.11.T; 4.22.K-M; 6.7.)

Measurements

Holotype male: L= 860 µm, onchiostyle= 46 µm, spicule length= 33 µm, a= 28, b= 7.3, T= 65.

Males: L= 616-890 µm (770 µm; 690 µm; 730 µm; 686 µm; 731 µm; 652 µm ± 35; -), onchiostyle= 32-53 µm (45 µm; 44 µm; 41 µm; 45 µm; 49 µm; 36.6 µm ± 2.3; -), spicule length= 22-37 µm (34 µm; 31 µm; 32 µm; 28 µm; 29 µm; 24.2 µm ± 1.3; -), gubernaculum= 8-18 µm, a= 17.6-33, b= 4.2.-11.5, c= 47-92.

Females: L=552-919 µm (750 µm; 670 µm; 720 µm; 659 µm; 811 µm; 638 µm ± 49; 710 µm), onchiostyle= 36-56 µm (46 µm; 41 µm; 41 µm; 44 µm; 48.2 µm; 37.4 µm ± 1.2; 45.4 µm), a=17.3-28, b=3.9-12.4, V=51-62% (56%; 56%; 55%; 55%; 56.5%; 55.1% ± 1.0; 56.7%) (Hooper, 1963: type population and Dutch population; Mancini *et al.*, 1979; Roca and Lamberti, 1984: 2 populations; Peneva, 1988; Coiro *et al.*, 1988).

Description

Males with three ventromedian cervical papillae anterior to S-E pore; anterior cervical papilla clearly within onchiostyle region, second papilla at base of onchiostyle. One lateral cervical pore on both sides just behind base onchiostyle. Sperm cells large; nucleus large, sausage-shaped 2x8-9.5 µm. Spicules, head rarely marked, shaft proximally wide, gradually tapers to small constriction about mid-spicule, then shaft wider again and gradually tapering to distal end; narrow mid-spicule with a few bristles. Gubernaculum usually positioned between spicules, posteriorly with distinct dorsal keel. Three ventromedian precloacal supplements present, posterior supplement just anterior to retracted spicules.

Females with one pair of postadvulvar lateral body pores in vulva region. Vulva pore-like. Vaginal sclerotizations oval-shaped, oblique or almost parallel to vaginal lumen. Vagina more or less rhomboid-shaped, internal differentiation present, well-developed constrictor muscles, divided in two groups.

Males and females in about equal numbers. Pharynx offset or with small overlapping of ventrosublateral pharyngeal glands.

Remarks

Loof (1973) observed one male specimen out of 36 with only two ventromedian precloacal supplements.

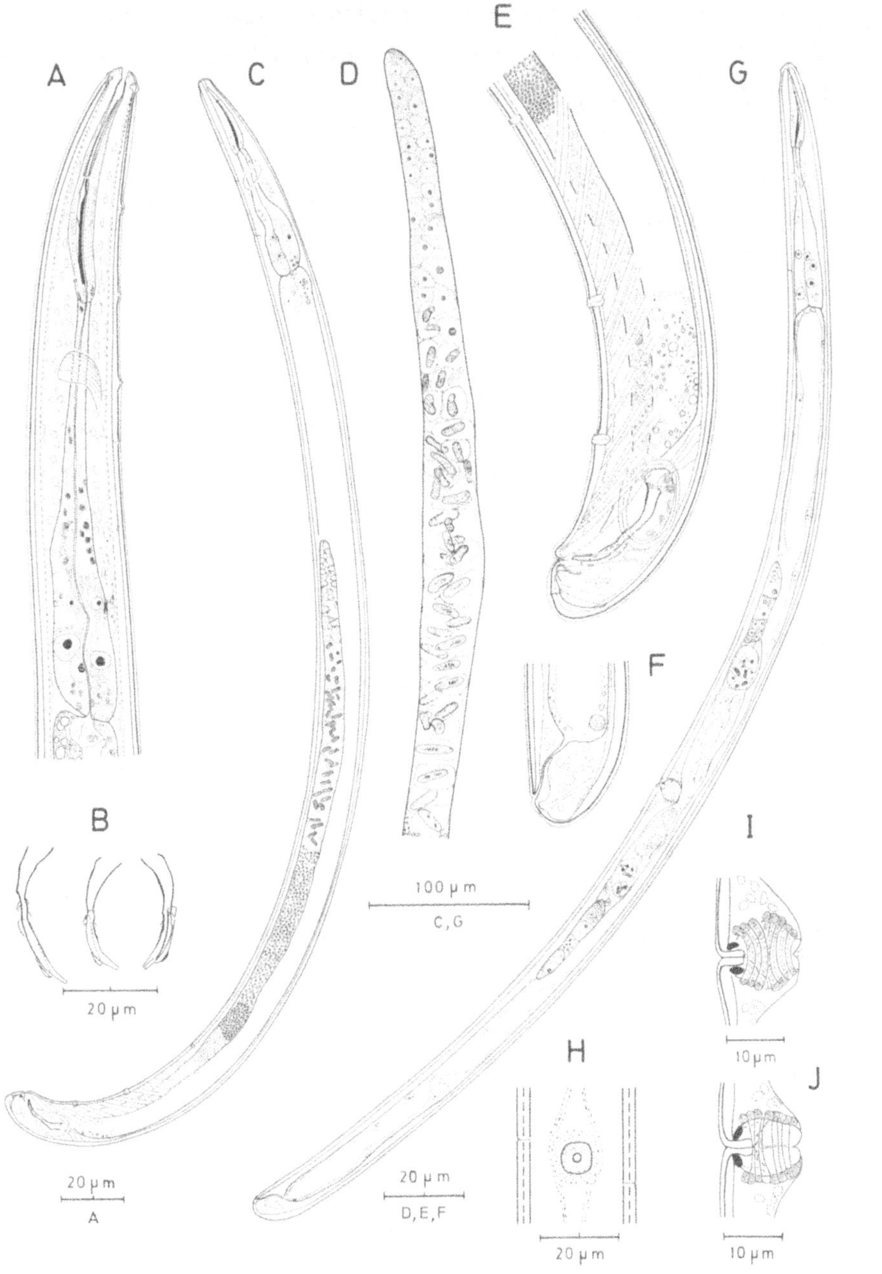

Fig. 6.7. *Trichodorus viruliferus*. Male: (A) anterior body region, (B) copulatory apparatus (after Decraemer, 1980), (C) total view, (D) testis, (E) posterior body region. Female: (F) tail, (G) total view, (H) ventral view vulva (redrawn after Hooper, 1964), (I,J) vaginal region (After Decraemer, 1991. Courtesy Marcel Dekker Inc.)

A bivulval specimen has been described from a population from a vineyard inthe province of Verona, Italy. The vulvae were apparently functional, interconnected and fused with the adjacent uteri (Coiro *et al.*, 1988).

Type host and locality

In light sandy soil around roots of wheat following grass, in a field at Roydon near Kings Lym, Norfolk, England.

Differential Diagnosis

T. viruliferus males are similar to those of *T. lusitanicus*, *T. azorensis*, *T. beirensis* and *T. velatus*, all characterized by spicules with a constriction at about mid-shaft and the presence of bristles. It can be distinguished from these species by the degree of narrowing, length of the indented zone and by a more slender spicule shaft. Females differ mainly in the shape of vagina and vaginal sclerotizations.

Distribution and Host

T. viruliferus is known mainly from western Europe (England, the Netherlands, Belgium, West Germany) and is common in Poland, Italy, Bulgaria and the southeast of France. It was also reported from Florida, U.S.A. (Mead, 1982). Hosts are wheat, rye, barley, potato, apple and pea (Symalla, 1972; van Hoof, 1964; Hooper, 1976). Other plant associations are sugar beet, grass, grape vine, maize (Ritter in van Hoof *et al.*, 1966); olive, pear, peach, tomato, artichoke, pepper, poplar, hazelnut, oak, orange, fig, lemon, walnut, Bermuda grass and *Taraxacum* sp. (Roca & Lamberti, 1984; Peneva, 1988 and white pine (Mancini *et al.*, 1979).

Biology

T. viruliferus most frequently occurs in sand and sandy-loam soils but occasionally can be found in loam soils. Usually most of the population is found at depths between 10 and 40 cm, and the remainder at 40-60 cm. General seasonal fluctuations in the total population are reflected in minimum densities during summer/autumn and maximum densities during winter/spring (Seddon, 1973; Noffsinger, 1984). Males and females are usually present in about equal numbers. On apple roots the life cycle was completed in about 45 days with reproduction occurring throughout the year, even in winter at 5°C at 20 cm depth. Oocyte development seems closely linked with the food source (Pitcher &

McNamara, 1970).

Host-Parasite Relationship

T. viruliferus causes direct damage through its feeding, such as root browning, stunting, and the occasional swelling of the root tips. In England *T. viruliferus* is a vector of PEBV (Gibbs and Harrison, 1964). It is also a vector of TRV, causing spraing in potato tubers (van Hoof, 1964, 1968). In fallow soil a viruliferous population can remain infective for 3 years (Symalla, 1972).

Control

The fumigants dichloropropane-dichloropropene or 1,3-dichloropropene give satisfactory control of docking disorder of sugar beet and spraing disease in potato tubers (Cooper and Thomas, 1971). Nemacur P prevented nematodes from transmitting TRV for a period of time although it did not reduce nematode numbers (Reepmeyer, 1973).

6.8. *Trichodorus cedarus* Yokoo, 1964 (Figs 3.2.D; 3.4 (8f, 9a); 3.6.P; 3.9 (8f, 9a); 4.4.A-I; 6.8.)

syn. *Trichodorus longistylus*, Yokoo, 1964

syn. *Trichodorus kurumeensis*, Yokoo, 1966

Measurements

Holotype female: L= 650 µm, onchiostyle= 60 µm, a= 14.4, b= 4.2, V= 60%.

Females: L= 494-897 µm (648 µm; 770 µm; 575 µm ± 36; 603 µm; 571 µm; 586 µm; 776 µm ± 70; 665 µm ± 30; 692 µm ± 28), onchiostyle= 45-70 µm (61.3 µm; 60 µm; 56.2 µm ± 2.7; 51 µm; 55 µm; 57 µm; 66 µm ± 2.8; 59 µm ± 1.6; 47 µm ± 1.3), a= 13.1-25, b= 3.6-6.2, V= 53-61% (57.9%; 57%; 58.3% ± 1.5; 56.4%; 57.2%; 56.9%; 58% ± 1.5; 59% ± 1.7; 57% ± 1.4), anterior end to S-E pore= 63-142 µm.

Males: L= 510-899 µm (573 µm; 710 µm; 603 µm ± 40; 554 µm; 599 µm; 600 µm; 790 µm ± 58; 660 µm ± 49; 710 µm ± 49), onchiostyle= 41-70 µm (59.7 µm; 60 µm; 55.8 µm ± 2.9; 51 µm; 54 µm; 57 µm; 67 µm ± 2.3; 58 µm ± 1.6; 47 µm ± 1.0), spicule length= 36-53 µm (43.5 µm; 43 µm; 40 µm ± 1.7; 38 µm; 41 µm; 42 µm; 44 µm ± 1.8; 42 µm ± 1.4; 41 µm ± 1.3), gubemaculum= 14-24 µm, anterior end to S-E pore= 56-142 µm, a= 11.9-27, b= 3.5-6.3, T= 51-81%.

(Yokoo, 1964; Mamiya, 1967; Shishida, 1979: 4 populations; Xu & Decraemer, 1995: 3 populations).

Description

Males with three ventromedian cervical papillae, posterior to onchiostyle and anterior to

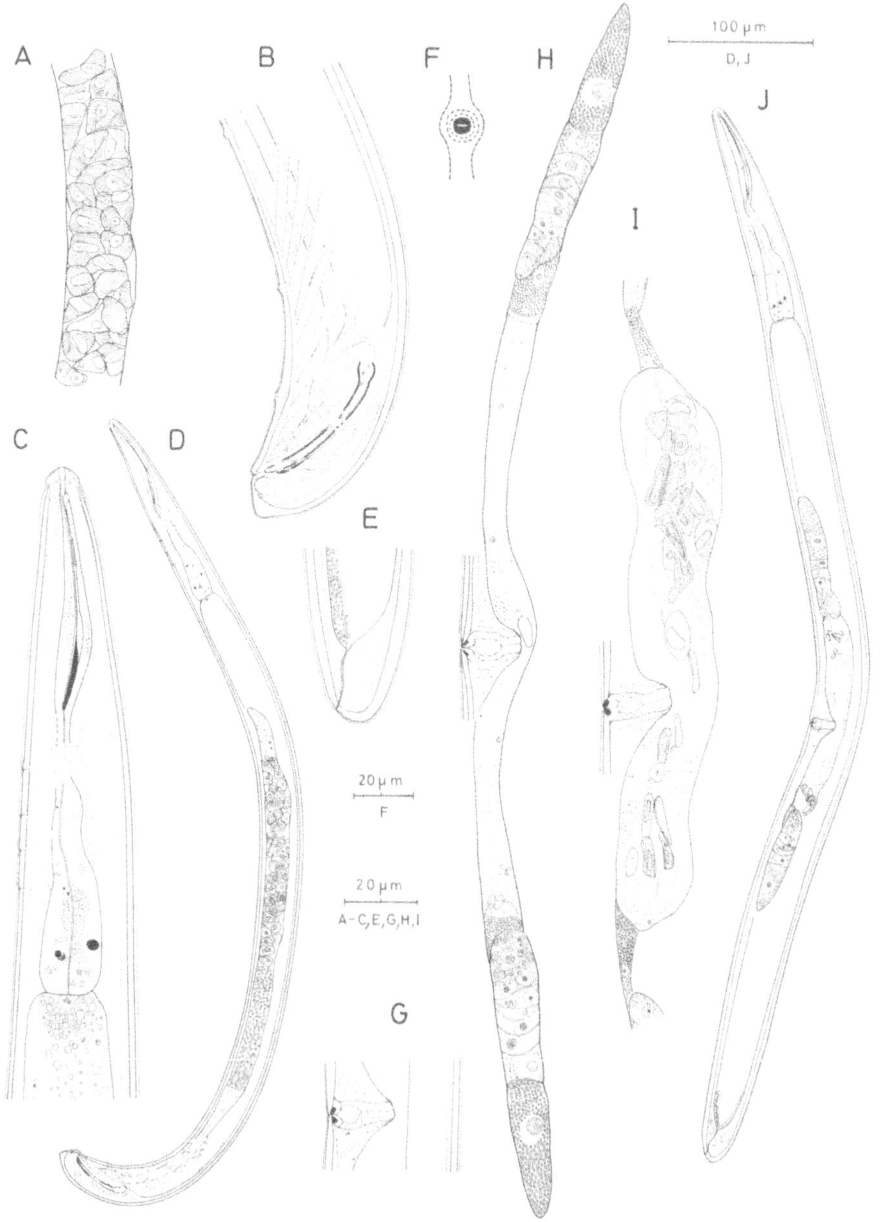

Fig. 6.8. *Trichodorus cedarus*. Male: (A) sperm, (B) posterior body region, (C) anterior body region, (D) total view. Female: (E) tail, (F) ventral view vulva (after Mamiya, 1967), (G) vaginal region, (H) reproductive system, (I) distal part of reproductive system, (J) total view. (After Decraemer, 1991. Courtesy Marcel Dekker Inc.)

S-E pore. One lateral cervical body pore on each side at level S-E pore or posterior cervical papilla. Sperm cells large, clearly fibrillar; nucleus large, oval-shaped, 3x6.5 µm. Spicules slightly ventrally curved; cephalated; spicule shaft with transverse striae except at extremities; rarely with some delicate bristles at midshaft. Gubernaculum almost straight, dorsal keel on posterior third. Three ventromedian precloacal supplements present, posterior supplement opposite midway of retracted spicules. Tail terminal cuticle more or less thickened, contour not evenly rounded.

Females with one pair of ventrosubmedian to ventrosublateral advulvar body pores (one-third to one-half body width posterior to vulva). Vulva a narrow transverse slit. Vaginal sclerotizations small, more or less triangular-shaped. Vagina pear-shaped. Tail terminal cuticle not thickened. Pharyngeal bulb offset in males and females.

Remarks

Shishida (1979) observed two males out of 263 with four ventromedian precloacal supplements. Number of ventromedian cervical papillae usually three; exceptionally varying: 1 (2 specimens), 2 (8 specimens) or 4 (1 specimen).

The Chinese populations exhibit high inter-population variability in onchiostyle length, indicating that onchiostyle length is of little diagnostic value for *T. cedarus* (Shishida, 1979; Xu & Decraemer, 1995).

Type host and locality

Cryptomeria japonica D. Don (Cedar), Asakura-Machi, Asakura-Gun, Fukuoka Prefecture, Japan.

Differential diagnosis

T. cedarus most closely resembles a recently described species from China: *T. paracedarus* in morphometric data, in males by three ventromedian cervical papillae between base onchiostyle and S-E pore, spicule shape, posterior precloacal supplement clearly within region of retracted spicules and in females by similar vaginal sclerotizations. It differs mainly in males by the curved posterior body end without bursa (straight posterior body end and bursa in *T. paracedarus*), a less pronounced thickening of the terminal tail cuticle; females are difficult to differentiate. *T. cedarus* females show some similarity with *T. borneoensis* in shape vaginal sclerotizations, one pair of postadvulvar body pores and a peach-shaped vagina, but differ in vulva shape (a transverse slit *vs* a pore).

Distribution and Host

T. cedarus has only been reported from the Far East, Japan, South Korea and China (Yokoo, 1964; Mamiya, 1969; Lee, 1976; Xu & Decraemer, 1995), except for one record from Florida (Mead, 1974; record not checked). It has a wide distribution and is associated with a wide variety of plants. The widespread occurrence in Japan makes it one of the most important plant parasitic nematodes in Japanese forest nurseries (Mamiya, 1969). Plant associations are *Cryptomeria japonica*, *Chamaecyparis obtusa*, *Larix leptolepis*, *Abies sachalinensis*, *A. homolepis*, *Picea jezoensis* v. *hondoensis*, *Pinus densiflora*, *P. sylvestris*, *P. strobus* and *P. resinosa* (Mamiya, 1969a). *Torreya nucifera*, *Pinus thunbergii*, *Carpinus tschonoskii*, *Quercus acutissima*, *Castanea crenata*, *Castaneopsis cuspidata*, *Zelkova serrata*, *Celtis sinensis*, *Cercidiphyllum japonicum*, *Magnolia obovata*, *Cinnamomum camphora*, *Neolitsea sericea*, *Kerria japonica*, *Prunus yedoensis*, *Rhus succedanea*, *Daphniphyllum macropodum*, *Camellia sinensis*, *C. japonica*, *Fatsia japonica*, *Cornus controversa*, *Rhododendron indicum* and *Callicarpa japonica* (Shishida, 1979) and soybean, cabbage, apple and barley (Lee, 1976). In China it was associated with Chinafir trees (*Cunninghamia lanceolata*), pear trees (*Pyrus pyrifolia*), peach trees (*Prunus persica*), apricot trees (*Prunus armeniaca*) and persimoon trees (*Diospyros kaki*) in Changrin, northern Zhejiang Province, with apple trees in Ganyu, northern Jiangsu Province and with peach trees in Wuxi, southern Jiangsu Province (Xu & Decraemer, 1995).

Biology

Yokoo (1964) considered *T. cedarus* to be the cause of stubby root in nurseries of black pine and cedar. No data are available on its life cycle.

Control

In a loamy soil moderately infested with *T. cedarus* on *Cryptomeria* seedlings the nematode was effectively controlled by applying a dichloropropane-dichloropropene mixture (Mamiya, 1969b).

CHAPTER 7

ECOLOGICAL ASPECTS: distribution and bionomics

Apart from Winfield & Cooke (1975) no recent compilation work exists on the ecology of trichodorid nematodes. The majority of "ecological" data are on the distribution of trichodorid species and include only restricted information on the influence of biotic and abiotic factors. This chapter is a synthesis of these separate data on the distribution of trichodorids in various regions, mainly Belgium and Great Britain where national surveys have been undertaken.

7.1. Distribution

7.1.1. GEOGRAPHICAL DISTRIBUTION

Trichodoridae are known from all over the world. They are widespread in Europe and North America, taking into account that those continents are also the most extensively studied.

The geographical distribution of trichodorid nematodes in several European countries was presented as a series of distribution maps in the European atlas of the Longidoridae and Trichodoridae (Alphey & Taylor, 1986). However, further information and new species have been described since the publication of the atlas. An emended list of the distribution of trichodorid species in Europe is given hereafter (Table 7.1).

So far, the monodelphic genera, *Monotrichodorus* and *Allotrichodorus* have been described only from the Neotropical region, with *Monotrichodorus* being widely recorded to the north and north west of the Amazon Basin (except for *M. proporifer* (= *M.m. monohystera)* from Para State, Brazil (Siddiqi, 1991)) and *Allotrichodorus* only from the south east of the Basin and restricted to Bahia State (Rashid *et al.*, 1986; Hunt, 1993). They have been found in soil around the roots of several crops (banana, citrus, sugarcane,

Table 7.1. Number of *Trichodorus* and *Paratrichodorus* species in Europe (*introduced)

	Belgium	Bulgaria	Denmark	Finland	France	Germany	Great Britain	Ireland	Italy	The Netherlands	Norway	Poland	Portugal	Roumania	Russia	Slovac Republic	Spain	Sweden	Switserland
species	10	6	4	1	10	14	13	2	12	11	3	9	12	4	4	3	8	6	3
T. aquitanensis	-	-	-	-	+	-	-	-	-	-	-	-	-	-	-	-	-	-	-
T. azorensis	-	-	-	-	-	-	-	-	-	-	-	-	+	-	-	-	-	-	-
T. beirensis	-	-	-	-	-	-	-	-	-	-	-	-	+	-	-	-	-	-	-
*T. cedarus**	-	-	-	-	-	-	-	-	-	-	-	-	-	-	-	-	+	-	-
T. cylindricus	+	-	-	-	+	+	+	-	-	+	-	+	-	-	-	-	+	-	+
T. giennensis	-	-	-	-	-	-	-	-	-	-	-	-	-	-	-	-	+	-	-
T. hooperi	-	-	-	-	-	+	+	-	-	-	-	-	-	-	-	-	-	-	-
T. lucitanicus	-	-	-	-	-	-	-	-	-	-	-	-	+	-	-	-	-	-	-
T. primitivus	+	+	+	-	+	+	+	+	+	+	+	+	+	+	+	+	+	+	-
T. similis	+	+	+	-	+	+	+	-	+	+	+	+	-	+	+	+	-	+	-
T. sparsus	+	+	+	-	-	+	+	-	+	+	-	+	-	-	+	-	-	+	+
T. taylori	-	-	-	-	-	-	-	-	+	-	-	-	-	-	-	-	-	-	-
T. variopapillatus	+	-	-	-	-	+	+	-	+	+	-	+	-	-	-	-	-	-	-
T. velatus	+	+	-	-	+	+	+	-	-	-	-	+	-	-	-	-	-	-	-
T. viruliferus	+	+	-	-	+	+	+	-	+	+	-	+	-	-	-	-	+	+	-
*P. acutus**	-	-	-	-	-	-	-	-	-	-	-	-	+	-	-	-	+?	-	-
*P. allius**	-	-	-	-	-	-	-	-	+	-	-	-	+	-	-	-	-	-	-
P. anemones	-	-	-	-	+	-	+	-	-	+	-	-	+	-	-	-	-	-	-
P. hispanus	-	-	-	-	-	-	-	-	-	-	-	-	-	-	-	-	+	-	-
P. macrostylus	-	-	-	-	-	-	-	-	-	-	-	-	-	+	-	-	-	-	-
*P. minor**	-	-	-	-	-	+	-	-	+	-	-	-	+	-	-	-	+	-	-
P. nanus	+	-	-	-	+	+	+	-	+	+	-	-	+	-	-	-	-	-	-
P. pachydermus	+	+	+	+	+	+	+	+	+	+	+	+	+	+	+	+	-	+	+
P. porosus	-	-	-	-	-	-	-	-	-	-	-	-	+	-	-	-	-	-	-
*P. renifer**	-	-	-	-	-	+	+	-	-	+	-	-	-	-	-	-	-	-	-
P. teres	+	-	-	-	+	+	+	-	+	+	-	+	+	-	-	-	-	+	-
P. tunisiensis	-	-	-	-	-	-	-	-	+	-	-	-	-	-	-	-	-	-	-
P. weischeri	-	-	-	-	-	+	-	-	-	-	-	-	-	-	-	-	-	-	-

oilpalm, cocoa, coffee, etc.) and also in soil from primary forest (*M. m. monohystera*, *M. m. vangundyi*, *A. campanullatus*, resp. in Venezuela, Panama and Brazil). Their association with native plants supports the theory that the monodelphic trichodorids are endemic to South and Central America. *A. westindicus*, the aberrant form of the genus (see chapter on taxonomy) has a somewhat different distribution compared with the other monodelphic species. It was originally described from Trinidad, West Indies and further recorded from Martinique (Baujard & Germani, 1985) and Brazil (Rashid *et al.*, 1986). The didelphic trichodorids reported from Central and South America may have been introduced with non-indigenous plants (Rashid *et al.*, 1986). They mainly belong to the genus *Paratrichodorus*, rarely to *Trichodorus* (Tables 7.2, 7.3). They have been found associated with agricultural crops e.g. *P. minor* with sugar-cane, rice, cotton, banana, cocoa; *P. porosus* with sugar-cane and *Trichodorus* sp. on tomato (Campos & Sturhan, 1987).

Some of the didelphic species only occur in certain geographic areas and are absent or rare in others. *Trichodorus* is a cosmopolitan genus with hitherto 47 species described. It occurs predominantly in the more temperate regions (Table 7.2) whilst the genus *Paratrichodorus* with 31 species (Table 7.3) is present mainly in tropical and subtropical regions (Luc, Sikora & Bridge, 1990). Only 39% of *Trichodorus* species have been described from former Gondwanaland countries. From this it can be assumed that the genus probably originated in Laurasia, with main speciation in Europe (see Table 7.2). About 75% of *Paratrichodorus* species have been described from former Gondwanaland countries. Because information on species distribution in this area is scanty and correlated with the occurrence of nematologists interested in the group (e.g. in India, South Africa), speciation cannot be deduced from the present species distribution.

Bernard (1992a) provided an account of the present distribution of *T. californicus* (California, Alaska, Colorado, South Dakota, Wisconsin) in relation to its historical biogeography, past glacial events and morphological differences observed between populations. *T. californicus* appears most common in taiga and coniferous rhizospheres, and its distribution is highly correlated with non-glaciation. He concluded that the central Alaskan population probably has been isolated from the Californian type population for many thousands of years because of morphological differences observed between males of both populations. Other reports of the occurrence of *T. californicus* e.g. from Wisconsin, which was totally glaciated, can be only explained by human activity.

Part of the present distribution of *Trichodorus* and *Paratrichodorus* species is probably due to introductions with non-indigenous plants. Reports from the 'Florida Department of Agriculture and Consumer Services', published during the last few years,

Table 7.2.: List of *Trichodorus* species reported from the continents:

Europe	**N- America**	**Africa**	**South-Central America**	**Asia**
T. aquitanensis	*T. aequalis*	*T. coomansi*	*T. primitivus**	*T. aequalis**
T. azorensis	*T. borneoensis**	*T. eburneus*	*T. cedarus**	*T. borai*
T. beirensis	*T. californicus*	*T. kilianae*	*T. obscurus**	*T. borneoensis*
T. cylindricus	*T. carlingi*	*T. magnus*		*T. cedarus*
T. hooperi	*T. cedarus**	*T. parorientalis*	**Oceania**	*T. complexus*
T. lusitanicus	*T. cylindricus**	*T. petrusalberti*	*T. cottieri*	*T. nanjingensis*
T. primitivus	*T. dilatatus*	*T. philipi*	*T. primitivus**	*T. orientalis*
T. similis	*T. elefjohnsoni*	*T. rinae*		*T. pakistanensis*
T. sparsus	*T. elegans*	*T. sanniae*		*T. persicus*
T. taylori	*T. intermedius*	*T. vandenbergae*		*T. primitivus**
T. variopapillatus	*T. obscurus*			*T. tricaulatus*
T. velatus	*T. obtusus*			*T. yokooi*
T. viruliferus	*T. pakistanensis**			*T. minzi*
T. giennensis	*T. paucisetosus*			*T. hooperi**
T. cedarus?	*T. primitivus**			*T. obscurus**
	*T. similis**			*T. paracedarus*
	*T. velatus**			
	*T. viruliferus**			

contain more and more references to species, previously only reported from Europe for example, being intercepted with planting material or from sites in the U.S.A. The first record of *Trichodorus velatus* in Florida, U.S.A. dates from 1989 when it was found in samples from the rhizosphere of *Quercus* sp. (oak) and was probably introduced by man (McGowan, 1989a). Some species such as *T. primitivus* (common in Europe) are occasionally reported from U.S.A. and New Zealand. Other examples of species possibly introduced to a continent are indicated with an * in Tables 7.2, 7.3.

Table 7.3.: List of *Paratrichodorus* species reported from the continents:

Europe	**N - America**	**Africa**	**South-Central America**	**Asia**
*P. allius**	*P. acutus**	*P. acutus*	*P. anthurii*	*P. acaudatus*
P. anemones	*P. allius*	*P. alleni*	*P. minor**	*P. allius*
*P. acutus**	*P. anemones**	*P. allius*	*P. porosus**	*P. delhiensis*
P. hispanus	*P. atlanticus*	*P. catharinae*	*P. allius**	*P. faisalabadensis*
P. macrostylus	*P. grandis*	*P. lobatus*	*P. lobatus**	*P. minor*
*P. minor**	*P. minor*	*P. meyeri*		*P. mirzai*
P. nanus	*P. mirzai**	*P. minor*		*P. pachydermus*
P. pachydermus	*P. nanus**	*P. nanus*	**Oceania**	*P. paramirzai*
P. porosus	*P. pachydermus*	*P. porosus*	*P. acutus*	*P. paraporosus*
P. renifer	*P. porosus*	*P. renifer*	*P. lobatus*	*P. porosus*
P. teres	*P. renifer**	*P. rhodesiensis*	*P. minor*	*P. psidii*
P. tunisiensis	*P. teres**	*P. sacchari*	*P. orrae*	*P. renifer**
P. weischeri	*P. anthurii**	*P. teres*	*P. porosus*	
	*P. catharinae**	*P. tunisiensis*	*P. queenslandensis*	
			P. renifer	

Currently *Trichodorus* species have not been recorded from Australia and only the genus *Paratrichodorus* is reported from this country (Decraemer & Reay, 1991). Data for South and Central America are restricted to a few species (see before). *P.*

minor and *P. porosus* appear to be cosmopolitan species. However, *P. minor* is not indigenous to South and Central America (Rashid *et al.*, 1986) and is only occasionally recorded from Europe (Sicily, Belgium, The Netherlands, Portugal, Sweden, Switzerland and West-Germany (Decraemer, 1991)). *P. pachydermus* is widespread only in the northern hemisphere whereas other species e.g. *P. catharinae*, *P. meyeri* and *P. sacchari* seem to be endemic only in South Africa (De Waele & Kilian, 1992).

7.1.2 HORIZONTAL AND VERTICAL DISTRIBUTION

Distribution of Trichodoridae may be irregular both horizontally and vertically with the horizontal distribution being more difficult to predict than the vertical distribution; besides a stronger border-colonization may be possible (Pietler, 1977). Although the horizontal distribution has been assumed to have a random distribution, it has been shown that most plant-parasitic nematodes have an aggregated or contiguous distribution which may be described statistically e.g. by Taylor's Power Law (Boag & Topham, 1984). Boag *et al.* (1987) studied the horizontal distribution of virus vector nematodes and found that trichodorid nematodes aggregate more than longidorid nematodes. Indeed trichodorid nematodes can aggregate in large numbers around the growing points of host root systems (Pitcher, 1967). A knowledge of the horizontal distribution of nematodes can be used to help devise optimum sampling procedures (Boag *et al.*, 1992).

Trichodoridae are rarely numerous in the surface layers, except after periods of prolonged rainfall. They appear susceptible to desiccation but are able to avoid such stress by migrating vertically for considerable distances for example to deeper, moister soil layers in summer; in autumn, they move up through the soil to recolonise the upper soil layers (Rössner, 1972; Decraemer, 1980b).

Trichodorids appear to be susceptible to mechanical disturbance (Bor & Kuiper, 1966) and are usually found below the depth of cultivation (Cooke, 1973). This susceptibility is much lower in water logged soils (Bor & Kuiper, 1966).

The vertical distribution of some common trichodorid species is summarized in Fig.7.1, depths with highest population densities being indicated by coarse punctuation. This information is based mainly upon data from the literature. Boag (1981) recorded no correlation between the vertical distribution of *P. pachydermus* and root distribution of the host plant, *Sitka* spruce. Also, *P. pachydermus* could withstand both drought and waterlogged conditions.

7.2. Biotope preference

7.2.1. SOIL TEXTURE, MOISTURE

The particle size and crumb structure of soils are of fundamental importance to soil inhabiting nematodes. Pore space and moisture holding capacity are probably the most important features of a soil that determine whether or not it is a suitable habitat for nematodes. Clay soils have a texture that tends to prevent free movement of air and water. Loams aerate and drain well but the clay fraction (8- 15 %) assumes that the soils can withstand short periods of drought. Sandy soils are composed of relatively large particles which gives an open texture so that they are free-draining and subject to rapid drying during periods of drought. Trichodorids in general are not found in fine textured soils that contain much silt or clay. They have a preference for coarser textured soils (Winfield & Cooke, 1975; Decraemer *et al.*, 1979; De Pelsmaeker *et al.*, 1985). In Great Britain, for example, they were frequently found in the sandy loams and loamy sand soils, but not in any clay or silt soils (Alphey & Boag, 1976). Similarly a national survey in Belgium, showed that 40% of all samples with a sand fraction higher than 90% were infested with trichodorids compared with 6% for samples with a sand fraction lower than 80%. Furthermore, all populations of *T. sparsus* and *P. nanus* and most populations of *T. similis*, *T. cylindricus*, *P. pachydermus* and *T. viruliferus* were associated with soils with a sand fraction higher than 90% and a silt fraction smaller than 10%. However, 44% of soils with a silt fraction of 20% or higher contained *T. primitivus* (De Waele & Coomans, 1991). *T. primitivus* seems to be an exception to the general soil type distribution pattern of trichodorid nematodes, by occurring also in clay soils (Seinhorst, 1963; Cooper, 1971). In South Africa trichodorids are as numerous, although a little less widespread, in clays and sandy clays as in sandy soils (Spaull, 1981). However, in sandy soils, large populations of *Paratrichodorus* were frequently associated with poorly growing sugar-cane with typical stubby root symptoms.

Trichodorids are, among all plant-parasitic nematodes, amongst the most susceptible to desiccation. Movement of *Trichodorus* was studied in different moisture regimes in sandy soils. Movement was greatest when soil pores were half-full of water, least in waterlogged or dry soil. The fact that trichodorid nematodes are almost entirely restricted to free-draining sandy soils means that they are likely to encounter very dry

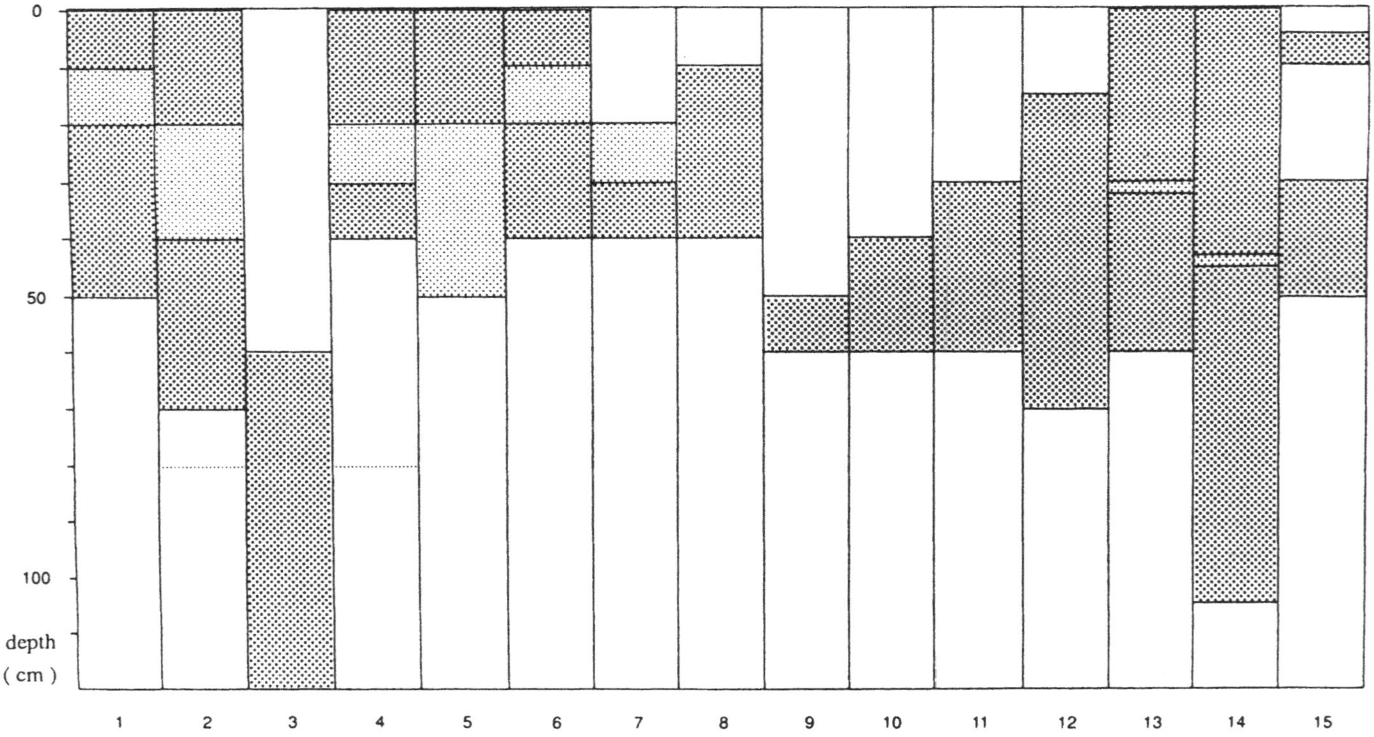

Fig. 7.1. Vertical distribution of *Trichodorus* and *Paratrichodorus* species. (1) *T. cylindricus*, under potato; (2-3) *T. primitivus*, under potato (2) and under clover after a long period of drought (3); (4) *T. similis*, under potato, rarely to a depth of 80cm; (5) *T. proximus* (= *T. obtusus*) under potato; (6-9) *T. viruliferus*, under potato (different authors); (10-12) *P. pachydermus*, under potato (10-11), in summer (11), in absence of plant roots (12); (13) *P. minor*; (14) *P. porosus*, under sugar cane; (15) *P; teres* (after Carneiro *et al.*, 1982; Cooke, 1984; De Pelsmaecker, 1987, De Pelsmaecker *et al.*, 1985; De Waele *et al.*, 1985; Harrison & Smart, 1975; Kuiper, 1977; Brodie, 1976 in Noffsinger, 1984; Pietler, 1977; Richter, 1969).

soil conditions, so it is not surprising that trichodorid numbers in the upper layer of the soil are correlated with rainfall. The increased survival and activity of *Trichodorus* in conditions of high water content are reflected in the close correlation between the area of sugar beet in England affected by Docking disorder and the total rainfall in May, the month in which most damage to the seedling sugar beet takes place (Cooke & Draycott, 1971). Similarly, numbers of *P. porosus* in sugar-cane fields in Brazil were positively correlated with rainfall and soil moisture. However, in the Philippines, numbers of *T. borneoensis* were correlated with rainfall in only one of four cane fields investigated (Estioko & Reyes, 1984).

7.2.2. pH

In Belgium, in general, relatively more populations (30%) of trichodorids occur in soils with a pH lower than 5.5. e.g. *T. similis*, and *P. pachydermus* most frequently occurs in a slightly acid soil, although *T. similis* may show some variability (pH 5.1 to 6.9) according to the soil texture (sand or sandy loam)(De Pelsmaeker *et al.*, 1985).
About 40 % of *T. primitivus*, *T. variopapillatus*, *T. cylindricus* and *T. velatus* populations were found in soils with a pH higher than 7 (De Waele & Coomans, 1991).

In Britain, only 6.5% of the soils with a pH lower than 5.5 and 12.4% of the soils with a pH between 5.5 and 6.4 were infested with trichodorids, whereas 16.9% of the soils with a pH of 6.5 or higher contained at least one trichodorid species (Alphey & Boag, 1976).

T. primitivus shows a relatively high tolerance for soil acidity; it also has a higher tolerance for copper in the soil than the other *Trichodorus* species (Alphey & Boag, 1976).

7.3. Species associations

Topham *et al.* (1985) made a comparison between plant-parasitic nematode species associations in Great Britain and in Belgium. Prior to the Quaternary period Great Britain and Belgium formed part of the same land mass and may therefore be expected initially to have had similar nematofaunas and recent surveys confirm this within the Longidoridae and Trichodoridae. Although within the Longidoridae the most frequently occurring species were the same in both countries, this was not so within the Trichodoridae. This difference may be explained by differences in the availability of suitable habitats. Comparison of patterns of species associations between Belgium and Great Britain showed that they were broadly similar except for *T. viruliferus* and *P. teres*. In Belgium, the degree of species association was greatest between *T. viruliferus*, *T. similis*, and *P. pachydermus* and then, by decreasing levels of similarity, *T. cylindricus*, *T. primitivus*, *P. teres* and *T. velatus*. In Great Britain, the most common associations were between *T. cylindricus* and *P. teres*. They were most commonly associated with *T. similis*, *P. pachydermus*, *T. primitivus*, *T. velatus* and less with *T. viruliferus*.

7.4. Host-parasite relationship

7.4.1. ASSOCIATIONS WITH VEGETATION TYPE

Trichodorids are polyphagous migratory ectoparasites of the roots of perennial and woody plants. The wide host range of polyphagous trichodorids can be best illustrated by the more or less cosmopolitan species *P. minor* which has been found associated with over 100 plant species (economic crops, grasses, broad-leaf plants, weeds) (Noffsinger, 1984).

In temperate climatic areas

National surveys in Belgium showed that in general, relatively more trichodorid populations were associated with grasses at road sides, deciduous woodland and arable crops than with pastures or coniferous woodland (De Waele & Coomans, 1991). In contrast to Belgium, arable crops in Great Britain supported more trichodorids (63% against 21%) than any other vegetation type (Alphey & Boag, 1976). For more extensive data in temperate agriculture see Evans *et al.* (1993).

In subtropical and tropical climatic areas

Although the genus *Paratrichodorus* is distributed worldwide it is more common in tropical and subtropical regions. The genera *Monotrichodorus* and *Allotrichodorus* are restricted to Central and South America. Useful information and more extensive bibliographies can be found in Luc *et al.* (1990) and chapter 6 on economically important species.

7.5. Feeding behavior

Trichodorid nematodes have a feeding behavior almost unique among plant parasitic nematodes by (1) the continuous thrusting of the onchiostyle throughout all phases of a complete feeding cycle, (2) the production of a feeding tube as an essential suction tool for food ingestion and by (3) possessing an onchiostyle with a completely solid tip at its anterior end.

The feeding mechanism of trichodorid nematodes was analyzed for *T. similis* with the aid of cinematography (Wyss,1972). The complete feeding process, lasting only a few minutes, can be divided into five phases: (1) exploration (lip-rubbing along the cell wall), (2) penetration (puncturing of the cell wall), (3) salivation, (4) food ingestion and (5) withdrawal from the feeding site. The feeding tubes are initiated in a specialized region of the feeding apparatus, where the cuticular lining contains three strengthening rods. In the exploration phase, the strengthening rods remain positioned behind the oral aperture until a satisfactory site is located. Then, the strengthened region is protruded so that the tips of the rods appear to touch the surface of the cell wall to be penetrated (Fig. 7.2). Once contact is made, the onchiostyle is immediately repeatedly trusted at the cell wall at a rapid rate. If the nematode stops thrusting before perforation of the cell wall, it draws back the strengthening rods and leaves the cell; a small tube, 1-2 µm long, can be seen adhering to the cell wall. If penetration occurs and feeding follows, a feeding tube is formed that is as long as the strengthening rods. The feeding tube is formed by secretions, released by the nematode through the pharyngeal lumen (probably produced in the pharyngeal glands). These secretions harden rapidly around that part of the lumen where the lining is stiffened by the strengthening rods. These secretions accumulate more thickly at the penetration site, spread out over the inner lining of the cell wall and form a plug, which effectively anchors the tubes (Fig. 7.2). After penetration, the thrust rate of the spear increases, the cytoplasm of the cell flows to the site of cell perforation together with the cell nucleus. The transition towards food ingestion is initiated by several deeper thrust of the stylet. In the last phase of cytoplasm concentration in the attacked plant cell, food ingestion occurs at a high rate and later becomes slower. When the feeding process is completed, the remaining cytoplasm beyond the feeding tube becomes coagulated; the nucleus is disorganized (Wyss, 1971).

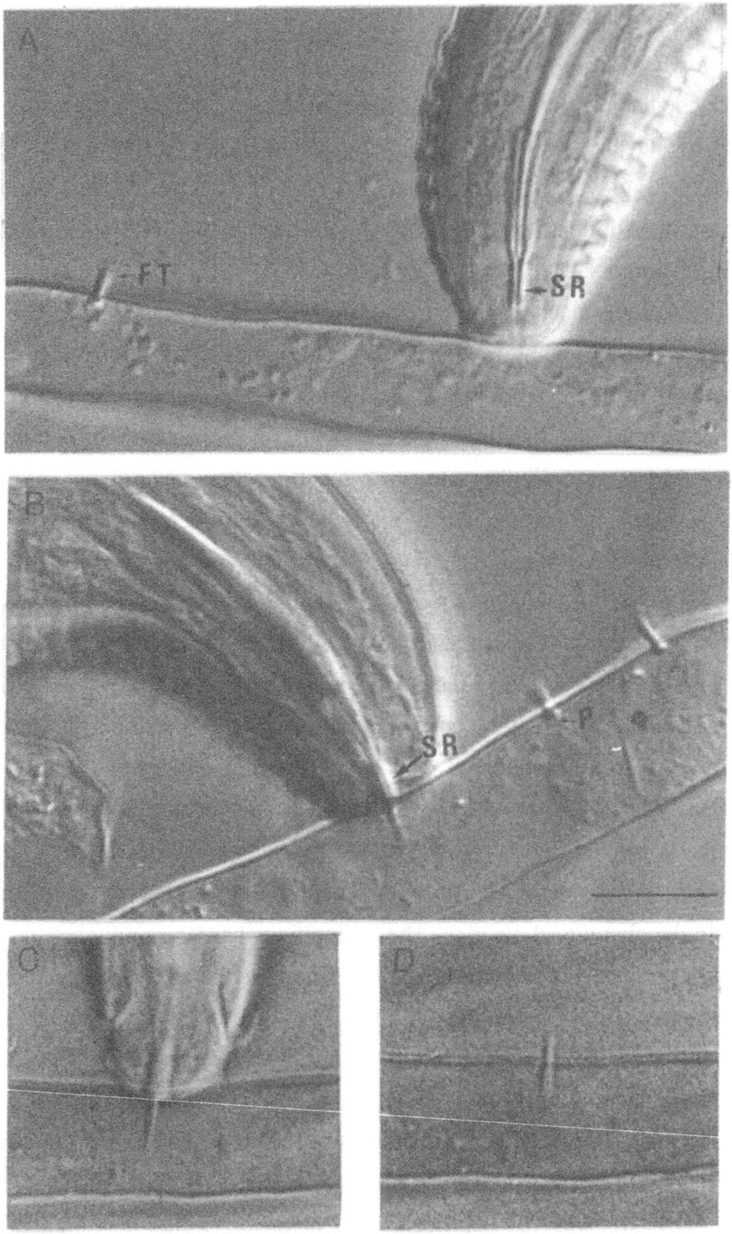

Fig. 7.2. *T. similis* (female) feeding on root hairs of *Nicotiana tabacum* L. var. 'Samsun' seedlings in agar culture. (A) Local exploration, with strengthening rods (SR) of pharyngeal lumen still behind oral aperture. Left: former feeding site with feeding tube (FT) and plug (P). (B) Perforation of root hair with previous feeding sites. Anterior end of strengthening rods now in close contact with cell wall. (C) Onchiostyle penetrating into empty root hair. (D) Same root hair after nematode's departure (after Wyss *et al.*, 1979 with courtesy of *Nematologica*). Scale: 10 µm

7.5. Reproduction

Reproduction is usually regarded as being amphimictic or parthenogenetic (see chapter 3 on morphology). The life cycle of the bisexual species *T. viruliferus* on apple roots takes about 45 days from egg to adult at 15-20°C. Breeding occurs throughout the year, but the level of breeding activity (indicated by number of oocytes produced) is much higher in the rhizosphere samples than in the random samples, especially during spring and summer. Oocyte development seems closely linked with food source (Pitcher, 1967; Pitcher & McNamara, 1970). Similarly, *T. sparsus* on *Prunus mahaleb* completes its life cycle in 40-42 days at 27°C (Coiro & Sasanelli, 1994).

The embryogenesis of *P. minor*, *P. porosus* and *P. teres* was studied under controlled conditions. For details see chapter 3 on general morphology and chapter 6 on economically important species.

CHAPTER 8

TRICHODORIDS AS VIRUS VECTOR NEMATODES

8.1. Historical background

Although nematodes had long been suspected as possible vectors of plant viruses, the first records of trichodorid species shown to be vectors of tobacco rattle virus date from 1960.

At the end of last century, a disease in Dutch tobacco fields was descriptively called "rattle disease" because of the sound caused by the wind playing with the dried-out infected leaves (van Hoof, 1972). Behrens (1899) was the first to describe this soil-borne disease in tobacco. Later, this disease was recognized as being caused by tobacco rattle tobravirus (Quanjer, 1947).

In 1960, Sol *et al.* (1960) demonstrated that *Paratrichodorus pachydermus* could transmit tobacco rattle virus (TRV). Subsequently pea early-browning virus (PEBV) and pepper ringspot virus (PRV) have been reported to be transmitted by trichodorid nematodes (van Hoof, 1962 and Salomao, 1973 respectively).

8.2. Viruses transmitted by Trichodoridae

The three viruses transmitted by trichodorid nematodes belong to the tobravirus group, formerly called netuviruses (*ne*matode, *tu*bular). They are characterized by their rod shaped particles. Tobravirus isolates fall into two main classes: (1) M-type which produce long and short nucleoprotein particles containing respectively genome RNA1 and genome

RNA2, and (2) NM-type which produce RNA1 but not RNA2 which codes for the coat protein. Because coat protein is not produced, infection by NM-type isolates cannot be detected by serological means and the virus is not transmissible by trichodorids (Robinson, 1989).

Tobacco rattle virus is widespread in Europe and has also been recorded from North America, New Zealand and Japan (Ploeg, Brown & Robinson, 1989). With more than 400 plants species being susceptible to infection with TRV, this virus has the widest known host range of all plant viruses.

Crops in which TRV causes serious damage include potato ("spraing" disease in the tuber, stem mottle in the aerial parts), tobacco ("rattle"-disease), lettuce (mosaic patterns in the leaves), in *Capsicum* - peppers (mosaic patterns), bulbous ornamental crops ("notched-leaf" in *Gladiolus*, color breaking and leaf distortion in tulip), and a range of other crops (Robinson & Harrison, 1989).

TRV is known to have many variants, which differ in their serological properties, particle length and /or symptomatology (Harrison & Robinson, 1986; Robinson & Harrison, 1989). Therefore, establishing TRV as the agent causing a disease in a crop either on the basis of symptom development in inoculated test plants, or by electron microscopic examination of infected sap, or by serological tests (e.g. ELISA) may not always be conclusive. Nucleic acid hybridization techniques using probes complementary to parts of the highly conserved RNA-1 genome is at present the most reliable method for identifying TRV. Such a probe can be used in dot-blot tests which are practicable for routine diagnosis (Robinson, 1989; Robinson & Harrison, 1989).

The occurrence of TRV usually reflects the distribution of the vector nematodes, although the virus can occur without the vector being present in the field as a result of using vegetatively propagated infected plant material (e.g. potato tubers, flower bulbs). Primary infection (=nematode transmitted) in a field can usually be distinguished from secondary infection (=infected planting material) because of the patch like distribution of the former compared with the random distribution of the latter (Asjes, 1989; Cremer & Schenk, 1967).

Pepper ringspot virus, formerly known as the TRV strain CAM, an isolate from South America, classified as the sole representative of TRV serotype III, has thus far only been reported from Brazil (Harrison & Robinson, 1986).

Pea early-browning virus which is less frequently found than TRV has been recorded only from Europe and North Africa; it is mainly important as the cause of the disease in pea crops (Harrison & Robinson, 1986). Within PEBV, serological differences between isolates are known to occur. PEBV can be distinguished from TRV by inoculation of a differential host plant (e.g. *Phaseolus vulgaris*) or by nucleic acid hybridization techniques using probes specific for PEBV RNA-1 sequences (Robinson & Harrison, 1989).

8.3. Virus-vector associations

After the initial report of virus transmission by a trichodorid nematode (Sol *et al.*, 1960), much research was initiated to investigate the possible role of other trichodorids as virus vectors. Trudgill *et al.* (1983) established criteria which should be fulfilled in order to consider longidorid nematode species as being virus vectors: (1) the nematode and virus must be fully identified and characterized, (2) bait plant tissue must be shown to be infected with the virus, and (3) the nematode must be shown to be the only possible vector of the virus.

Brown *et al.* (1989) modified the criteria of Trudgill *et al.* (1983) and studied the data available on trichodorid vector-virus associations. They concluded that only 12 out of 40 reported trichodorid and tobravirus associations contained adequate evidence to support the virus and vector association. At present, the following trichodorid species are considered valid vectors of tobraviruses: *P. allius, P. anemones, P. minor, P. nanus, P. pachydermus, P. teres, P. tunisiensis, T. cylindricus, T. primitivus, T. similis, T. viruliferus* (Brown *et al.*, 1989; Brown, 1995; Ploeg & Brown, 1995)

Brown, Ploeg & Robinson (1989) developed a system which enables the ability of an individual *Paratrichodorus* or *Trichodorus* nematode to transmit a given TRV isolate to be assessed. The system is applicable for bait testing of field soil and thus identifying associations between nematodes and tobraviruses from natural biotopes.

It is now widely accepted that trichodorid nematodes are the natural vectors of TRV, however, it has also been suggested that transmission may occur in the absence of trichodorid vector nematodes e.g. through soil water and root contact (Sol, 1961, 1963; Fritzsche *et al.*, 1985).

8.4. Plant virus transmission and retention

8.4.1. MECHANISM OF TRANSMISSION

Feeding and Acquisition

Transmission of a virus by a nematode consists of a complex series of events: acquisition (uptake of virus particles from infected plants), retention (retention of the infected particles within the vector nematode), and subsequent release of the virus (inoculation by the vector to another plant cell, and infection of the receptor plant by the virus) (Harrison *et al.*, 1974).

The nature of the mechanism involved in the specific retention of TRV by its trichodorid vectors is largely unknown. Trichodorid nematodes presumably acquire virus when the cytoplasmic contents of an infected cell are ingested. However, inoculation leading to virus infection is likely to occur when feeding is interrupted and the cell not damaged so severely as to prevent the resumption of cytoplasmic streaming (Harrison, 1977). The nature of trichodorid feeding also suggests that viruses may more readily be acquired than inoculated.

TRV was first observed *in situ* in its vector *P. pachydermus* by Taylor & Robertson (1970), who found TRV particles throughout the length of the pharyngeal lumen.

Transmission

Taylor & Robertson (1970, 1978) envisaged the transmission process as a mechanical contamination of the nematode stylet involving specific adsorption of virus particles on the cuticular lining of the pharyngeal lumen. Subsequent release of the virus would occur under the influence of a pH change caused by the saliva flow produced by the nematode during the initial stages of the feeding process.

In recent research, the lining of the pharynx of *P. pachydermus*, which is the site of retention for TRV, was stained for carbohydrates. The staining appeared continuous over the whole of the lining and therefore virus retention might possibly involve recognition between carbohydrate moieties present there and complementary lectin-like molecules on the protein coat of the virus (Robertson & Henry, 1986). Nematode transmissibility of the virus may well depend on characteristics of the virus coat protein (Brown *et al.*, 1989).

Juveniles as well as adults can transmit viruses. However, the virus is not transmitted through the egg or molting. This is due to the attachment of the virus particles to the cuticularized parts of the feeding apparatus which are shed during molting (Taylor & Robertson, 1974).

8.4.2. SPECIFICITY AND EFFICIENCY OF TRANSMISSION

Van Hoof (1968) reported a population of *P. pachydermus* transmitting an isolate of TRV originating from the same locality as the nematodes, but failed to transmit several isolates of TRV from other localities. Another example of some specificity of transmission between trichodorids and tobraviruses was demonstrated by *P. anemones* and *T. primitivus* from England, both transmitting English isolates of PEBV but not a Dutch isolate (Harrison, 1966, 1968). Further support for the suggestion that specific associations exist between trichodorids and tobraviruses was obtained with results from transmission experiments in which TRV isolates with similar serological characteristics were transmitted by the same populations of the species from different geographical origins (Brown & Ploeg, 1989; Brown *et al.*, 1989; Ploeg *et al.*, 1989). Brown, Ploeg & Robinson (1989) studied the transmission of tobraviruses by populations of *(Para)Trichodorus* from Britain, The Netherlands, and Scandinavia and the results were very suggestive, although not conclusive, that each tobravirus serotype has a single *Trichodorus* or *Paratrichodorus* species as its vector.

Infected vector nematodes can be very efficient in the transmission of virus to healthy plants: e.g. a single viruliferous *P. pachydermus* male successively infected three *Nicotiana rustica* seedlings with TRV within a period of four days (van Hoof, 1964). The relatively low transmission rates frequently obtained when testing viruliferous populations may result from only one species out of several present transmitting virus, and/or a low abundance of virus sources in the field (Harrison & Robinson, 1986).

CHAPTER 9

AGRICULTURAL IMPORTANCE AND CONTROL

9.1. Direct damage

Since *T. christiei* (now *P. minor*), the first trichodorid recognized as damaging plants, many species of Trichodoridae have been recorded as plant pathogens, but some species are considered more economically important than others. *P. minor* for example, appears to be fairly cosmopolitan and a pest of many crops. It is considered to be the most economically important species of stubby root nematodes in the USA. Typical above-ground symptoms of infected plants are retarded growth, chlorotic and wilted foliage, and sensitivity to drought (McGowan, 1983; Christie, 1959; Heyns, 1975). The amount of damage to the root system varied with the crop attacked, and further appeared greatly influenced by season, soil type and fertility (Jones *et al.*, 1969; Cooke, 1973). Often, the crop can compensate for initial damage and yield almost normally, but this was not the case in tests with wheat and barley (Spaull, 1980).

In an underground laboratory constructed for the study of the growth of fruit-tree roots up to 160 cm depth, Pitcher (1967) studied the host-parasite relations of *T. viruliferus* on apple roots. The most striking feature was the massing of nematodes feeding at the elongation zone of young extending roots i.e. 1-3 mm behind the root tip. The feeding of large numbers of nematode upon epidermal and cortical cells led to a marked decline in root growth usually, accompanied by a fall in nematode numbers and tendency for the attack to be transferred to the apical meristem. Once root growth had ceased most of the nematodes dispersed into the soil.

Direct feeding of trichodorid nematodes can cause severe damage e.g.'Docking disorder' of sugar beet in Great Britain resulting from feeding of *T. cylindricus* and *P. anemones* on seedlings, and characterized by patches of stunted plants in the crop in the early stage of growth (Whitehead & Hooper, 1970). *P. porosus* was responsible for extensive damage to the root system of *Camellia* (Barriga, 1965) and for "black scurf-like syndrome" of Chinese yam, *Dioscorea opposita* in Japan, resulting in blacking, cracking and corkiness of the tuber tips. (Nishizawa, 1973). The disease increased with successive planting of yam.

Direct damage is mainly apparent in seedlings and is highly depended upon the mobility of the nematode which is strongly determined by pore space and moisture capacity.

9.2. Damage caused by trichodorid nematodes transmitting plant viruses

Several other trichodorid nematodes are known to be responsible for stubby root symptoms on a variety of plants, but major pest status results from the known ability of trichodorids to transmit tobraviruses.

About half of the species of *Trichodorus* and *Paratrichodorus* in Europe, are vectors for plant viruses; this percentage is much lower on a worldwide scale (Ploeg & Brown, 1995).

The literature on trichodorids and virus transmission is voluminous (Harrison *et al.*, 1974; Taylor, 1978; Brown *et al.*, 1989; Brown & Ploeg, 1995). The following examples are given to illustrate the variety of diseases caused by trichodorid nematodes transmitting tobraviruses. *T. primitivus* is a vector of TRV, causing spraing in potato tubers (Hooper & Siddiqi, 1972; De Pelsmaeker *et al.*, 1985) and stem mottle of potato (van Hoof, 1980). It also transmits PEBV, which damages pea crops and infects other plants such as lucerne, cucumber, tobacco, and sugar beet (Gibbs & Harrison, 1964; Harrison, 1964). *T. similis* is a vector of TRV, causing spraing disease in potato (van Hoof, 1968, 1972; Eriksson, 1974; De Pelsmaeker, 1988) and notched leaf symptoms in *Gladiolus* (Cremer & Schenk, 1967).

9.3. Control

9.3.1. DISPERSION OF INFESTATION

Usually the first indication of nematode-transmitted virus infection in a crop is the appearance of patches, frequently oblong in shape of diseased plants. An additional characteristic is the persistence of an infection in the soil for long periods of time.

Probably one of the most important and common ways to spread an infestation is by the distribution and use of infected planting material. For example, the services of Plant Protection of Martinique intercepted bonzais of *Pinus pentaphylla* imported from Japan which were heavily infested with *Rotylenchus robustus* and *Trichodorus cedarus* (Cadet & Van den Berg, 1992). Ornamental bulbs, for example, are effective disseminators of TRV.

Soil adhering to farm machinery, to transplants, and to the feet of man and animals as well as nematodes transported by the wind may be responsible for the spread of virus vectors (Boag, 1985).

The roots of weed plants e.g. *Stellaria media*, *Solanum nigrum*, *Capsella bursa-pastoris* and *Senecia vulgaris* are frequently found to be virus infected (van Hoof, 1963; Noordham, 1956; own observations). Infected weeds and root pieces remaining after the harvest of an infected crop can provide an overwintering reservoir of viruses (Cooper & Harrison, 1973).

Seeds are known to carry the TRV and PEBV (Lister & Murant, 1967). Also virus retained in the vector trichodorids may easily survive winter fallow between susceptible crops.

9.3.2. CONTROL

Crop rotation, as a non-chemical control strategy, in general offers little prospect of practical prevention of damage since TRV and its polyphagous trichodorid nematode vectors have a wide host range. Though, knowledge of the precropping history of a field is important on a limited scale. Crops as for example tulip and seed-potatoes are more sensitive for virus, others as e.g. wheat, onions and sugar beet are more sensible for direct damage (Zoon, pers. comm.). Some crops such as oil radish and *Tagetes patula* appear to be poor hosts for trichodorids (Kuiper, 1977). *Crotalaria spectabilis* has been shown to be a non-host of the nematode and when used as a cover crop will reduce nematode densities (Rhoades, 1964). *Asparagus officinalis* var. *altilis* L. has also been shown to be resistant to nematode attack as the plant induces the production of highly toxic glycoside when the nematodes begin to feed (Rohde & Jenkins, 1957). Green manures as Italian rye may decrease the yield, while organic fertilizers as e.g. chicken manure and compost provide the same effects as soil disinfection and increase the yield.

Trichodorid vector nematodes are susceptible to physical damage and desiccation during cultivations and repeated intensive soil cultivation may affect population size. **Tillage prior to planting** may give good results (Kuiper, 1977). However, McSorley & Gallaher (1993) observed in tropical corn, that tillage had little effect on densities of most plant-parasitic nematodes, including *P. minor*.

Control of virus vector nematodes by **chemical nematicides** gives satisfactory results. Large amounts of methylbromide, D-D, or chloropicrin can kill a high percentage of the trichodorid population in the soil (Taylor, 1978). Systemic nematicides, such as aldicarb, fenamiphos, and oxamyl, do not kill the nematodes but do control virus spread. Fumigant and non-fumigant nematicides are effective in reducing initial damage and in giving crops an initial advantage over the nematode. However, it has been shown that nematode populations build-up quickly (Perry, 1953). Some of the carbamate and phosphate non-fumigant nematicides exhibit longer durations of control than the fumigants (Rhoades, 1967, 1968). With sugar-cane in subtropical and tropical climatic areas, both Martin (1967) and Harris (1975) noted that the number of trichodorids declined sharply after soil was treated with fumigant nematicide, but that after 3 or 4 months the populations increased dramatically to levels in excess of those present before fumigation. In Puerto Rico, trichodorids increased from non-detectable levels to 6000 per dm^3 of soil 9 months after fumigation, (Roman, 1967). A similar but smaller response was recorded in Burkina Faso (Cadet, 1979).

Chemical control measures and the use of partly tolerant or **resistant cultivars** of potatoes are currently the main control measures for TRV in Britain. Nematicides are frequently used by growers for extremely sensitive cultivars for processing and routine application of expensive nematicides which eliminates any possible chance of infection and consequent loss of revenue. With pressures on reduced profit margins and increasing environmental concern on the use of nematicides this method of control must be viewed as short term. This will place increasing reliance on plant resistance. Potato plant breeders attempt to breed new cultivars with low levels of sensitivity to TRV but they require reliable testing of large numbers of potato genotypes in a glasshouse test, a source of heritable variation and an understanding of how such variation is inherited (Dale, 1989). The problem of nematode-transmitted viruses can be avoided by planting non-susceptible crops on infected land.

Conventional nematicides are expensive, toxic substances and their continued use in agriculture is in question. In the United States too, chemical fumigants have been widely marketed for the control of soilborne pests. Environmental concerns have created social and legislative pressure to remove many agricultural pesticides (e.g. methylbromide) from the market. Soil solarization is an example of a new approach in the search for alternative, nonchemical management of soilborne pests. Chelleni *et al.* (1993) found that **soil solarization** reduced populations of *P. minor* on tomato, and achieved similar results as fumigation with a 67:33 mixture of methylbromide and chloropicrin (448kg/ha). This method is unlikely to have any practical relevance in Europe due to lower temperatures and the ability of trichodorids to migrate vertically downwards to escape from the deleterious effects of drought and for high temperatures.

The search for alternative control strategies has led to the need for more basic information on the ecology of nematodes to aid the development of new environmentally sensitive control methods. Further measures to eliminate TRV from nematode populations, or to eliminate the nematode vectors are required to reduce the use of soil disinfectants. One obvious area for further research is the mechanism by which TRV survives in soil from year to year. Reducing nematode populations or freeing them from virus, either through the culture of non-susceptible crops or cultivars may be an option to solve the complexity of TRV infestations on the long term.

REFERENCES

Ahmad, W., 1989. Taxonomy of the superfamily Trichodoroidea (Nematoda: Triplonchida). in Jairajpuri, M.S. (Es.) *Nematode Pest Identification.* Department of Zoology, A.M.U.,: 246-273.

Allen, M.W., 1957. A review of the nematode genus *Trichodorus* with descriptions of ten new species. *Nematologica*, 2: 32-62.

Allen, M.W., 1960. The genera *Xiphinema, Longidorus* and *Trichodorus.* In: Sasser, J.N. & Jenkins, W.R. (Eds). *Nematology. Fundamentals and Recent Advances with Emphasis on Plant Parasitic and Soil Forms.* The University of North Carolina Press, Chapel Hill, chapter 24: 227-228.

Almeida, M.A., T., De Waele, D., Santos, M.S.N. De & Sturhan, D., 1989. Species of *Trichodorus* (Nematoda: Trichodoridae) from Portugal. *Revue de Nématologie,* 12: 219-233.

Almeida, M.A., T., Santos, M.S.N. De, 1988. Trichodoridae nematodes in Portugal. *Nematologica*, 34: 252-253.

Alphey, T.J.W., 1978. Oxamyl sprays for the control of potato spraing disease caused by nematode transmitted tobacco rattle virus. *Annals of Applied Biology,* 88: 75-80.

Alphey, T.J.W. & Boag, B., 1976. Distribution of trichodorid nematodes in Great Britain. *Annals of applied Biology,* 84: 371-381.

Alphey, T.J.W., Cooper, J.I. & Harrison, B.D., 1975. Systematic nematicides for the control of trichodorid nematodes and of potato spraing disease caused by tobacco rattle virus. *Plant Pathology,* 24: 117-121.

Alphey, T.J.W. & Taylor, C.E., 1986. *European Atlas of the Longidoridae and Trichodoridae.* Scottish Crop Research Institute and European Science Foundation. 123 pp.

Andersen, H.J., 1972. Migrerende rodnematoder hos baelgplanter. *Tidsskrift for Planteavl.*, 76: 559-569.

Andrassy, I., 1968. The scientific results of the Hungarian soil zoological expedition to the Brazzaville-Congo. 31. Nematoden aus Grundwasser. *Annales Universitatis Scientiarum, Budapest,* 9-10: 3-26.

Andrassy, I., 1973. 100 neue Nematodenarten in der ungarischen Fauna. *Opuscula Zoologica, Budapest,* 11: 7-48.

Andrassy, I., 1976. *Evolution as a basis for the systematization of nematodes.* London, San Francisco & Melbourne, Pitman Publishing, 288p.

Andrassy, I., 1989. Six new species from South America. *Acta Zoologica Hungarica*, 35: 1-16.

Anonymous, 1982. Bureau of Nematology. *Tri-ology,* 21 (11): 11-12.

Anonymous, 1986. Bureau of Nematology. *Tri-ology,* 25 (2): 5-6.

Anonymous, 1987a. Bureau of Nematology. *Tri-ology,* 26 (2): 7.

Anonymous, 1987b. Bureau of Nematology. *Tri-ology,* 26 (8): 7.

Anonymous, 1988a. Bureau of Nematology. *Tri-ology,* 27 (2): 6.

Anonymous, 1988b. Bureau of Nematology. *Tri-ology,* 27 (5): 4.

Anonymous, 1988c. Bureau of Nematology. *Tri-ology,* 27 (8): 3.

Anonymous, 1988d. Bureau of Nematology. *Tri-ology,* 27 (11): 2.

Anonymous, 1989a. Bureau of Nematology. *Tri-ology,* 28 (5): 2-3.

Anonymous, 1989b. Bureau of Nematology. *Tri-ology,* 28 (8): 8.

Arias, M. & Bello, A., 1988. Status of Trichodoridae in Spain. *Nematologica* 34:254.

Arias, M. & Andres, M.F., 1989. Virus-vector nematodes in cereals and fruit crops in Spain. *Bulletin OEPP/EPPO,* 19: 625-632.

Arias-Delgado, M., Jimenez-Millan, F. & Lopez-Pedregal, J.M., 1965. Tres nuevas especies de nematodos posibles fitoparasitos en suelos espanoles. *Publicaciones del Instituto de biologia aplicada, Barcelona,* 38: 47-58.

Arias, M. & Roca, F., 1986. *Trichodorus castellanensis,* Arias Delgado, Jimenez Millan et Lopez Pedregal, a junior synonym of *Trichodorus primitivus* (de Man). *Nematologia mediterranea,* 14: 279-281.

Asjes, C.J., 1989. Tobacco rattle virus and other soil-borne viruses in ornamental bulb culture in The Netherlands. *Aspects of applied Biology,* 22: 83-86.

Ayala, A. & Allen, M., 1968. Transmission of the Californian tobacco rattle virus by three species of the nematode genus *Trichodorus. Journal of Agriculture of the University of Puerto Rico,* 52: 101-125.

Ayala, A., Allen, M.W. & Noffsinger, E.M., 1970. Host range, biology, and factors affecting survival and reproduction of the stubby root nematode. *Journal of Agriculture of the University of Puerto Rico,*

64: 341-369.

Baldwin, J., 1977. Bureau of Nematology. *Tri-ology*, 16 (6): 2.

Banck, A., 1977. Inventrering av rotnematoder i matlök. *Växtskyddsnotiser*, 41: 111-114.

Banck, A., 1988. The vertical distribution of trichodorids in a Swedish barley field. *Nematologica*, 34: 255-256.

Barriga, R., 1965. Parasitism of camellias by some nematode species. Abstract. *Nematologica*, 11: 34.

Baujard, P., 1980. *Trichodorus aquitanensis* n.sp. et données nouvelles sur *Paratrichodorus (Atlantadorus) pachydermus* (Seinhorst, 1954) Siddiqi, 1974 (Nematoda: Trichodoridae). *Revue de Nématologie*, 3: 21-27.

Baujard, P., 1983. Observations sur les Trichodoridae Thorne, 1935 (Nematoda) de l'Afrique de l'Ouest. *Revue de Nématologie*, 8: 35-39.

Baujard, P. & Germani, G., 1985. Description de *Monotrichodorus sacchari* n.sp., *Paratrichodorus anthurii* n.sp. et d'une population de *Paratrichodorus westindicus* Rodriguez-M, Sher & Siddiqi, 1978 (Nematoda: Trichodoridae). *Revue de Nématologie*, 8: 35-39.

Behrens, J., 1899. Die Mauche (Mauke) des Tabaks. *Landwirtschaftlichen Versuchsstationen*, 52: 101.

Benson, D. & Barker, K., 1977. Efficacy of 1,2-dibromo-3-chloropropane and aldicarb for control of parasitic nematodes in the root zone of American boxwood, and Japanese and Chinese hollies. *Nematologica*, 9: 263-264.

Bernard, E.C., 1992a. Terrestrial Nematodes of Alaska I. Trichodoridae (Nemata). *Journal of Nematology*, 24: 67-77.

Bernard, E.C., 1992b. *Trichodorus elefjohnsoni* n.sp. (Nemata: Trichodoridae) from Undisturbed Appalachian Forest. *Journal of Nematology*, 24: 78-83.

Bird, A.F. & Bird, J., 1991. *The structure of Nematodes*. Academic Press, Inc., San Diego, New York, 316pp.

Bird, G.W., 1967. *Trichodorus acutus* n.sp. (Nematoda: Diphtherophoroidea) and a discussion of allometry. *Canadian Journal of Zoology*, 45: 1201-1204.

Bird, G.W., 1971. Digestive system of *Trichodorus porosus*. *Journal of Nematology*, 3: 50-57.

Bird, G.W., Goodman, R.M. & Mai, W.F., 1968. Observations on the embryogenesis of *Trichodorus christiei* (Nematodea: Diphtherophoroidea). *Canadian Journal of Zoology*, 46: 292-293.

Bird, G.W. & Mai, W.F., 1965. Plant species in relation to morphometric variations of the New York population of *Trichodorus christiei*. (Abstract). *Nematologica*, 11: 34.

Bird, G.W. & Mai, W.F., 1968. Morphometric and allometric variations of *Trichodorus christiei*. *Nematologica*, 13 (1967): 617-632.

Boag, B., 1978. Nematodes in Scottish forest nurseries. *Annals of Applied Biology*, 88: 279-286.

Boag, B., 1981. Observations on the population dynamics and vertical distribution of trichodorid nematodes in a Scottish forest nursery. *Annals of applied Biology*, 98: 463-469.

Boag, B.,1985. The localised spread of virus-vector nematodes adhering to farm machinery. *Nematologica*, 31: 234-235.

Boag, B., Brown, D.J.F. & Topham, P.B., 1987. Vertical and horizontal distribution of virus-vector nematodes and implications for sampling procedures. *Nematologica*, 33: 83-96.

Boag, B., Neilson, R. & Brown, D.J.F., 1992. Seminar: Nematode sampling and prediction. *Nematologica*, 38: 459-465.

Boag, B. & Topham, P.B., 1984. Aggregation of plant parasitic nematodes and Taylor's Power Law. *Nematologica*, 30: 348-357.

Bongers, T., 1988. *De Nematoden van Nederland*. Stichting Uitgeverij Koninklijke Nederlandse Natuurhistorische Vereniging, Utrecht: 408 pp.

Bor, N.A. & Kuiper, K., 1966. Gevoeligheid van *Trichodorus teres* en *T. pachydermus* voor uitwendige invloeden. *Mededelingen Rijksfaculteit Landbouwwetenschappen, Gent*, 31: 609-616.

Brodie, B.B., 1976. Vertical distribution of three nematode species in relation to certain soil properties. *Journal of Nematology*, 8: 243-247.

Brodie, B.B., Good, J.M. & Adams, W.E., 1969. Population dynamics of plant nematodes in cultivated soil: effect of sod-based rotations in Cecil sandy loam. *Journal of Nematology*, 1: 309-312.

Brown, D.F.J., 1995. Transmission of Viruses by Nematodes. In: Santos, M.S. De, O. Abrantes, I.M. De,

Brown, D.J.F. & Lemlos, R.J.V.C. (Eds). *An introduction to virus-vector nematodes and their associated viruses*. Chapter 12. Coimbra University.

Brown, D.F.J. & Ploeg, A.T., 1989. Specificity of transmission of tobravirus variants by their *(Para)Trichodorus* nematodes. *Journal of Nematology*, 21: 553.

Brown, D.J.F., Ploeg, A.T. & Robinson, D.J., 1989. A review of reported associations between *Trichodorus* and *Paratrichodorus* species (Nematoda: Trichodoridae) and tobraviruses with a description of laboratory methods for examining virus transmission by trichodorids. *Revue de Nématologie*, 12: 235-241.

Brown, E.B. & Sykes, G.B., 1973. Control of tobacco rattle virus (spraing) in potatoes. *Annals of Applied Biology*, 75: 462-464.

Brzeski, M. & Szczygiel, A., 1974. (Nematodes of the genus *Trichodorus* Cobb (Nematoda: Trichodoridae) in Poland.) *Roczniki Nauk Rolniczych , seria E*, 4: 59-66.

Burrows, P.R., Kerry, B.R. & Perry, R.N., 1994. New approaches to plant-parasitic nematode control. *Journal of Zoology*, London, 232: 341-346.

Cadet, P. 1979. Traitements nématicides sur canne à sucre en Haute-Volta. *Congrès Lutte contre Insectes Milieu tropical. Marseille, 13-16 mars, 1979*: 201-217.

Cadet, P. & Van den Berg, E., 1992. Les nématodes phytoparasites de la montagne pelée à la Martinique. *Compte rendu de la Société de biogéographie*, 68: 89-104.

Campos, V.P. & Sturhan, D., 1987. (Occurrence and distribution of plant parasitic nematodes on vegetables in Minas Gerais State, Brazil.). *Nematologica Brasileira*, 11: 153-158.

Cameiro, R.G., Gomes, Cameiro, R. Gomes & Monteiro, A.R., 1982. Distribuiçao vertical de quatro especies de nematoides parasitos da cana de açucar, em relaçao a certas propriedades do solo. *Sociedade Brasileira de Nematologia*, Piracicaba, SP, Brasil, 6: 117-132.

Chaves, E., 1984. *Observations on plant parasitic nematodes from Argentina*. Ph. D. Dissertation, Rijkuniversiteit Gent.

Chellemi, D.O., Olsen, S.M., Scott, J.W., Mitchell, D.J. & McSorley, R., 1993. Reduction of phytoparasitic nematodes on tomato by soil solarization and genotype. *Supplement of Journal of Nematology*, 25 (4S): 800-805.

Chèvres-Roman, R., Gross, H.D. & Sasser, J.N., 1971. The influence of selected nematode species and number of consecutive plantings of corn and sorghum on forage production, chemical composition of plant and soil, and water use efficiency. *Nematropica*, 1: 4041-4046.

Choleva, B., 1988. Nematodes of family Trichodoridae in Bulgaria. (Abstract). *Nematologica*, 34: 261-262.

Christie, J.R., 1959. *Plant parasitic nematodes: their bionomics and control*. University of Florida, Gainesville: 256pp.

Christie, J.R. & Perry, U.G., 1951. A root disease of plants caused by a nematode of the genus *Trichodorus*. *Science*, 113 (2937): 491-493.

Christie, J.R., Perry, U.G. & Wilson, J.W., 1950. Control of nematodes injurious to vegetable crops. *Report of the Florida agricultural Experimental Station*, 4950: 144-145.

Clark, W.C., 1961. A revised classification of the order Enoplida (Nematoda). *New Zealand Journal of Science*, 4: 123-150.

Clark, W.C., 1963. A new species of *Trichodorus* (Nematoda: Enoplida) from Westland, New Zealand. *New Zealand Journal of Science*, 6: 414-417.

Cobb, N.A., 1913. New nematode genera found inhabiting fresh water and non-brackish soils. *Journal of the Washington Academy of Sciences*, 9: 217-343.

Cobb, N.A., 1920. One hundred new nemas. *Contributions to a Science of Nematology*, 9: 217-343.

Coiro, M.I., Alphey, T.J. & Agostinelli, A., 1988. A bivulval female *Trichodorus viruliferus*. *Nematologia Mediterranea*, 16: 239-240.

Coiro, M.I., Alphey, T.J.W. & Agostinelli, A., 1989. Distribution of trichodorids in the vineyards of the province of Trento (Northeastern Italy). *Nematologia mediterranea*, 17: 45-53.

Coiro, M.I. & Sasanelli, N., 1994. The life cycle and reproductive potential of individual *Trichodorus parsus* (Nematoda) on S. Lucie cherry. *Nematologia Mediterranea*, 22: 233-236.

Colbran, R.C., 1956. Studies of plant and soil nematodes. 1. Two new species from Queensland. *Queensland*

Journal of Agricultural Science, 13: 123-126.

Colbran, R.C., 1964. Studies of plant and soil nematodes. 7. Queensland records of the order Tylenchida and the genera *Trichodorus* and *Xiphinema*. *Queensland Journal of Agricultural Science*, 21: 77-123.

Colbran, R.C., 1965. Studies of plant and soil nematodes. 9. *Trichodorus lobatus* n.sp. (Nematoda: Trichodoridae), a stubby-root nematode associated with citrus and peach trees. *Queensland Journal of Agricultural Science*, 22: 273-276.

Cooke, D.A., 1973. The effect of plant parasitic nematodes, rainfall and other factors on Docking disorder of sugar beet. *Plant Pathology*, 22: 161-170.

Cooke, D.A., 1976. Economics of control of Docking disorder of sugar beet. *Proceedings of the Association of Applied Biologists*, 84: 451-455.

Cooke, D.A., 1984. The effect of soil-applied granular pesticides and irrigation on the yield of sugar beet in fields infested with *Trichodorus viruliferus*. *Annals of applied Biology*, 105: 253-261.

Cooke, D.A. & Draycott, A.P., 1971. The effects of soil fumigation and nitrogen fertilizers on nematodes and sugar beet in sandy soils. *Annals of applied Biology*, 69: 253-264.

Coolen, W.A., Decraemer,W., Hendrickx, G.J. & De Meyer, P., 1980. Voorkomen en verspreiding van virusoverdragende nematoden in de West- en Oostvlaamse pootaardappelenverbouwende gebieden. *Landbouwtijdschrift*, 33: 333-337.

Coomans, A. & De Grisse, A., 1981. Sensory Structures. In: Zuckermann, B.M. (Ed.) *Plant parasitic Nematodes*. III: 127-174.

Coomans, A. & Loof, P.A.A., 1970. Morphology and taxonomy of Bathyodontina (Dorylaimida). *Nematologica*, 16: 180-196.

Cooper, J.I., 1971. The distribution in Scotland of tobacco rattle virus and its nematode vectors in relation to soil type. *Plant Pathology*, 20: 51-58.

Cooper, J.I. & Harrison, H., 1973. The role of weed hosts and the distribution and activity of vector nematodes in the ecology of tobacco rattle virus. *Annals of applied Biology*, 73: 53-66.

Cooper, J.I. & Thomas, P.R., 1971. Chemical treatment of soil to prevent transmission of tobacco rattle virus to potatoes by *Trichodorus* spp.. *Annals of applied Biology*, 69:23-34.

Cotton, J. & Hooper, D., 1991. Two new records of nematodes associated with azaleas in England: *Paratrichodorus renifer* Siddiqi, *Tylenchorhynchus claytoni* Steiner. *Plant Pathology*, 40: 308-310.

Craveiro, M.T. & Santos, M.S.N. De A., 1984. Studies on Trichodoridae found in Portugal. In: *Proceedings of the First International Congress of Nematology. Guelph, Ontario, Canada*. 19 p Gainesville, Florida U.S.A.

Cremer, M.C. & Kooistra, G., 1964. Investigations on notched leaf ('kartelblad') of *Gladiolus* and its relation to rattle virus. *Nematologica*, 10: 69-70.

Cremer, M.G. & Schenk, P.K., 1967. Notched leaf in *Gladiolus* spp., caused by viruses of the tobacco rattle virus group. *Netherlands Journal of Plant Pathology*, 73: 33-48.

Dale, M.F.B., 1989. Advances in breeding for Tobacco Rattle Virus sensitivity in potatoes. *Aspects of applied Biology. Roots and the soil environment*, 22: 67-72.

Decraemer, W., 1979. Differentiation of juvenile specimens to species level in Trichodoridae (Nematoda) from fields of seed potatoes in Belgium. *Mededelingen van de Faculteit Landbouwwetenschappen*, Rijksuniversiteit Gent, 44: 273-287.

Decraemer, W., 1980a. Systematics of the Trichodoridae (Nematoda) with keys to their species. *Revue de Nématologie*, 3: 81-99.

Decraemer, W., 1980b. Distribution of Trichodoridae in a strongly infected field. *Mededelingen van de Faculteit Landbouwwetenschappen*, Rijksuniversiteit, Gent, 45: 783-793.

Decraemer, W., 1986. Morphological comparison between the monodelphic genera *Monotrichodorus* and *Allotrichodorus* with some remarks on their relation to both didelphic genera of the Trichodoridae (Nematoda). *Nematologica*, 32: 256-264.

Decraemer, W., 1988. Morphometric variability and value of the characters used for specific identification in *Trichodorus* Cobb, 1913. *Bulletin Koninklijk Belgisch Instituut voor Natuurwetenschappen, Biologie*, 58: 29-44.

Decraemer, W. 1989a. Identification of trichodorids. In: *Nematode identification and expert-system technology*,

R. Fortuner (Ed.). New York, Plenum Publication Corporation: 157-170.

Decraemer, W., 1989b. Morphologic variability and value of the characters used for species identification in *Paratrichodorus* Siddiqi, 1974 (Nematoda: Trichodoridae). *Nematologica*, 35: 37-61.

Decraemer, W., 1991. Stubby root and virus vector nematodes, *Trichodorus*, *Paratrichodorus*, *Allotrichodorus* and *Monotrichodorus*. In *Manual of Agricultural Nematology*. Ed. Nickle, W.R. Marcel Dekker, Publisher: 587-625.

Decraemer, W. & Almeida, M.A.T. Identification of *Trichodorus*/ *Paratrichodorus* species (Diphtherophorina). In: Santos, M.S. DE, O. Abrantes, I.M. DE, Brown, D.J.F. & Lemlos, R.J.V.C. (Eds). *An introduction to virus-vector nematodes and their associated viruses*. Chapter 8. Coimbra University.

Decraemer, W. & Chaves, E., 1988. Morphology of the reproductive system and structure of sperm cells in some species of the genus *Paratrichodorus* with uncommon males (Nematoda: Diphtherophorina). *Revue de Nématologie*, 11: 405-409.

Decraemer, W. & Cheng, H., 1994. On the occurrence of an inner stylet in adult trichodorids and additional information on *Trichodorus nanjingensis* Liu & Cheng, 1990. *Fundamental and applied Nematology*, 17: 503-507.

Decraemer, W., Coolen, W. & Hendrickx, G.J., 1979. Evaluation of extraction methods for a survey of Trichodoridae in fields of seed potatoes. *Nematologica*, 25: 494-495.

Decraemer, W. & De Waele, D., 1981. Taxonomic value of the position of oesophageal gland nuclei and of oesophageal gland overlap in the Trichodoridae (Diphtherophorina). *Nematologica*, 27: 82-94.

Decraemer, W. & Kilian, S., 1992. Description of *Trichodorus parorientalis* n.sp. (Nemata: Diphtherophoroidea) from South Africa. *Fundamental and applied Nematology*, 15: 539-543.

Decraemer, W. & Marais, M., 1993. Two new species of *Trichodorus* from South Africa, with a note on *T. petrusalberti* (Nemata: Trichodoridae). *Fundamental and applied Nematology*, 16: 211-214.

Decraemer, W., Roca, F., Castillo, P., Pena-Santiago, R. & Gomez-Barcina, A., 1993. Trichodoridae from southern Spain, with description of *Trichodorus giennensis* n.sp. (Nemata: Trichodoridae). *Fundamental and applied Nematology*, 16: 407-416.

Decraemer, W. & Reay, F., 1991. Trichodorid nematodes from Australia with description of two new species from native vegetation. *Australasian Plant Pathology*, 20: 52-66.

De Grisse, A., 1969. Redescription ou Modifications de quelques Techniques utilisées dans l'étude des Nématodes Phytosanitaires. *Mededelingen van de Rijksfaculteit Landbouwwetenschappen, Gent*, 34: 351-369.

De Ley, P., Van De Velde, M.C., Mounport, D., Baujard, P. & Coomans, A., 1995. Ultrastructure of the stoma in Cephalobidae, Panagrolaimidae and Rhabditidae, with proposal of a revised stoma terminology in Rhabditida (Nematoda). *Nematologica*, 41: 153-182.

De Maeseneer, J., 1962. Relation of *Trichodorus christiei* to Dieback of *Rhododendron* and *Azalea*. (Abstract). *Nematologica*, 7: 13.

De Pelsmaeker, M. 1988. Het belang en de controle van virusvectornematoden en grondvirussen in de aardappel-, de hop- en de aardbeienteelt. I.W.O.N.L.-I.R.S.I.A., Brussel, (1987) 93pp.

De Pelsmaeker, M., Calus, A. & Coomans, A., 1985. Vertical distribution of *Trichodorus* and TVR. *Mededelingen van de Faculteit Landbouwwetenschappen*, Rijksuniversiteit Gent, 50/3a: 769-775.

De Waele, D., 1986. Taxonomical remarks concerning *Trichodorus proximus* Allen, *T. eburneus* De Waele & Carbonell and *T. coomansi* De Waele & Carbonell (Nematoda: Trichodoridae). *Nematologica*, 32: 115-117.

De Waele, D., 1988. *Trichodorus petrusalberti* n.sp. (Nematoda: Trichodoridae) from rice with additional notes on the morphology of *T. sanniae* and *T. rinae*. *Journal of Nematology*, 20: 85-90.

De Waele, D., Alphey, T.J.W. & Barbez, D., 1985. *Paratrichodorus pachydermus*. *C.I.H. Descriptions of Plant-parasitic Nematodes*, 8, No 112.

De Waele, D. & Carbonell, E., 1983. Two new species of *Trichodorus* (Nematoda: Diphtherophorina) from Africa. *Nematologica*, 28: 387-397.

De Waele, D. & Cohn, E., 1992. Trichodoridae from Israel, with a description of *Trichodorus minzi* n.sp. *Fundamental and applied Nematology*, 15: 201-207.

De Waele, D. & Coomans, A., 1983. Distribution of Longidoridae and Trichodoridae. In: Alphey, T.J.W. (Ed.).

Atlas of Plant Parasitic Nematodes of Belgium. Scottish Crop Research Institute, 42pp.

De Waele, D. & Coomans, A., 1991. Occurrence and ecology of trichodorid nematodes in Belgium. *Revue de Nématologie*, 14: 127-132.

De Waele, D. & Hashim, Z., 1984. *Trichodorus orientalis* n.sp. (Nematoda: Trichodoridae) from Jordan and Iran. *Systematic Parasitology*, 6: 63-67.

De Waele, D. & Kilian, S., 1992. *Trichodorus vandenbergae* n.sp. and *Paratrichodorus meyeri* n.sp. (Nematoda: Trichodoridae) from South Africa, with notes on *P. teres*, *P. catharinae* and *P. sacchari*. *Fundamental and applied Nematology*, 15: 539-543.

De Waele, D., Mancini, G., Roca, F. & Lamberti, F., 1982. *Trichodorus taylori* n.sp. (Nematoda: Dorylaimida) from Italy. *Nematologia Mediterranea*, 10: 27-37.

De Waele, D., Meyer, A.J. & Van Mieghem, A.P., 1990. *Trichodorus philipi* n.sp. from South Africa, with notes on *Paratrichodorus lobatus* and *P. acutus*. *Journal of Nematology*, 22: 200-206.

De Waele, D. & Sturhan, D., 1987. *Trichodorus persicus* n.sp. (Nematoda: Trichodoridae) from Iran. *Systematic Parasitology*, 10: 79-83.

Edward, J.C. & Misra, S.L., 1970. Two new species of *Trichodorus* from Uttar Pradesh, India. *The Allahabad Farmer*, 44: 167-171.

Eriksson, K.B., 1974. Virus-transmitting nematodes. *Växtskyddsnotiser*, 38: 43-51.

Eroshenko, A.S. & Teplyakov, A.A., 1975. A new species of ectoparasitic nematode in *Trichodorus* from coniferous forest in the Far East. (in Russian). *Parazitology*, 9: 545-547.

Esser, R.P., 1971. A compendium of the genus *Trichodorus* (Dorylaimoidea: Diphtherophoridae). *Soil and Crop Science Society of Florida. Proceedings*, 31: 244-253.

Esser, R.P. (Ed.), 1990a. Bureau of Nematology. *Tri-ology*, 29 (7): 5.

Esser, R.P. (Ed.), 1990b. Bureau of Nematology. *Tri-ology*, 29 (9): 1.

Esser, R.P. (Ed.), 1991a. Bureau of Nematology. *Tri-ology*, 30 (3): 2.

Esser, R.P. (Ed.), 1991b. Bureau of Nematology. *Tri-ology*, 30 (5): 7.

Esser, R.P. (Ed.), 1993. Bureau of Nematology. *Tri-ology*, 32 (4): 9.

Estioko, R.V. & Reyes,T.T., 1984. Population dynamics of plant-parasitic nematodes associated with sugarcane in Negros Occidental in relation to soil type and weather pattern. *Proceedings of the Philippine Sugar Technologists' Association*, 31: 235-252.

Evans, K., Trudgill, D.L. & Webster, J.M., 1993. *Plant Parasitic Nematodes in Temperate Agriculture*. CAB International, University Press, Cambridge, 648 pp.

Fortuner, R. & Wong,Y., 1984. Review of the genus *Helicotylenchus* Steiner, 1945. 1: A computer program for identification of the species. *Revue de Nématologie*, 7: 385-392.

Fritzsche, R., Kegler, H., Decker, H., Barchend, G. & Thiele, S., 1985. Einflusz von nematoden aus der Gruppe der Dorylaimiden mit kurzem Mundstachel auf die Infektion von Pflanzen mit dem bodenbürtigen Tabakrattle-Virus (Tobacco rattle virus). *Archiv fur Phytopathologie und Pflanzenschutz.*, 21: 259-264.

Geraert, E., Grootaert, P. & Decraemer, W., 1980. Structure of the female reproductive system in some Dorylaimida and Enoplida (Nematoda). *Nematologica*, 26 (1980): 255-271.

Gibbs, A.J. & Harrison, B.D., 1964. A form of pea early-browning virus found in Britain. *Annals of applied Biology*, 54: 1-11.

Gomez-Barcina, A. & Castillo, P., 1988. Resencia de *Trichodorus cylindricus* Hooper, 1962, en Espana, y algunas notas sobre su distribucion. *Revista iberica Parasitologia*, 48: 289-293.

Goodey, T., 1951. *Soil and freshwater nematodes*. London, Methuen, 390pp.

Gooris, J. & D'Herde, C., 1972. *A method for the quantitative extraction of eggs and second stages juveniles of* Meloidogyne *spp. from soil*. Publication State Nematology and Entomology Research Station, Merelbeke, Belgium: 1-36.

Gügerli, P., 1977. Untersuchungen über die groszräumige und lokale Verbreitung des Tabaksrattlevirus (TRV) und seiner Vektoren in der Schweiz. *Phytopathologische Zeitschrift*, 89: 1-24.

Hafez, S., Raski, D. & Lownsberry, B., 1981. Effect of *Paratrichodorus christiei* on Thompson seedless grape. *Revue de Nématologie*, 4: 115-120.

Harris, R.H.G., 1975. Studies on nematode populations in sugarcane profiles. *Proceedings of the South African*

Sugar Technologists' Association, 49: 164-170.

Harrison, B.D., 1964. Specific nematode vectors for serologically distinctive forms of raspberry and tomato black ring viruses. *Virology*, 22: 544-550.

Harrison,B.D., 1966. Further studies on a British form of pea early-browning virus. *Annals of applied Biology*, 57: 121-129.

Harrison, B.D., 1968. Pea early-browning virus (PEBV). *Annual Report for 1967. Rothamsted Experimental Station, England*: 115.

Harrison, B.D., 1977. Ecology and control of viruses with soil-inhabiting vectors. *Annual Review of Phytopathology*, 15: 331-360.

Harrison, B.D., Robertson, W.M. & Taylor, C.E.,1974. Specificity of retention and transmission of viruses by nematodes. *Journal of Nematology*, 6: 155-164.

Harrison, B.D. & Robinson, D.J., 1986. Tobraviruses. In: Van Regenmortel, M.H.V. & Fraenckel-Conrat, H. (Eds). *The Plant Viruses. Vol. 2. The rod-shaped plant viruses. Plenum Press. London and New York: 339-369.*

Harrison, B.D. & Smart, G.C.Q. jr, 1975. Vertical distribution of *Trichodorus christiei* and *Trichodorus proximus* relative to soil moisture. *Journal of Nematology*, 7: 324.

Heathcote, G.D., 1973. Nematode-transmitted Viruses of sugar beet in England, 1965-1972. *Plant Pathology*, 22: 156-160.

Heyns, J., 1975. *Paratrichodorus christiei. C.I.H. Descriptions of Plant-Parasitic Nematodes*, 5, No 69.

Hirumi, H., 1970. Fine structure of the feeding apparatus and digestive tract of *Trichodorus christiei. Journal of Parasitology*, 56: 147.

Hirumi, H., Chen, T.A., Lee, K.L. & Maramorosch, K., 1968. Ultrastructure of the feeding apparatus of the nematode *Trichodorus christiei. Journal of Ultrastructure Research*, 24: 434-453.

Hirumi, H. & Hung, C.L., 1969. Ultrastructure of intestinal epithelium in a plant parasitic nematode, *Trichodorus christiei. Journal of Nematology*, 1: 292.

Hirumi, H., Raski, D.J. & Jones, N.O., 1969. Ultrastructure of somatic muscles of *T. christie* and *Longidorus elongatus*: comparative cytology of two plant-parasitic nematodes. Abstract. *Journal of Nematology*, 1: 291.

Hirumi, H., Raski, D.J. & Jones, N.A., 1970. Fine structure of sensory cilia in two plant-parasitic nematodes, *Trichodorus christiei* and *Longidorus elongatus*. (Abstract). *Journal of Parasitology*, 56: 148.

Hirumi, H., Raski, D.J. & Jones, N.A., 1971. Primitive muscle cells of nematodes: Morphological aspects of platymyarian and shallow coelomyarian muscles in two plant parasitic nematodes, *Trichodorus christiei* and *Longidorus elongatus. Journal of Ultrastructure Research*, 34: 517-543.

Hooper, D.J., 1962. Three new species of *Trichodorus* (Nematoda: Dorylaimoidea) and observations on *T. minor* Colbran, 1956. *Nematologica*, 7: 273-280.

Hooper, D.J., 1963. *Trichodorus viruliferus* n.sp. (Nematoda: Dorylaimida). *Nematologica*, 9: 200-204.

Hooper, D.J., 1972. Two new species of *Trichodorus* (Nematoda: Dorylaimida) from England. *Nematologica*, 18: 59-65.

Hooper, D.J., 1973. Nematodes. In: Gibbs, A.J. (Ed.) *Viruses and invertebrates*., Amsterdam, The Netherlands, North Holland Publishing Co: 212-228.

Hooper, D.J., 1975. Morphology of trichodorid nematodes. In: Lamberti, F., Taylor, C.E. & Seinhorst, J.W. (eds): *Nematode Vectors of Plant Viruses*. London, U.K. & New York, U.S.A., Plenum Press: 91-101.

Hooper, D.J., 1976. *Trichodorus viruliferus. C.I.H. Descriptions of Plant-Parasitic Nematodes*, 5, No 103.

Hooper, D.J., 1977. *Paratrichodorus (Nanidorus) minor. C.I.H. Descriptions of Plant-Parasitic Nematodes*, 7, No 103.

Hooper, D.J., Kuiper, K. & Loof, P.A.A., 1963. Observation on the identity of *Trichodorus teres* Hooper, 1962 and *T. flevensis* Kuiper & Loof, 1962. *Nematologica*, 9: 646.

Hooper, D.J. & Siddiqi, M.R.,1972. *Trichodorus similis C.I.H. Descriptions of Plant-parasitic Nematodes*., Set 1 (15): 3pp.

Hooper, D.J. & Southey, J.F., 1978. Plant-parasitic Dorylaimida: Morphology and Identification. In: Southey, J.F. (ed.). *Plant Nematology*. Ministry of Agriculture, Fisheries and Food, London, 207-231.

Hunt, D.J., 1993. *Aphelenchida, Longidoridae and Trichodoridae. Their Systematics and Bionomics*. CAB International, Wallingford, U.K., 352pp.

Ivanova, T.S., 1977. (Virus-carrying root nematodes of the suborder Diphtherophorina). *Opredeliteli Nematod. Rast. Nochvy Nasek.*, 4: 1-93.

Jairajpuri, M.S. & Ahmad, W., 1992. *Dorylaimida. Free-living, Predaceous and Plant-parasitic Nematodes*. E.J. Brill, Leiden, New York, Kobenhavn, Koln, 458pp.

Jana, A. & Baqri, Q., 1984. Nematodes from West Bengal (India) XIV. On the occurrence of ectoparasitic nematodes of Longidoroidea and Trichodoroidea (Dorylaimida), with remarks on the validity of genus *Siddiqia* (Longidoridae). *Bulletin of the zoological Survey of India*, 6: 75-79.

Jensen, H.J., 1963. *Trichodorus allius*, a new species of stubby-root nematode from Oregon (Nemata: Dorylaimoidea). *Proceedings of the helminthological Society of Washington*, 30: 157-159.

Johnson, A. & Chalfaut, R., 1972. Control of nematodes and corn earworm on sweet corn. *Journal of Nematology*, 4: 227-228.

Johnson, A.W., Dowler, C.C. & Hauser, E.W., 1975. Crop rotation and herbicide effects on population densities of plant-parasitic nematodes. *Journal of Nematology*, 7: 158-168.

Jones, F.G.W., Larbey, D.W. & Parrot, D.M., 1969. The influence of soil structure and moisture on nematodes, especially *Xiphinema, Longidorus, Trichodorus* and *Heterodera* spp. *Soil Biology & Biochemistry*, 1: 153-165.

Katalan-Gateva, S., 1985. Species of Trichodoridae (Nematoda: Dorylaimida) Family in Bulgaria. *Annuaire de l'Université de Sofia "Kliment Ohridski"*, Zoologie, 76: 79-81.

Khan, T.H., Jairajpuri, M.S. & Ahmad, W., 1989. Descriptions of some new and known species of dorylaim nematodes. *Nematologica*, 35: 419-437.

Khan, E., Saha, M. & Lal, M., 1993. A new species of *Atlantadorus* Siddiqi, 1974 (Nematoda: Trichodoroidea) from Delhi. *Annals of Plant Protection Sciences*, 1: 15-17.

Khera, S. & Chaturvedi, Y., 1977. Nematodes from tea plantations of Dehra Dun, India. *Records of the Zoological Survey of India*, 72: 125-152.

Knobloch, N. & Bird, G.W., 1981. Plant parasitic nematodes of Michigan: with special reference to the genera of the Tylenchorhynchinae (Nematoda). *Michigan State University Agriculture Experiment Station Research Report*, 419: 35pp.

Komuro, Y., Yoshino, M. & Ichinohe, M., 1970. Tobacco rattle virus isolated from aster showing ringspot syndrome and its transmission by *Trichodorus minor* Colbran. (In Japanese). *Annals of the Phytopathological Society, Japan*, 36: 17-36.

Krebill, R.G., Barker, K.R. & Patton, R.F., 1967. Plant-parasitic nematodes of jack and red pine stands in Wisconsin. *Nematologica*, 13: 33-42.

Kuiper, K., 1977. Introductie en vestiging van planteparasitaire aaltjes in nieuwe polders, in het bijzonder van *Trichodorus teres*. *Verslagen en Mededelingen van de Plantenziektenkundige Dienst te Wageningen*, 136: 193-200.

Kuiper, K. & Loof, P.A.A., 1962. *Trichodorus flevensis* n.sp. (Nematoda: Enoplida), a plant nematode from a new polder soil. *Verslagen en Mededelingen van de Plantenziektenkundige Dienst te Wageningen*, 136: 193-200.

Kuiper, K. & Loof, P.A.A., 1964. Observations on some *Trichodorus* species. (Abstract). *Nematologica*, 10: 77.

Lee, Y., 1976. Two genera of Trichodoridae (Trichodoroidea: Nematoda) new to Korea. *Korean Journal of Plant Protection*, 15: 75-78.

Lehman, P. (Ed.), 1990. Bureau of Nematology. *Tri-ology*, 29 (6): 1.

Lister, R.M. & Murant, A.F.,1967. Seed transmission of nematode-borne viruses. *Annals of applied Biologists*, 59: 49-62.

Liu, L. & Ayala, A., 1970. Pathogenicity of *Fusarium moniliforme* and *F. roseum* and their interaction with *Trichodorus christiei* on sugarcane in Puerto Rico. Abstract. *Phytopathology*, 60: 15.

Liu, R. & Cheng, H., 1990. Occurrence of trichodorid species (Nemata: Trichodoridae) in China. *Journal of Nanjing Agricultural University*, 13: 50-54.

Loof, P.A.A., 1961. The nematode collection of Dr. J.G. de Man I. *Beaufortia*, 8: 169-254.

Loof, P.A.A., 1964. Free-living and plant parasitic nematodes from Venezuela. *Nematologica*, 10: 201-300.

Loof, P.A.A., 1965. *Trichodorus anemones* n.sp. with a note on *T. teres* Hooper, 1962 (Nematoda: Enoplida). *Verslagen en Mededelingen van de Plantenziektenkundige Dienst te Wageningen*, 142: 132-136.

Loof, P.A.A., 1973. Taxonomy of *Trichodorus aequalis*-complex (Diphtherophorina). *Nematologica*, 19: 49-62.

Loof, P.A.A., 1975. Taxonomy of Trichodoridae. In: Lamberti, F., Taylor, C.E. & Seinhorst, J.W. (Eds). *Nematode Vectors of Plant Viruses*. , London & New York, Plenum Press: 103-127.

Loof, P.A.A., 1994. Book Review of : *Aphelenchida, Longidoridae and Trichodoridae: their systematics and bionomics*. by Hunt, D.J., 1993, Wallingford, UK, CAB International. *Helminthological Abstracts*, 63: 37.

Lordello, L.G.E. & Zamith, A.P.L., 1958. Nota sôbre o gênero *Trichodorus* Cobb, 1913 com descriçao de *Trichodorus bucrius* n.sp. (Nematoda, Dorylaimoidea). *Anais da Academia brasileira de Ciencias*, 30: 103-105.

Luc,M., Sikora, R.A. & Bridge, J. (Eds), 1990. *Plant Parasitic Nematodes in Subtropical and Tropical Agriculture*. CAB International, Wallingford, UK, 629pp.

Malek, W.R., Jones, N.O. & Roggen, D.R., 1969. Effect of temperature on growth and reproduction of *Criconemoides curvatum* and *Trichodorus christiei*. (Abstract). *Nematologica*, 11: 41.

de Man, J.G., 1880. Die einheimischen, frei in der reinen Erde und im süssen Wasser lebende Nematoden. *Tijdschrift nederlandsche dierkundige Vereeniging*, 5: 1-104.

de Man, J.G., 1884. *Die frei in der reinen Erde und im süssen Wasser lebenden Nematoden der niederlandischen Fauna*. Leiden, Brill, 206pp.

Mamiya, Y., 1967. Descriptive notes on three species of *Trichodorus* (Dorylaimida: Trichodoridae) from forest Nurseries in Japan. *Applied Entomology and Zoology*, 2: 61-68.

Mamiya, Y., 1969a. Plant parasitic nematodes associated with coniferous seedlings in forest nurseries in eastern Japan. *Bulletin of the Government Forest Experiment Station*, 220: 121-132.

Mamiya, Y., 1969b. Effects of nematicides on nematode populations and growth of *cryptomeria* seedlings at a forest nursery. *Bulletin of the Government Forest Experiment Station*, 219: 95-119.

Mancini, G., Moretti, M., Cotroneo, A., Palenzona, M. & Ferrara, A.M., 1979. *Trichodorus aequalis* e *T. viruliferus* (Nematoda, trichodoridae) su semine di *Pinus strobus*. *Informatore Fitopatologico*, 29: 3-7.

Martin, F., 1967. Plant parasitic nematodes associated with sugarcane production in Rhodesia. *FAO Plan Protection Bulletin*, 15: 45-58.

McGlochon, N.E., Sasser, J.N. & Sherwood, R.T., 1961. Investigations of plant-parasitic nematodes associated with forage crops in North Carolina. *Technical Bulletin, North Carolina Agriculture Experiment Station*, 148.

McGowan, J.B. (Ed.), 1972. Bureau of Nematology. *Tri-ology*, 11 (9): 1.

McGowan, J.B. (Ed.), 1977. Bureau of Nematology. *Tri-ology*, 16 (12): 1.

McGowan, J.B. (Ed.), 1979. Bureau of Nematology. *Tri-ology*, 18 (6): 2.

McGowan, J.B., 1983. The stubby-root nematode, *Paratrichodorus christiei* (Allen, 1957) Siddiqi, 1974. *Nematology Circular*, Division of Plant Industrie, Florida Department of Agriculture and Consumer Services, no 97.

McGowan, J.B. (Ed.), 1986. Bureau of Nematology. *Tri-ology*, 25 (9): 1.

McGowan, J.B. (Ed.), 1987. Bureau of Nematology. *Tri-ology*, 26 (12): 2.

McGowan, J.B. (Ed.), 1988a. Bureau of Nematology. *Tri-ology*, 27 (3): 2.

McGowan, J.B. (Ed.), 1988b. Bureau of Nematology. *Tri-ology*, 27 (6): 1-2.

McGowan, J.B. (Ed.), 1989a. Bureau of Nematology. *Tri-ology*, 28 (3): 2.

McGowan, J.B. (Ed.), 1989b. Bureau of Nematology. *Tri-ology*, 28 (9): 1-2.

McSorley, R.M. & Gallaher, R.N., 1993. Effect of crop rotation and tillage on nematode densities in tropical corn. *Supplement to Journal of Nematology*, 25 (4S): 814-819.

Mead, F.W. (Ed.) 1971. Bureau of Nematology. *Tri-ology*, 10 (4): 9.

Mead, F.W. (Ed.) 1974. Bureau of Nematology. *Tri-ology*, 13 (7): 9.

Mead, F.W. (Ed.) 1979. Bureau of Nematology. *Tri-ology*, 18 (7): 7.

Mead, F.W. (Ed.) 1982. Bureau of Nematology. *Tri-ology*, 21 (1): 9.
Mead, F.W. (Ed.) 1984. Bureau of Nematology. *Tri-ology*, 23 (4): 3.
Mead, F.W. (Ed.) 1986a. Bureau of Nematology. *Tri-ology*, 25 (4): 4.
Mead, F.W. (Ed.) 1986b. Bureau of Nematology. *Tri-ology*, 25 (10): 5.
Mead, F.W. (Ed.) 1987a. Bureau of Nematology. *Tri-ology*, 26 (4): 5-6.
Mead, F.W. (Ed.) 1987b. Bureau of Nematology. *Tri-ology*, 26 (10): 4.
Mead, F.W. (Ed.) 1988. Bureau of Nematology. *Tri-ology*, 27 (10): 5.
Mead, F.W. (Ed.) 1989a. Bureau of Nematology. *Tri-ology*, 28 (1): 5.
Mead, F.W. (Ed.) 1989b. Bureau of Nematology. *Tri-ology*, 28 (7): 6.
Mead, F.W. (Ed.) 1990. Bureau of Nematology. *Tri-ology*, 29 (1): 3.
Micoletzky, H., 1922. Die freilebende Erd-Nematoden. *Archiv für Naturgeschichte, A*. 87: 1-650.
Miller, J.W. (Ed.), 1974. Bureau of Nematology. *Tri-ology*, 13 (8): 3.
Miller, J.W. (Ed.), 1977. Bureau of Nematology. *Tri-ology*, 16 (2): 6.
Miller, J.W. (Ed.), 1982. Bureau of Nematology. *Tri-ology*, 21 (2): 4.
Miller, J.W. (Ed.), 1984. Bureau of Nematology. *Tri-ology*, 23 (11): 3.
Morton, H.V. & Perry, V.G., 1968. Life cycle and reproductive potential of *Trichodorus christiei*. *Nematologica*, 14: 11.
Nasira, K. & Maqbool, M.A., 1994. Two new species of *Paratrichodorus* Siddiqi, 1974 (Nematoda: Trichodoridae) with observation on *P. mirzai* (Siddiqi, 1960) Siddiqi, 1974 and *P. renifer* Siddiqi, 1974 from Pakistan. *Fundamental and applied Nematology*, 17: 323-332.
Nishizawa, T.,1973. Pathogenicity of *Trichodorus porosus* to Chinese yam. *Japanese Journal of Nematology*, 3: 33-37.
Noffsinger, E.M.,1984. Adenophorea. In: *Distribution of plant-parasitic nematode species in North America*. A project of the Geographical Distribution Committee. Available from the Society of Nematologists: 20-29.
Noordham, D. 1956. Waardplanten en toetsplanten van het ratelvirus van de tabak. *Tijdschrift over Plantenziekten*, 62: 219-225.
Norton, D.C., Dunlap, D. & Williams, D.D., 1982. Plant-parasitic Nematodes in Iowa: Longidoridae and Trichodoridae. *Proceedings of the Iowa Academy of Sciences*, 89: 15-19.
Pena-Santiago, R., Castillo, D.C., Barcina, A.G. & Jimenez-Millan, F., 1987. Plant-parasitic nematodes associated with olive (*Olea europea* L.) in Jaén province, Spain. *Proc. 7th Congr. medit. phytopathol. Union, Granada, Spain*: 182-183.
Peneva, V., 1988. Nematodes of fam. Trichodoridae (Diphtherophorina) from forest nurseries in Bulgaria. *Bulgarian Academy of Sciences. Heminthology*, 26: 47-66.
Perry, V.G., 1953. The awl nematode, *Dolichodorus heterocephalus*, a devasting plant parasite. *Proceedings of the helminthological Society of Washington*, 20: 21-27.
Pietler, K., 1977. Zur Populationsdynamik virusübertragender Nematoden der Gattungen *Longidorus*, *Xiphinema* und *Trichodorus* unter Berücksichtigung der Standortverhältnisse und des Witterungsverlaufes. In 3 Vortragstagung zu Aktuellen Problemen der Phytonematologie am 2.6.1977 in Rostock. *Biologische Gesellschaft der deutschen Demokratischen Republik und Wilhem-Pieck-Universität*, Rostock: 70-80.
Pitcher, R.S., 1967. The host-parasite relations and ecology of *Trichodorus viruliferus* on apple roots, as observed from an underground laboratory. *Nematologica*, 13: 547-557.
Pitcher, R.S. & McNamara, D.G., 1970. The effect of nutrition and season of year on the reproduction of *Trichodorus viruliferus*. *Nematologica*, 16: 99-106.
Ploeg, A.T. & Brown, D.F.J., 1995. Trichodorid Nematodes and Their Associated Viruses. In: Santos, M.S. De, O. Abrantes, I.M. De, Brown, D.F.J. & Lemlos, R.J.V.C. (Eds). *An introduction to virus-vector nematodes and their associated viruses*. Chapter 3. Coimbra University.
Ploeg, A.T., Brown, D.J.F. & Robinson, D.J., 1989. Transmission of tobravirus by trichodorid nematodes. *Bulletin OEPP/EPPO Bulletin, 19: 605-610.*
Popovici, I., 1989. Distribution and dynamics of soil nematodes in mixed and spruce fir forest ecosystems. *Rev. Roum. Biol. Anim.*, 25: 171-179.

Quanjer, H.M., 1947. Bijdrage tot de kennis van de in Nederland voorkomende ziekten van tabak en van de tabaksteelt op kleigrond. *Plantenziekten*, 49: 37-51.

Rahman, M.F., Jairajpuri, M.S. & Ahmad, I., 1985. Two new species of the genus *Trichodorus* Cobb, 1913 (Nematoda: Dorylaimida) from India. *Revue de Nématologie*, 8: 103-107.

Rashid, F., De Waele, D & Coomans, A., 1986. Trichodoridae (Nematoda) from Brazil. *Nematologica*, 31 (1985): 289-320.

Raski, D.I., Jones, N.O. & Roggen, D.R., 1969. On the morphology and ultrastructure of the esophageal region of *Trichodorus allius* Jensen. *Proceedings of the helminthological Society of Washington*, 36: 106-118.

Rau, J., 1975. Das Vorkommen virusübertragender Nematoden in ungestörten Biotopen Niedersachsen. *Ph.D. Dissertation*, Osnabruck: 1-169.

Razjivin, A.A. & Penton, G., 1975. (A new species of the genus *Trichodorus* (Nematoda) from the rhizosphere of the sugar cane in Cuba). *Zoologiceskij Zhurnal*, 54: 1082-1083.

Reepmeyer, H., 1973. Untersuchungen zur Biologie einzelner *Trichodorus*-Arten und zu ihrer Bekämpfung. Abstract. *Mitteillungen aus der Biologischen Bundesanstalt für Land- und Forstwirtschaft*, 151: 301.

Rhoades, H.L., 1964. Effect of *Crotalaria spectabilis* and *Sesbania exaltata* on plant nematode populations and subsequent yield of snap beans and cabbage. *Proceedings Florida State Horticultural Society*, 77: 233-237.

Rhoades, H.L., 1965. Parasitism and pathogenicity of *Trichodorus proximus* to St Augustine grass. *Plant Disease Reporter*, 49: 259-262.

Rhoades, H.L., 1967. Efficacy of some experimental nematicides applied in-the-row on vegetables. *Proceedings Florida State Horticultural Society*, 80: 156-161.

Rhoades, H.L., 1968. Re-establishment of *Trichodorus christiei* subsequent soil fumigation in central Florida. *Plant Disease Reporter*, 52: 573-575.

Rhoades, H.L., 1969; Nematicide efficacy in controlling sting and stubby-root nematodes attacking onions in Central Florida. *Plant Disease Reporter*, 53: 728-730.

Rhoades, H.L., 1985. Comparison of fenamiphos and *Arthrobotrys amerospora* for controlling plant nematodes in central Florida. *Nematropica*, 15: 1-7.

Richter, E., 1969. Zur vertikalen Verteilung von Nematoden in einem Sandboden. *Nematologica*, 15: 44-54.

Ridings, E. (Ed.), 1980. Bureau of Nematology. *Tri-ology*, 19 (2): 11.

Riemann, F., 1972. *Kinonchulus sattleri* n.g.n.sp. (Enoplida, Tripyloidea), an aberrant freeliving nematode from the lower Amazonas. *Veröffentlichungen des Instituts für Meeresforschung in Bremerhaven*, 13: 317-326.

Robertson, W.M. & Henry, C.E., 1986. A possible role of carbohydrates in the retention of nematode-transmitted viruses. *Annual Report of the Scottish Crop Research Institute for 1985, Invergowrie, Scotland, 113.*

Robertson, W.M. & Wyss, U., 1983. Feeding Processes of Virus-Transmitting Nematodes. In: *Current topics in vector Research.* Vol.1. by Harris, K.F. (Ed.). Praeger Scientific.

Robinson, D.J., 1989. Tobacco rattle tobravirus: variation among strains and detection by cDNA probes. *Bulletin OEPP/EPPO Bulletin*, 19: 619-623.

Robinson, D.J. & Harrison, B.D., 1989. Tobacco rattle virus. *AAB Descriptions of Plant Viruses*, 346: 7pp.

Roca, F. & Arias, M., 1986. A new *Paratrichodorus* species (Nematoda: Trichodoridae) from Spain. *Nematologia Mediterranea*, 12: 181-185.

Roca, F. & Lamberti, F., 1984a. Trichodorids (Nematoda) from Italy. *Nematologia Mediterranea*, 12: 95-118.

Roca, F. & Lamberti, F., 1984b. Trichodorids (Nematoda) from Italy. *Nematologia Mediterranea*, 12: 132-141.

Rodriguez-M, R. & Bell, A.H., 1978. External morphology of the spicules of some Trichodoridae. *Journal of Nematology*, 10: 132-141.

Rodriguez-M., R., Sher, S.A. & Siddiqi, M.R., 1978. Systematics of the monodelphic species of Trichodoridae (Nematoda: Diphtherophorina) with descriptions of a new genus and four new species. *Journal of Nematology*, 10: 141-152.

Rohde, R.A. & Jenkins, W.R., 1957. Host range of a species of *Trichodorus* and its host-parasite relationships on tomato. *Phytopathology*, 47: 295-298.

Roman, J. 1967. Preliminary investigations on nematodes associated with sugarcane in Puerto Rico. *Proceedings of the international Society of Sugar Cane Technologists'*, 12: 1401-1407.

Rössner, J., 1969. Populationsentwicklung pflanzenparasitarer Nematoden unter Koniferen. *Mitteilungen aus der Biologischen Bundesanstalt für Land- und Forstwirtschaft*, 136: 50-58.

Rössner, J., 1972. Vertikalverteilung wandernder Wurzelnematoden im Boden in Abhängigkeit von Wassergehalt und Durchwurzelung. *Nematologica*, 18: 360-372.

Salomao, T.A., 1973. Soil transmission of artichoke yellow band virus. *2nd Congr. internaz. Studi Carciofo, Bari, Italy, 21-24 nov. 1973*, 831-854.

Sänger, H.L., 1961. Untersuchungen über schwer übertragbare Formen des rattle Virus. *Proceedings 4th Conference Potato Virus Diseases Braunschweig Germany*: 22-28.

Scotto La Massese, C., 1985. Distribution of Longidoridae, Xiphinemidae and Trichodoridae. In: Alphey, T.J.W. (Ed.) *Atlas of Plant Parasitic Nematodes of France*. Scottish Crop Research Institute, 43pp.

Seddon, J., 1973. School of Agriculture. Report 1972-1973. Loughborough, Leicestershire, United Kingdom. Plant Nematology: 146.

Seinhorst, J.W., 1954. On *Trichodorus pachydermus* n.sp. (Nematoda: Enoplida). *Journal of Helminthology*, 28: 111-114.

Seinhorst, J.W., 1963. A redescription of the male of *Trichodorus primitivus* (de Man), and the description of a new species *T. similis*. *Nematologica*, 9: 125-130.

Seinhorst, J.W., 1970. Replacement stylet in adult *Trichodorus nanus*. *Nematologica*, 16: 330.

Seinhorst, J.W. & Van Hoof, H.A., 1982. Distribution of Longidoridae, Trichodoridae and Xiphinemidae. In *Atlas of Plant Parasitic Nematodes of the Netherlands*. (Alphey, T.J.W., Ed.). Scottish Crop Research Institute, Dundee, Scotland: 33pp.

Sher, S.A., Bell, A.H. & Rodriguez-M, R., 1977. The face view of Trichodoridae. *Journal of Nematology*, 9: 254-256.

Shishida, Y., 1979. Studies of nematodes parasitic on woody plants. 1. Family Trichodoridae (Thorne, 1933) Clark, 1961. *Japanese Journal of Nematology*, 9: 28-44.

Siddiqi, M.R., 1960. Two new species of the genus *Trichodorus* (Nematoda: Dorylaimoidea) from India. *Proceedings of the helminthological Society of Washington*, 27: 22-27.

Siddiqi, M.R., 1962. *Trichodorus pakistanensis* n.sp. (Nematoda: Trichodoridae) with observations on *T. porosus* Allen, 1957, *T. mirzai* Siddiqi, 1960, and *T. minor* Colbran, 1956, from India. *Nematologica*, 8: 193-200.

Siddiqi, M.R., 1963. *Trichodorus* spp. (Nematoda: Trichodoridae) from Tunesia and Nicaragua. *Nematologica*, 9: 69-75.

Siddiqi, M.R., 1974. Systematics of the genus *Trichodorus* Cobb, 1913 (Nematoda: Dorylaimida), with descriptions of three new species. *Nematologica*, 19: 259-278.

Siddiqi, M.R., 1980a. On the generic status of *Atlantadorus* Siddiqi, 1974 and *Nanidorus* Siddiqi, 1974 (Nematoda: Trichodoridae). *Systematic Parasitology*, 1: 151-152.

Siddiqi, M.R., 1980b. The origin and phylogeny of the nematode orders Tylenchida Thorne, 1949 and Aphelenchida n.ord. *Helminthological Abstracts. Series B, Plant Nematology*, 49: 144-170.

Siddiqi, M.R., 1983. Phylogenetic relationships of soil nematode orders Dorylaimida, Mononchida, Triplonchida and Alaimida, with a revised classification of the subclass Enoplia. *Pakistan Journal of Nematology*, 1: 79-110.

Siddiqi, M.R., 1991. Studies on plant-parasitic nematode genus *Monotrichodorus*, with descriptions of three new species from South America and *Paratrichodorus paramirzai* sp.n. from India. *Afro-Asian Journal of Nematology*, 1: 180-191.

Siddiqi, M.R. & Brown, K.F., 1965. *Trichodorus rhodesiensis* and *Amphidelus trichurus*, two new nematode species from cultivated soils of Africa. *Proceedings of the helminthological Society of Washington*, 32: 239-242.

Siddiqui, I.A., Sher, S.A. & French, A.M., 1973. *Distribution of plant parasitic nematodes in California*. California Department of Food and Agriculture, Division of Plant Industry, Sacramento: 205pp.

Smolik, J.D. & Malek, R.B., 1972. *Tylenchorhynchus nudus* and other nematodes associated with Kentucky bluegrass turf in South Dakota. *Plant Disease Reporter*, 56: 898-900.

Sol, H.H., 1961. Grondvirussen. 1. Ratelvirus. *Annual Report for 1960. Instituut voor Plantenziektenkundig Onderzoek, Wageningen, The Netherlands*: 85-86.

Sol, H.H., 1963. Some data on the occurrence of rattle virus at various depths in the soil and on its transmission. *Netherlands Journal of Plant Pathology*, 69: 208-214.

Sol, H.H., van Heuven, J.C. & Seinhorst, J.W., 1960. Transmission of rattle virus and Atropa belladonna virus by nematodes. *Tijdschrift Plantenziekten*, 66: 228-231.

Sol, H.H. & Seinhorst, J.W., 1961. The transmission of rattle virus by *Trichodorus pachydermus*. *Tijdschrift Planteziekten*, 67: 307-309.

Spaull, A., 1980. Effect of *Paratrichodorus anemones* on growth of spring wheat and barley. *Nematologica*, 26: 163-169.

Spaull, V.W., 1981: Nematodes associated with sugarcane in South Africa. *Phytophylactica*, 13: 175-179.

Steudel, W., 1969. Untersuchungen zur Wirkung einiger Nematizide auf *Trichodorus* spp. *Mitteillungen Biologischen Bundesanstalt für Land- und Forstwirtschaft*, Berlin, Dahlem, 136: 77-86.

Sturhan, D., 1967. Vorkommen von *Trichodorus*-Arten in Westdeutschland. *Mitteilungen Biologischen Bundesanstalt für Land- und Forstwirtschaft*, 121: 146-151.

Sturhan, D., 1978. Untersuchungen über Vorkommen und Verbreitung pflanzenparasitärer Nematoden in der Bundesrepublik Deutschland. Annual report for 1977. *Biologischen Bundesanstalt für Land- und Forstwirtschaft*, Berlin und Braunschweig, part H: 68.

Sturhan, D., 1985. Ein neuer Phytonematode aus Deutschland: *Paratrichodorus weischeri* spec. nov. (Nematoda, Trichodoridae). *Mitteilungen aus der Biologischen Bundesanstalt für Land- und Forstwirtschaft.*, 226: 31-45.

Sturhan, D., 1989a. Hermaphroditism in *Paratrichodorus* species (Nemata: Dorylaimida). *Revue de Nématologie*, 12: 273-276.

Sturhan, D., 1989b. On the species status of *Paratrichodorus allius* (Jensen, 1963)) and *P. tansaniensis* Siddiqi, 1974 (Nematoda, Trichodoridae). *Nematologica*, 35: 62-68.

Symalla, H., 1972. Influence of different crops on TRV-infectivity and population density of *Trichodorus*. Abstract. *Summaries of the International Nematology Symposium of the European Society of Nematology, Reading*: 73-74.

Szczygiel, A., 1968. *Trichodorus sparsus* sp.n. (Nematoda, Trichodoridae). *Bulletin de l'Académie polonaise des Sciences*. Classe II, 16: 695-698.

Szczygiel, A. & Brzeski, M.W., 1985. Distribution of Longidoridae, Xiphinemidae and Trichodoridae. In: Alphey, T.J.W. (Ed.) *Atlas of Plant Parasitic Nematodes of Poland*. Scottish Crop Research Institute, 31pp.

Taylor, C.E., 1978. Plant-parasitic Dorylaimida: Biology and Virus Transmission. In : J.F. Southey (Ed.). *Plant Nematology*. HM Stationary Office, London: 232-243.

Taylor, C.E. & Robertson, W.M., 1970. Location of tobacco rattle virus in the nematode vector *Trichodorus pachydermus* Seinhorst. *Journal of General Virology*, 6: 179-182.

Taylor, C.E. & Robertson, W.M., 1974. 1974. Acquisition, Retention and Transmission of Viruses by Nematodes. In: Lamberti, F., Taylor, C.E. & Seinhorst, J.W. (Eds). *Nematode Vectors of Plant Viruses*. Plenum Press, London & New York: 253-276.

Taylor, C.E. & Robertson, W.M., 1978. Virus vector relationships and mechanics of transmission. *Proceedings of the American Phytopathological Society*, 4: 20-29.

Thorne, G., 1935. Notes on free-living and plant-parasitic nematodes. II. *Proceedings of the helminthological Society of Washington*, 2: 96-98.

Thorne, G., 1939. A monograph of the nematodes of the superfamily Dorylaimoidea. *Capita zoologica*, 8: 1-261.

Thorne, G., 1961. *Principles of Nematology*. McGraw-Hill, New York: 553pp.

Thorne, G., 1974. Nematodes of the Northern Great Plains. Part II Dorylaimoidea in part (Nemata: Adenophorea). *Technical Bulletin*, 41: 120pp.

Topham, P.B., Alphey, T.J.W., Boag, B. & De Waele, D., 1985. Comparison between plant-parasitic nematode species associations in Great Britain and in Belgium. *Nematologica*, 31: 458-467.

Trudgill, D.L., Brown, D.J.F. & McNamara, D.G., 1983. Methods and criteria for assessing the transmission of viruses by longidorid nematodes. *Revue de Nématologie*, 6: 133-141.

Van de Velde, M.C. & Coomans, A., 1991. The ultrastructure of the buccal cavity of the monhysterid nematodes *Geomonhystera disjuncta* and *Diplolaimella dievengatensis*. *Revue de Nématologie*, 14: 133-146.

van Hoof, H.A., 1962. *Trichodorus pachydermus* and *T. teres*, vectors of the early browning virus of peas. *Tijdschrift Plantenziekten*, 68: 391-396.

van Hoof, H.A., 1963. Het verband tussen het optreden van stengelbont (Ratelvirus), de besmetting van vangplanten en het voorkomen van *Trichodorus*-soorten. *Mededelingen LandbouwHogeschool en Opzoekingsstations van de Staat, Gent*, 28: 1001-1010.

van Hoof, H.A., 1964. Serial transmission of rattle virus by a single male of *Trichodorus pachydermus* Seinhorst. *Nematologica*, 10: 141-144.

van Hoof, H.A., 1968. Transmission of tobacco-rattle virus by *Trichodorus*-species. *Nematologica*, 14: 20-24.

van Hoof, H.A., 1970. Some observations on retention of tobacco rattle virus in nematodes. *Netherlands Journal of Plant Pathology*, 76: 329-330.

van Hoof, H.A., 1972. Soilborne viruses. In: *Viruses of potatoes and seed-potato production. De Bokx, J.A. (Ed.). Centre for Agricultural Publishing and Documentation. Wageningen, The Netherlands: 57-64.*

van Hoof, H.A., 1973. Bestrijding van het tabaksratelvirus bij tulp. *Bloembollencultuur*, 83: 1225.

van Hoof, H.A., 1980. Stem mottle. *Annual Report for 1979. Instituut voor Plantenziektekundig Onderzoek, Wageningen, The Netherlands*: 35.

van Hoof, H.A., Maat, D.Z. & Seinhorst, J.W., 1966. Viruses of the tobacco rattle virus group in northern Italy: their vectors and serological relationships. *Netherlands Journal of Plant Pathology*, 72: 253-258.

Vermeulen, W.J.J. & Heyns, J., 1983. Studies on Trichodoridae (Nematoda: Dorylaimida) from South Africa. *Phytophylactica*, 15: 17-34.

Vermeulen, W.J.J. & Heyns, J., 1984. Further studies on southern African Trichodoridae (Nematoda: Dorylaimida). *Phytophylactica*, 16: 301-305.

Vinciguerra, M.T. & Giannetto, L., 1987. Nematodi delle Faggete Italiane. *Animalia*, 14: 5-33.

Wallace, H., 1971. The influence of the density of nematode populations on plants. *Nematologica*, 17: 154-166.

Walkinshaw, C.H., Griffin, G.D. & Larson, R.H., 1961. *Trichodorus christiei* as a vector of potato corky ringspot (tobacco rattle) virus. *Phytopathology*, 51: 806-807.

Wang, S.H. & Wu, X.Y., 1991. Species and distribution of nematodes from the rhizosphere of grapes in China. In: Huang, K.X. (Ed.). *Thesis on the first Symposium of All-China Young-Middle Age's Scientistion Plant Protection*. Science & Technology Press, Beijing, China: 212-219.

Wasilewska, L., 1971. Nematodes in a young pine plantation in the Laski Forest Administration District of the Kampinos Forest. *Zesz. probl. Postop. Nauk. rohp.*, 121: 159-167.

Whitehead, A.G. & Hooper, D., 1970. Needle nematodes (*Longidorus* spp.) and stubby-root nematodes (*Trichodorus* spp.) harmful to sugar beet and other field crops in England. *Annals of applied Biology*, 65: 339-350.

Winfield, A.L. & Cooke, D.A., 1975. The ecology of *Trichodorus*. In Lamberti, F., Taylor, C.E. & Seinhorst, J.W. (Eds) *Nematode Vectors of Plant Viruses*., Plenum Press, London & New York, 309-341.

Wyss, U., 1970. Zur Bestimmung von *Trichodorus*-Arten. *Pflanzenkrankheiten und Pflanzenschutz*, 77: 506-514.

Wyss, U., 1971. De Mechanismus der Nahrungsaufnahme bei *Trichodorus similis*. *Nematologica*, 17: 508-518.

Wyss, U., 1972. *Trichodorus similis* (Nematoda) Saugen an Wurzeln von Sämlingen (Rübsen). *Encyclopaedia cinematographica*, Göttingen, 1-12.

Wyss, U., 1974. *Trichodorus similis*. *C.I.H. Descriptions of Plant-parasitic nematodes*, 4, No 59.

Wyss, U., 1979. On the formation and ultrastructure of feeding tubes produced by trichodorid nematodes. *Nematologica*, 25: 385-390.

Xu, J. & Decraemer, W., 1995. *Trichodorus* species from China, with a description of *T. paracedarus* n.sp. (Nemata: Trichodoridae). *Fundamental and applied Nematology*, in press.

Ye, W., 1995. *Paratrichodorus renifer* Siddiqi, 1974 found in China associated with imported Azalea. *Afro-Asian Journal of Nematology*, 5: 1-2.

Yeates, G.W., 1967. Studies of nematodes from dune sands. 4. Diphtherophoroidea. *New Zealand Journal of Science*, 10: 322-329.

Yeates, G.W., 1969. The status of *Trichodorus clarki* Yeates, 1967, *Ereptonema inflatum* Yeates, 1967 and *Mononchoides potohikus* Yeates, 1969. *Nematologica*, 15: 430.

Yokoo, T., 1964. On the stubby root nematodes from the Western Japan. *Agricultural Bulletin of the Saga University*, 23: 1-6.

Yokoo, T., 1966. On a new stubby root nematode (*Trichodorus kurumeensis* n.sp.) from Kyushu, Japan. *Agricultural Bulletin of the Saga University*, 23: 1-6.

INDEX

ABBREVIATIONS

a: body length divided by greatest body width
b: body length divided by pharynx length (= to junction with intestine)
c: body length divided by tail length
CP: ventromedian cervical papilla
CP_1: anterior ventromedian cervical papilla
L: body length
onch: onchiostyle length
PEBV: Pea early-browning virus
PRV: Pepper ringspot virus
SP_1: posterior precloacal supplement
SP_2: second precloacal supplement
T: distance from cloaca to anterior part of testis as a percentage of total body length
TRV: Tobacco rattle virus
V: distance of vulva from anterior end as a percentage of total body length
a, b, c, T, V: indices of de Man.

The manufacturer's authorised representative in the EU is Springer Nature Customer Service Centre GmbH, Europaplatz 3, 69115 Heidelberg, Germany. If you have any concerns regarding our products, please contact ProductSafety@springernature.com

Printed and bound by CPI Group (UK) Ltd, Croydon, CR0 4YY
30/09/2026
02225436-0011